Postgraduate Physiology

Recent Advances

Postgraduate Physiology

Recent Advances

RK Marya MBBS MD PhD

Former
Professor and Head
Department of Physiology
Pt BD Sharma Postgraduate Institute of Medical Sciences
Rohtak, Haryana

Professor and Head
Unit of Physiology
AIMST University
Bedong, Malaysia

CBS Publishers & Distributors Pvt Ltd

New Delhi • Bengaluru • Chennai • Kochi • Kolkata • Mumbai
Bhopal • Bhubaneswar • Hyderabad • Jharkhand • Nagpur • Patna • Pune • Uttarakhand • Dhaka (Bangladesh)

Postgraduate Physiology
Recent Advances

ISBN: 978-93-88902-95-3

Copyright © Authur and Publisher

First Edition: 2020

Published by Satish Kumar Jain and produced by Varun Jain for

CBS Publishers & Distributors Pvt Ltd
4819/XI Prahlad Street, 24 Ansari Road, Daryaganj, New Delhi 110 002, India.
Ph: 23289259, 23266861, 23266867   Fax: 011-23243014   Website: www.cbspd.com
e-mail: delhi@cbspd.com; cbspubs@airtelmail.in.
Corporate Office: 204 FIE, Industrial Area, Patparganj, Delhi 110 092
Ph: 4934 4934   Fax: 4934 4935   e-mail: publishing@cbspd.com; publicity@cbspd.com

Branches

- **Bengaluru:** Seema House 2975, 17th Cross, K.R. Road,
  Banasankari 2nd Stage, Bengaluru 560 070, Karnataka
  Ph: +91-80-26771678/79   Fax: +91-80-26771680   e-mail: bangalore@cbspd.com
- **Chennai:** 7, Subbaraya Street, Shenoy Nagar, Chennai 600 030, Tamil Nadu
  Ph: +91-44-26680620, 26681266   Fax: +91-44-42032115   e-mail: chennai@cbspd.com
- **Kochi:** 42/1325, 1326, Power House Road, Opposite KSEB Power House,
  Ernakulam 682 018, Kochi, Kerala
  Ph: +91-484-4059061-65   Fax: +91-484-4059065   e-mail: kochi@cbspd.com
- **Kolkata:** 6/B, Ground Floor, Rameswar Shaw Road, Kolkata 700 014, West Bengal
  Ph: +91-33-22891126, 22891127, 22891128   e-mail: kolkata@cbspd.com
- **Mumbai:** 83-C, Dr E Moses Road, Worli, Mumbai 400018, Maharashtra
  Ph: +91-22-24902340/41   Fax: +91-22-24902342   e-mail: mumbai@cbspd.com

Representatives

- **Bhopal** 0-8319310552
- **Bhubaneswar** 0-9911037372
- **Hyderabad** 0-9885175004
- **Jharkhand** 0-9811541605
- **Nagpur** 0-9421945513
- **Patna** 0-9334159340
- **Pune** 0-9623451994
- **Uttarakhand** 0-9716462459
- **Dhaka (Bangladesh)** 01912-003485

Printed at Mudrak Noida, UP, India

Preface

In the last two decades, the number of biomedical research publications has increased exponentially. Now, information is available on the mechanism of many physiological phenomenon that were obscure a decade earlier. In ever-increasing information explosion, postgraduate students in physiology cannot depend merely on undergraduate textbooks. A postgraduate student in physiology, preparing for his/her future role as a teacher and research worker, has to attain a higher degree of depth and proficiency in the subject. Moreover, for this role, the postgraduate student needs to develop critical thinking skill and an ability to analyse and interpret the observed experimental data. The textbooks of physiology alone are not sufficient for this purpose. Though excellent review articles are available to fulfil this requirement, but with the limited time at their disposal, postgraduate students cannot be expected to read 50–100 page review articles with a couple of hundred references each, and make comprehensive and concise summaries of the contents. The author of this book has attempted to fulfil this lacuna in postgraduate medical education in physiology.

Basically, each chapter in the book is the latest review article on the subject, but it is presented in a textbook format. Information culled from scores of review articles and original articles has been distilled and condensed in approximately 10-page write up. Each chapter ends with approximately 20–25 most important references. The references authenticate the information given in the text, and also serve as a ready source for further study in still greater depth, if one desires. Besides postgraduate students, members of faculty in physiology department would also find this book a simple path for updating their knowledge of the subject.

The book has the hallmark of the author's previous books in physiology, namely conciseness and simplicity of presentation. A large number of simple illustrations help to clarify the text.

Some of the topics covered in book are

- Exercise hyperpnea.
- Molecular mechanism of smooth muscle contraction.
- Non-motor functions of basal ganglia.
- Thermoregulatory pathways.
- Neurobiology of sleep-wakefulness.
- Neurobiology of circadian rhythms.
- Executive functions of prefrontal cortex.
- Kisspeptins.
- Natriuretic peptides.

I am proud of my association with Mr Satish Kumar Jain for over 25 years and I am grateful for the publication of my numerous books by CBS Publishers & Distributors. Mr Satish Jain's support was available all along. I also thank Mr YN Arjuna and his team for the beautiful layout of the book.

I would thankfully welcome any feedback on the book.

RK Marya
e-mail: proframeshmarya@gmail.com

Contents

Preface   v

1. Erythropoiesis   1

2. Hemostasis   11

3. The Spleen   20

4. Regulation of Coronary Blood Flow   26

5. Mechanisms of Increased Cardiac Output and
   Skeletal Muscle Hyperemia during Exercise   35

6. Blood Pressure Regulation during Exercise   44

7. Exercise Hyperpnea   51

8. Enteric Nervous System   56

9. Gastrointestinal Hormones   64

10. Lower Esophageal Sphincter   75

11. Exocrine Pancreas   82

12. Molecular Basis of Smooth Muscle Contraction and Relaxation   92

13. Neuroglia   103

14. Motor and Nonmotor Functions of the Basal Ganglia   112

15. Thermoregulation   122

16. Appetite Regulation and Weight Control   132

17. Sleep-Wakeful Cycles   143

18. Circadian Rhythms   152

19. The Blood–Brain Barrier   162

20. The Executive Functions of Prefrontal Cortex   171

21. Growth Hormone   176

22. Islets of Langerhans — 183

23. Vitamin D Hormone (Dihydroxycholecalciferol) — 193

24. Endocrine Functions of Bone — 202

25. Blood–Testis Barrier — 208

26. Kisspeptins — 217

27. Parturition — 225

28. Prolactin — 233

29. Juxtaglomerular Apparatus — 241

30. Natriuretic Peptides — 252

Index — *261*

Chapter 1

Erythropoiesis

INTRODUCTION

In postnatal life, bone marrow is the only normal site for erythropoiesis. After bone marrow transplantation, long-term hemopoiesis is established only in the bone marrow. The morphology of various stages of development of blood cells in the bone marrow is well-known for almost a century.[1] During the last few decades, the interest in stem cell research has produced an explosive information on the initial stages of developing pluripotent stem cells, and growth factors involved at these stages. In addition, many advances have been reported in the physiology of erythropoietin, especially the site of production in the kidney and regulation of its secretion. In erythropoiesis, iron metabolism has a crucial role. In the last few decades, many significant advances have been reported in physiology and pathophysiology of iron metabolism. In this chapter, recent advances in physiology of erythropoietin and iron metabolism have been discussed.

In humans, the lifespan of RBCs is about 120 days. Under normal conditions, approximately 1% of RBCs are synthesized each day but RBC production can increase substantially during times of acute or chronic stress, such as acute hemorrhage or hemolysis.

As mentioned earlier, during the last few decades, the interest in stem cell research has produced an explosive information on the initial stages of developing pluripotent stem cells, and growth factors. The nomenclature of the early stages of hemopoiesis, such as colony-forming unit-erythroid or granulocyte, monocyte, etc. is based on the behavior of the precursor cells in culture, not on any specific morphological or biochemical difference (Fig. 1.1). Each stage responds to a particular growth factor(s). The details of growth factors involved at this stage of hemopoiesis are of interest for scientists involved in stem cell research.

In the bone marrow, hematopoietic cells are formed continuously from a small population of pluripotent stem cells that generate progenitors committed to one or a few hematopoietic lineages (Fig. 1.1). In the erythroid lineage, the earliest committed progenitors identified *ex vivo* are the slowly proliferating burst-forming unit-erythroid (BFU-E). Early BFU-E cells divide and further differentiate through the mature BFU-E stage into rapidly dividing colony-forming unit-erythroid (CFU-E).[2] CFU-E progenitors divide 3 to 5 times over 2 to 3 days as they differentiate and undergo many substantial changes, including a decrease in cell size, chromatin condensation, and hemoglobinization, leading up to their enucleation and expulsion of other organelles[1]. During this differentiation process, the

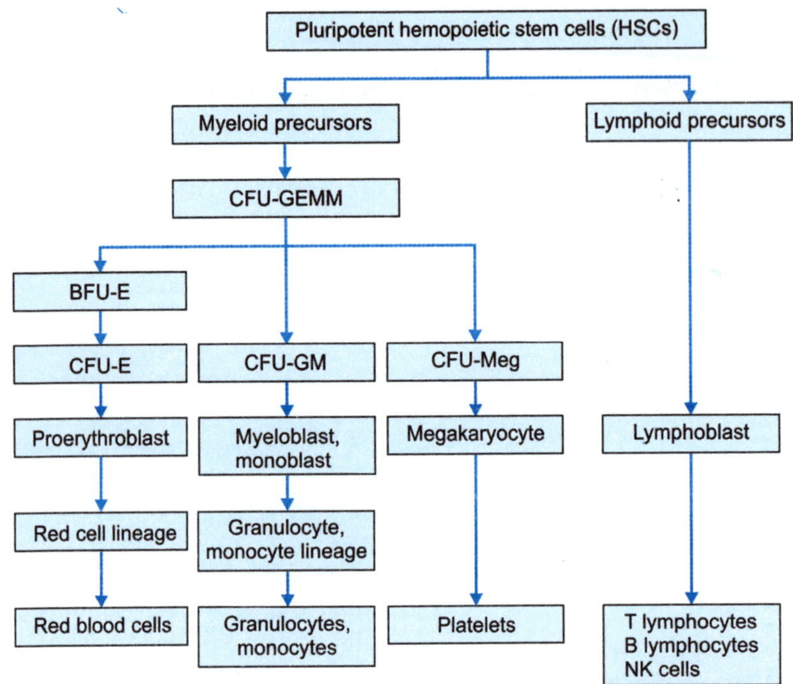

Fig. 1.1: Initial stages of hemopoiesis. CU-GEMM: Colony-forming unit-granulocytes-erythrocytes, monocytes, megakaryocytes; BFU-E: Burst-forming units-erythroid; CFU-E: Colony-forming units-erythroid; CFU-GM: Colony-forming units-granulocytes, monocytes; CFU-Meg: Colony-forming units-megakaryocytes

cells became progressively sensitive to erythropoietin (EPO) due to the appearance of the EPO receptors on these cells. In adult erythropoiesis, the CFU-E stage seems to be the one most responsive to environmental influence. The main targets of EPO are represented by bone marrow erythroid cells expressing EPOR, particularly, late erythroid progenitors (CFU-E) and proerythroblasts, where the EPOR is maximally expressed (about 1000 receptors per cell). EPO and its receptor are not required for the commitment of hemopoietic stem cell to the erythroid lineage (Fig. 1.2). In addition to EPO, BFU-E cells respond to many hormones, stem cell factor (SCF), insulin-like growth factor 1 (IGF-1), glucocorticoids (GCs), IL-3, and IL-6. Besides production of EPO in the kidney, it is not entirely known which cells in the bone marrow produce other regulatory cytokines.

Since, the most important function of red blood cells is transport of oxygen, it is appro-

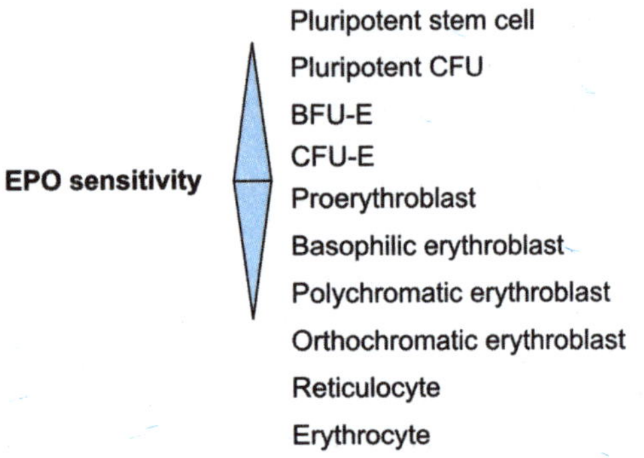

Fig.1.2: EPO-sensitive stages of erythropoiesis

priate that the body maintains normal oxygen carrying capacity by hypoxic regulation of erythropoiesis. Any decrease in oxygen carrying capacity results in proportionate increase in secretion of a glycoprotein, erythropoietin (EPO) by the kidney (Fig. 1.3). Thus, both hypoxic and anemic type of hypoxia regulates red cell count and hemoglobin concentration of blood.

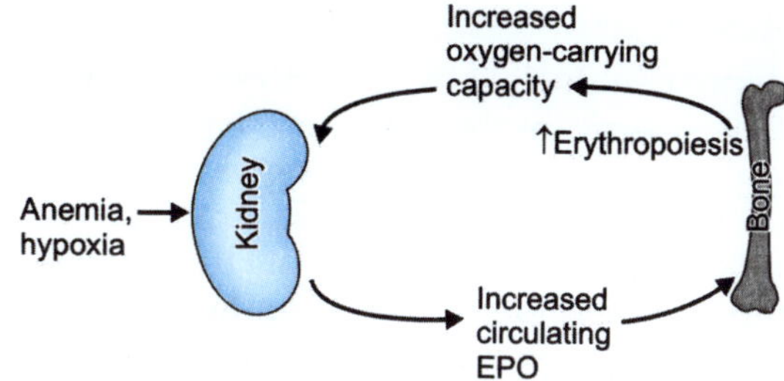

Fig.1.3: Oxygen carrying capacity and EPO feedback

ERYTHROPOIETIN (EPO)

The history of erythropoietin begins in 1878 when Bert noticed higher hemoglobin concentrations in animals living at an altitude of 4,000 m than low landers.[3] Bert considered it to be an inherited effect. This concept was disproved by the French anatomist François-Gilbert Viault who had traveled to the highland of Peru (about 4500 m above sea level). After 23 days of his stay at high altitude, he found his red cell count to increase from 5 million/dl to 8 million/dl. His companions, which included a doctor, few miners, a dog and a rooster, also showed the same reaction.[4] Viault concluded that increased red cell production was a result of an acute physiologic response to high altitude rather than an inherited condition.

The idea of hormonal regulation of erythropoiesis was first formulated by Carnot, and Deflandre in 1906.[5] These investigators had subjected rabbits to a bloodletting (30 ml), and took another blood sample on the next day and injected the serum in normal rabbits. The concentration of red blood cells in the recipients increased by 20–40% within 1–2 days. Carnot and Deflandre concluded that the serum contained a hematopoietic factor, which they called hemopoietin. The more specific name 'erythropoietin' was introduced by Bonsdorff and Jalavisto in 1948.[6]

Leon Jacobson and his colleagues reported in 1957 that nephrectomised rats subjected to hypoxic stress fail to show the expected increase in the plasma EPO level.[7] Subsequently, low plasma EPO activity in anemic patients with chronic kidney disease was also repor-

ted.[8,9] Human EPO was first purified in 1977 from 2500 liters of urine from patients suffering from aplastic anemia.[10]

Normally, low plasma concentrations of EPO are sufficient to replace the aged red cells destroyed by the macrophages. However, plasma EPO levels increase dramatically in hypoxia or after an acute blood loss. The glycoprotein hormone erythropoietin regulates the proliferation and differentiation of erythroid progenitor cells, and therefore controls the homeostasis of blood oxygen carrying capacity (Fig. 1.3). The normal range of EPO levels in human plasma is approximately 5–25 mU/ml, but EPO levels can be increased 100- to 1000-fold in response to hypoxia or blood loss. In healthy individuals and patients with various types of anemia (e.g. caused by blood loss, hemolysis, iron deficiency, aplastic bone marrow or nutritional deficiencies), EPO levels are inversely correlated with hematocrit and hemoglobin levels and reflect the reciprocal relationship between oxygen supply to the tissues and EPO production rate.

EPO levels are disproportionately low in anemic patients with chronic disorders such as rheumatoid arthritis, AIDS and cancer. The mechanism of anemia in these disorders is complex, but inhibition of EPO production and erythroid progenitor proliferation by inflammatory cytokines, such as IL-1 and TNF, is thought to be a major cause.

EPO levels with patients suffering from polycythemia vera are either low or within the normal range. In contrast, secondary polycythemia is caused by EPO overproduction, which results in increased proliferation and maturation of normal erythroid progenitor cells. EPO overproduction is a normal physiological response to decreases in arterial oxygen saturation caused by respiratory or cardiac disorders or low oxygen tension at high altitudes.

Site of Production

Recent evidence proves that EPO is produced by peritubular interstitial fibroblasts (Fig. 1.4), and not by renal tubular epithelial cells or

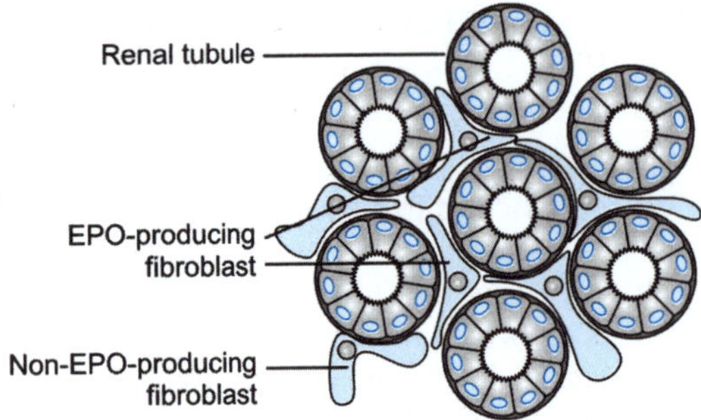

Fig. 1.4: EPO-producing fibroblasts in renal cortex

peritubular endothelial cells as reported earlier. Renal EPO-producing cells (REPC) can be typically found in the renal cortex (predominantly juxtamedullary region) and outer medulla.[11] However, not all peritubular fibroblast cells are capable of EPO production. It is reported that <20% of renal fibroblasts produce EPO.[12] While the kidney is the primary physiologic source of EPO synthesis in adults, the liver is the main site of EPO production during embryonic development. However, in adults, the liver retains its ability to produce EPO in response to moderate/severe hypoxia.[13]

Regulation of EPO Secretion

The primary function of EPO is to maintain the blood hemoglobin concentration in the normal range during steady-state conditions and to hasten red cell production after hemorrhage. The concentration of circulating EPO increases exponentially with decreasing hemoglobin levels in uncomplicated anemia (absence of renal disease or inflammation). However, from the early days of EPO research, it has been clear that the controlled variable is not the concentration of erythrocytes or of hemoglobin in blood. Instead, *feedback regulation is based on the tissue O_2 pressure (tissue pO_2)*, which depends on the hemoglobin concentration, the arterial pO_2, the O_2 affinity of the hemoglobin and the rate of blood flow. The kidney is very appropriate for controlling EPO production because the pO_2 in the renal cortex is little affected by the rate of blood flow.[14] EPO production is regulated by hypoxia-inducible factors discussed below.

Hypoxia-inducible Factors

Hypoxia-inducible factors (HIFs) are heterodimeric proteins that regulate the physiologic response to hypoxia by increased EPO production as well as improved uptake and utilization of iron. Discovery of HIFs filled the gap in our understanding on how anemia and/or hypoxia may lead to augmentation of erythropoiesis.[15, 16]

HIFs stimulate production of the EPO in the kidney. HIFs also stimulate intestinal iron uptake by repressing production of hepcidin in the liver.[17] HIFs also activate hepatic synthesis of transferrin, the major plasma protein responsible for transporting iron from the intestine to the bone marrow. HIFs are also involved in the utilization of iron from senescent red blood cells via macrophages.

This way, HIFs increase supply of iron from both absorption from duodenum and from the recycling of senescent red blood cells.[18,19] Thus, HIFs directly regulate the expression of EPO, hepcidin and transferrin, involving 5 different organs (kidney, liver, intestine, blood, and bone marrow) to control erythropoiesis (Fig. 1.5).[20]

HIFs and Hepcidin

Hepcidin, a small peptide synthesized in the liver, is a key regulator of iron absorption and homeostasis. Hepcidin is secreted by the hepatocytes in response to iron loading. It acts by binding to transport protein ferroportin and promotes its degradation. Thus, hepcidin, released from the liver when body iron stores are full, blocks further intestinal absorption of iron. The iron release from iron stores (hepatocytes and macrophages) is also blocked by hepcidin, decreasing iron utilization in bone marrow. Hepcidin expression is decreased by iron deficiency and, hypoxia.[19] Role of hepcidin in iron metabolism is discussed in greater detail later in this chapter.

Possible Role of Angiotensin II in Erythropoiesis

The signal to produce more erythrocytes following hemorrhage, besides production of

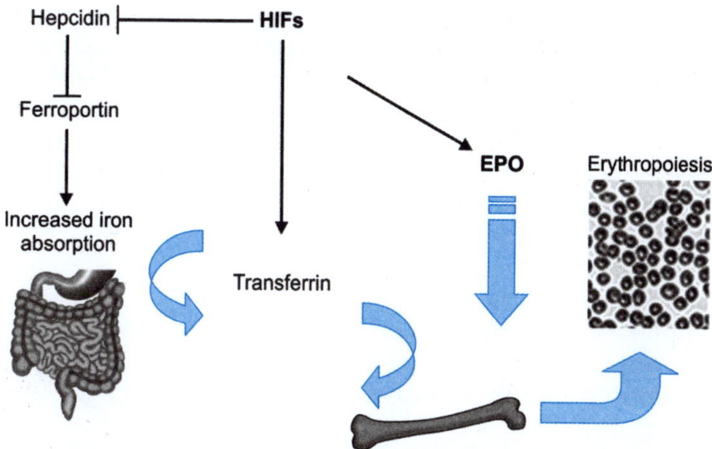

Fig. 1.5: Role of HIFs in erythropoiesis

HIFs, is apparently linked to the renin-angiotensin system. Angiotensin II is believed to stimulate erythropoiesis by two means. First, angiotensin II increases EPO production. Second, angiotensin II is a growth factor for the myeloid erythrocytic progenitors.[21] A feedback regulation of red cell mass and blood volume by means of EPO and angiotensin II seems to exist. Treatment of healthy men with recombinant human EPO produces an increase in red cell mass. However, the increase in hematocrit is accompanied by a decrease in plasma volume, which is probably due to a downregulation of the renin-angiotensin-aldosterone system and results in a constancy of the blood volume. Thus, EPO treatment in healthy humans raises hemoglobin by two mechanisms—an increase in red cell mass, and a decrease in plasma volume.[22]

IRON METABOLISM

Iron metabolism is an integral component of erythropoiesis. Since, many facts about iron metabolism are available in undergraduate textbooks, only recently discovered aspects of its metabolism would be highlighted.

Intestinal Absorption

The predominant forms of iron in the human diet are:

i. heme iron (ferrous iron), present in meat, fish, liver, etc.

ii. Inorganic iron (ferric iron), present as iron salts in eggs and vegetarian food such as green leafy vegetables, beans, cereals, etc.

The intestine can absorb 10–20% of dietary heme iron as compared to 0–10% absorption of non-heme or inorganic iron. The acid environment of the stomach and exposure to digestive enzymes cause a partial release of these iron forms from the food. Heme and non-heme iron appear to be absorbed by separate mechanisms (Fig. 1.6).

Despite the importance of heme iron as a dietary source of iron, a little is known about its transport and metabolism in the enterocyte.

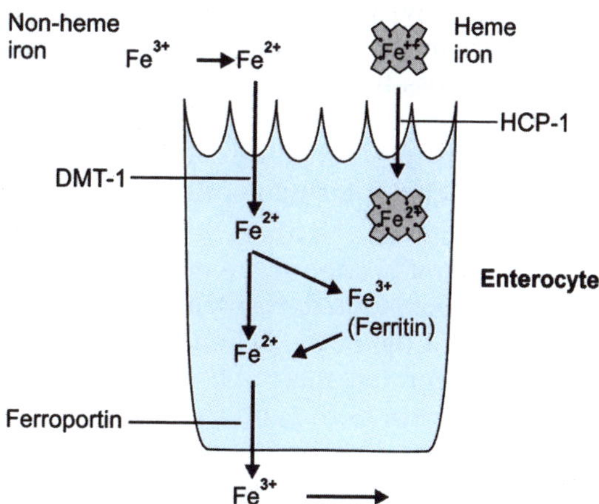

Fig. 1.6: Mechanism of iron absorption in the small intestinal enterocyte. DMT-1: Divalent metal transporter 1; HCP-1: Heme carrier protein 1

A carrier protein, called heme carrier protein-1 (HCP-1), seems to be involved in transfer of heme (Fe^{2+}) from the lumen of the intestine into the enterocyte. In contrast, the transport of inorganic iron has been studied in detail for several decades. Before, non-heme iron can be absorbed, it needs to be converted from ferric (Fe^{3+}) to ferrous (Fe^{2+}) state by a brush border enzyme, ferric reductase present in the upper small intestine. In the enterocyte, the most important apical uptake transporter of inorganic iron is the divalent metal transporter 1 (DMT-1) which imports ferrous iron as well as several other divalent metals but not trivalent (ferric) iron. DMT-1 is expressed on the brush-border membrane of duodenal enterocytes. In the cytosol of the intestinal epithelial cell, Fe^{2+} can take one of the two possible routes (Fig. 1.6):

1. It may cross the basolateral border of the epithelial cell into the bloodstream. The efflux of iron across the basolateral membrane and into the circulation is mediated by the iron transport protein ferroportin. In addition to ferroportin, the basolateral efflux of iron from enterocytes requires another transport protein called hephaestin. Although the exact role of hephaestin has not been defined, its ability to oxidize ferrous to ferric iron is predicted to be important for its function.

2. Some cytosolic Fe^{2+} is converted to Fe^{3+} and combines with a storage protein called apoferritin to constitute ferritin. Ferritin cannot cross the basolateral border of the enterocytes. Within a few days, this form of iron is lost from the body along with the desquamated intestinal epithelial cells.

The fraction of absorbed iron that crosses the basolateral border of the enterocytes and that becomes ferritin varies with the status of iron stores of the body. When the body iron stores are saturated, most of the absorbed iron becomes ferritin in the enterocytes and does not enter the body. In iron deficiency states, a larger percentage of absorbed iron enters blood circulation and only small amount becomes ferritin and lost from the body. This regulatory control over intestinal absorption

(*mucosal block hypothesis*) is critically important for prevention of body iron overload, since once iron enters the body, it cannot be excreted (Fig. 1.7). That is why in iron deficiency states, parenteral administration of iron should be undertaken with outmost care and only when absolutely necessary.

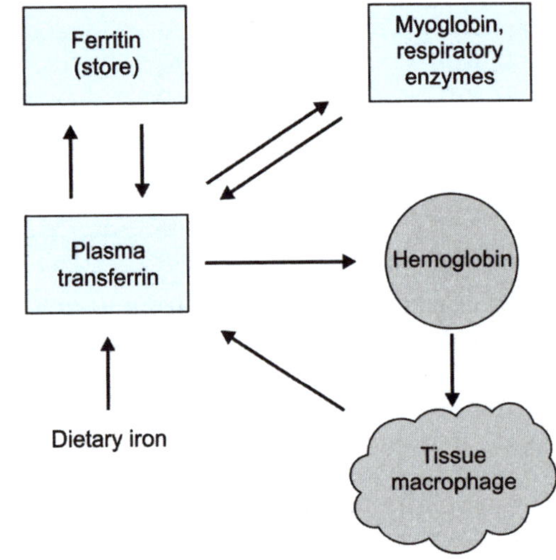

Fig. 1.7: Iron metabolism in the body

The mucosal block theory was initially proposed by Hahn et al.[23] However, it failed to explain the mechanism by which the status of iron stores in the body is communicated to the enterocytes. Recently advances in the knowledge about iron metabolism include the handling of iron in the duodenal enterocytes and macrophages, and regulatory role of hepcidin in export of iron from the enterocytes as well as the macrophages.

Iron Transport in Plasma

Iron is transported in plasma in combination with a transport protein called transferrin. Iron is distributed to tissues through blood plasma which contains only 2–4 mg iron, bound to transferrin. Of all cells, erythrocytes have the highest concentration of iron. Blood contains about 2–2.5 g iron, i.e. about 50% of total body iron. Most plasma iron is derived from aged erythrocytes that are recycled by macrophages in the spleen and other organs.

About 15–25 mg of all erythrocyte iron is recycled everyday, although smaller amounts of iron from other cell types are also recovered by macrophages.[24]

Iron Stores

Iron is stored mainly in the parenchymal cells and macrophages of liver, and in macrophages of bone marrow and spleen. In these tissues, iron, in ferric state, combines with a storage protein called apoferritin to form ferritin. From the iron stores, iron is released into plasma and utilized for formation of hemoglobin, myoglobin, respiratory enzymes, etc. (Fig. 1.7).

Excessive ferritin accumulation within iron storage cells results in formation of large aggregates of ferritin called hemosiderin. Iron in both ferritin and hemosiderin states can be released from the storage tissues, although it is less readily available from hemosiderin. Under light microscope, hemosiderin can be seen in the cytoplasm as yellow to brown granules. Ferritin can be identified only under electron microscope.

Iron Import and Export in Tissue Macrophages

An analogous sequence of events takes place in macrophage lysosomes when they phagocytose senescent erythrocytes. The hydrolytic environment of the phagolysosome digests the erythrocyte and its hemoglobin, releasing heme, which is then degraded by the enzyme, heme oxygenase 1 (HO-1), freeing iron for cytoplasmic storage or export to blood plasma. Normally most iron is exported to plasma across the macrophage cell membrane.

Iron Export

The sole known mammalian iron exporter is ferroportin. It is expressed at all sites involved in iron transfer to plasma, i.e. the basolateral membranes of duodenal enterocytes, the membranes of macrophages, and the sinusoidal surfaces of hepatocytes.[25] Cellular iron export is dependent on a copper-containing ferroxidase, hephaestin[26] that uses molecular oxygen to oxidize ferrous to ferric iron. Hephaestin is a transmembrane protein expressed predominantly in enterocyte. Hephaestin facilitates ferroportin-mediated Fe^{2+} efflux by oxidizing iron to its ferric form Fe^{3+}, allowing its uptake by transferrin.

HORMONAL CONTROL OF IRON HOMEOSTASIS BY HEPCIDIN

Hepcidin

Iron absorption is increased during periods of iron deficiency, whereas iron absorption is decreased by iron overload. These observations have led to the expectation that one or more systemically acting hormones regulate the major flows of iron and are in turn regulated by iron. The iron-regulatory hormone hepcidin is a 25-amino acid peptide. Hepcidin, synthesized and secreted by hepatocytes, circulates in blood plasma mostly free except for weak binding to albumin and α_2-macroglobulin.[27]

Mechanism of Action of Hepcidin

Hepcidin acts by controlling the membrane concentration of ferroportin, the sole known cellular iron exporter.[27] Hepcidin is located on the basolateral surface of gut enterocytes and the plasma membrane of monocyte macrophages. Hepcidin binds directly to ferroportin, inducing the endocytosis of ferroportin and its proteolysis in lysosomes (Fig. 1.8). Inhibition of ferroportin prevents iron from being exported and the iron is sequestered in the cells. Injection of synthetic hepcidin into mice elicits a profound decrease in serum iron concentrations within 1 hour, and the hypoferremia persists for many hours (Fig. 1.9).[28, 29] Hepcidin deficiency in mice or humans causes hyperabsorption of iron and iron overload in many organs including the liver, pancreas, and the heart. These effects of hepcidin excess or deficiency are evidence of the fundamental role of hepcidin in the control of iron absorption and the release of recycled iron from macrophages.

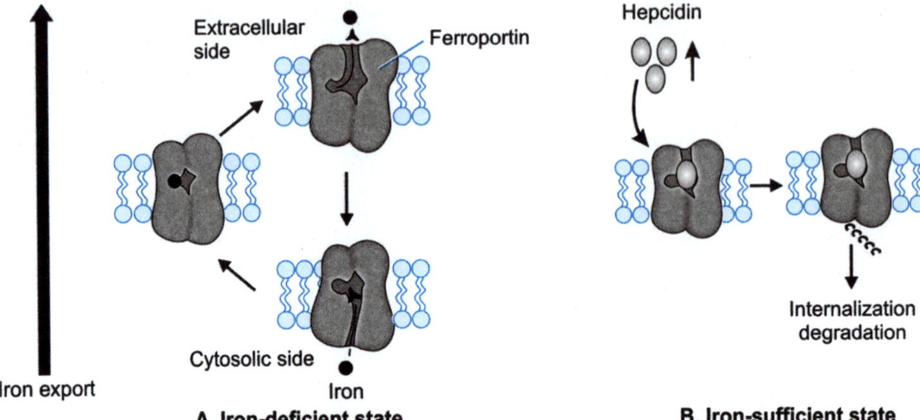

Fig. 1.8: Action of ferroportin in iron export (A), and action of hepcidin on ferroportin internalization and degradation (B)

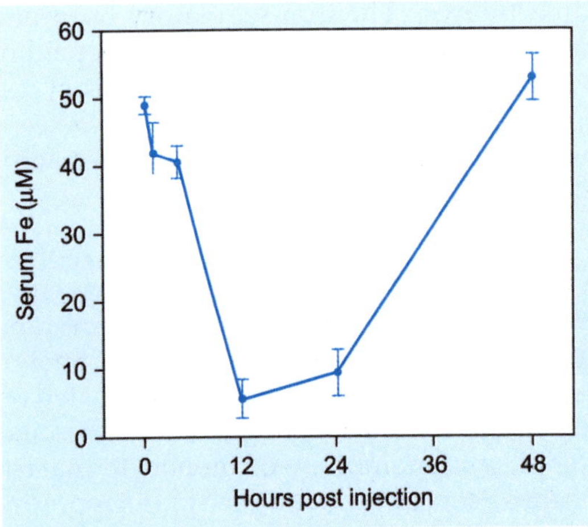

Fig. 1.9: Serum iron levels after injection of synthetic hepcidin in mice

hepcidin levels were observed to rise dramatically in response to transient increases in plasma iron with a delay of ~8 hours.[30] On the other hand, the relatively narrow distributions of estimated body iron stores in men and women on varied diets suggest that body stores could regulate hepcidin independently of the short-term effects of plasma iron concentrations. In humans, observations that support hepcidin regulation by iron stores include the correlation between hepcidin mRNA and iron stores in human liver biopsies and the strong correlation between serum ferritin, a recognized marker of iron stores, and serum hepcidin.[31] Hepcidin regulation by plasma iron and by tissue iron stores appears to operate on different time scales (hours *vs* days), and this could allow the two regulators to function in parallel.

Dual Regulation of Hepcidin by Extracellular Iron and Iron Stores

The relative stability of plasma iron concentrations despite rapid turnover of iron suggests feedback regulation of hepcidin by plasma iron. Experimentally, regulation of hepcidin by plasma iron concentrations was detected in human volunteers given small doses of iron sufficient to raise plasma iron concentrations transiently but too small to contribute significantly to iron stores. Serum

Regulation of Hepcidin by Erythropoiesis

Intestinal iron absorption is greatly increased in response to hemorrhage or erythropoietin leading to the hypothesis that an *erythroid regulator* modulates intestinal iron absorption, assuring adequate supply of iron when needed for accelerated erythropoiesis. Patients with ineffective erythropoiesis (e.g. in β-thalassemia), whose erythroid precursor populations are greatly expanded but fail to mature into functional erythrocytes, also have increased intestinal iron absorption, despite often severe systemic iron overload. Although blood transfusions given for severe anemia (e.g. in β-thalassemia major) contribute to the lethal iron overload in ineffective erythropoiesis, many patients with less severe anemia (exemplified by β-thalassemia intermedia) receive few or no transfusions but still become severely iron-overloaded.[31]

After the discovery of hepcidin, the erythroid regulator concept was modified from that of a direct regulator of iron absorption to a regulator of hepcidin. Hypoxia and erythropoietin were initially thought to regulate hepcidin directly, but the preponderance of data now supports a model in which the bone marrow produces a hepcidin suppressor, in response to erythropoietin.[32] A similar suppressive substance has been postulated in anemias with ineffective erythropoiesis where hepcidin is decreased despite iron overload even in the absence of transfusions. Identification of the physiological and pathological erythroid regulators of hepcidin is an important priority for future studies.[24]

In some chronic inflammatory disorders, the patients suffer from microcytic hypochromic anemia, even though the body is not deficient in iron (anemia of chronic diseases). This paradox was solved by the knowledge that hepcidin secretion by the liver is increased by chronic inflammations. Hence, even though body iron stores are full, iron is not exported from macrophages as well as tissue stores for hemoglobin synthesis in the bone marrow (Fig. 1.10).

GENETIC DISORDERS OF THE HEPCIDIN–FERROPORTIN SYSTEM

Mutations in the genes encoding hepcidin, its various regulators, or its molecular target ferroportin may manifest as disorders of iron metabolism. Hepcidin deficiency or ferroportin resistance to the endocytic effect of hepcidin results in hereditary hemochromatosis, a group of diseases characterized by systemic iron overload due to hyperabsorption of dietary iron, with subsequent injury to iron-overloaded tissues. Depending on the age of onset and the severity of the disease, the destructive process may affect the liver, causing cirrhosis and liver cancer; the heart, leading to heart failure; and endocrine glands, where the effects are wide-ranging, including delayed growth and sexual development in the juvenile forms of the disease and diabetes mellitus in the juvenile and adult forms. The age of onset and the rate of disease progression correlate roughly with the severity of the hepcidin deficiency.[24]

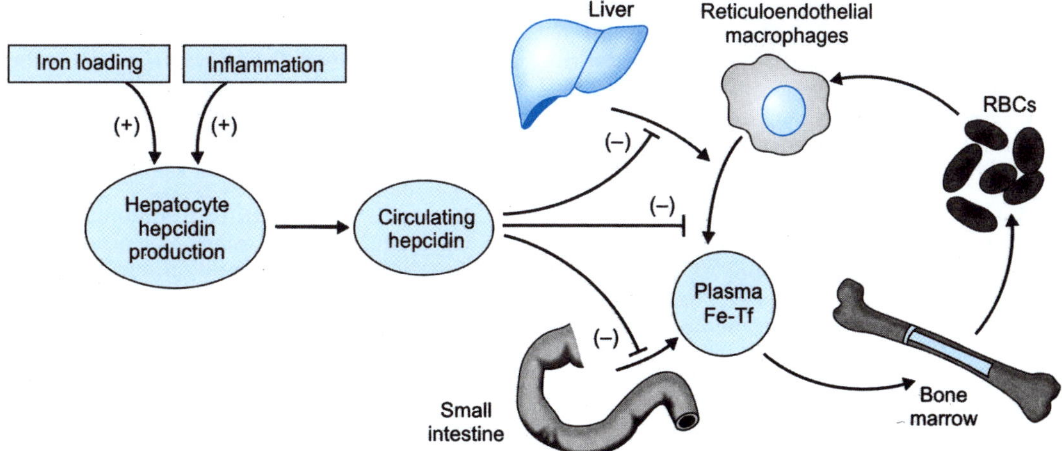

Fig. 1.10: Consequences of hepatic production of hepcidin in iron loading and chronic inflammation (Fe-Tf: Iron transferrin)

References

1. Fawcett DW, Jensh RP. In Bloom and Fawcett: Concise Histology (1st ed). Oxford: Arnold Publishers; 1997.
2. Wu H, Liu X, Jaenisch R, et al. Generation of committed erythroid BFU-E and CFU-E progenitors does not require erythropoietin or the erythropoietin receptor. Cell. 1995; 83: 59–67.
3. Bert P. La Pression Barometrique. Recherches de Physiol-gie Experimentale Paris: Libraire de l'Acade'mie de Medicine, 1878.
4. Viault F. Sur l'augmentation considerable du nombredesglobules rouges dans le sang chez les habitants des hautsplateaux de l'Amerique du Sud. C R Acad Sci Paris. 1890; 111: 917–18.
5. Carnot P, Deflandre C. Sur l'activitehemo-poietiqueduserum au cours de la regeneration du sang. C R Acad Sci Paris. 1906; 143: 384–86.
6. Bonsdorff E, Jalavisto E. A humoral mechanism in anoxic erythrocytosis. Acta Physiol Scand. 1948; 16: 150–70.
7. Jacobson LO, Goldwasser E, Fried W, et al. Role of the kidney in erythropoiesis. Nature.1957; 179: 633–44.
8. Gurney CW, Goldwasser E, Pan C. Studies on erythropoiesis. IV Erythropoietin in human plasma. J Lab Clin Med. 1957; 50: 534–40.
9. Gallagher NI, McCarthy JM, Hart KT, et al. Evaluation of plasma erythropoietic-stimulating factors in uremic patients. Blood. 1959; 14: 662–67.
10. Miyake T, Kung KH, Goldwasser, E. Purification of human erythropoietin. J Biol Chem. 1977; 252: 5558–556.
11. Bachmann S, Le Hir M, Eckardt KU.Co-localization of erythropoietin mRNA and ecto-52-nucleotidase immunoreactivity in peritubular cells of rat renal cortex indicates that fibroblasts produce erythropoietin. *J Histochem Cytochem.* 1993; 41: 335–41.
12. Maxwell PH, Osmond MK, Pugh CW, et al. Identification of the renal erythropoietin-producing cells using transgenic mice. Kidney Int. 1993; 44: 1149–162.
13. Fried W. The liver as a source of extrarenal erythropoietin production. Blood. 1972; 40: 671–77.
14. Erslev AJ, Caro J, Besarab A. Why the kidney? Nephron. 1985; 41: 213–16.
15. Solak Y, Cetiner M, Siriopol, D, et al. Novel Masters of Erythropoiesis: Hypoxia-inducible factors and recent advances in anemia of renal disease. Blood Purif. 2016; 42: 160–67.
16. Hasse VH. Regulation of erythropoiesis by hypoxia-inducible factors. Blood Reviews. 2013; 27: 41–53.
17. Andrews NC. Forging a field: the golden age of iron biology. Blood. 2008; 112: 219–30.
18. Peyssonnaux C, Zinkernagel A, Schuepbach, RA et al. Regulation of iron homeostasis by the hypoxia-inducible transcription factors (HIFs). J Clin Invest. 2007; 117: 1926–932.
19. Nicolas G, Chauvet C, Viatte L, et al. The gene encoding the iron regulatory peptide hepcidin is regulated by anemia, hypoxia, and inflammation. J Clin Invest. 2002; 110: 1037–1044.
20. Semenza GL. Involvement of oxygen-sensing pathways in physiologic and pathologic erythropoiesis. Blood. 2009; 114: 2015–19.
21. Dunn A, Lo V, Donnely S. The role of kidney in blood volume regulation: the kidney as a regulator of the haematocrit. Am J Med Sci. 2007; 334: 65–71.
22. Lundby C, Thompson JJ, Boushel R. Erythropoietin treatment elevates hemoglobin concentration by increasing red cell volume and depressing plasma volume. J Physiol. 2007; 578: 309–14.
23. Hahn PF, Bale WF, Ross JF, et al. Radioactive iron absorption by gastrointestinal tract: influence of anemia, anoxia, and antecedent feeding distribution in growing dogs. J Exp Med. 1943; 78: 169–88.
24. Ganz T. Systemic iron homeostasis. Physiol Rev. 2013; 93: 1721–41.
25. Canonne-Hergaux F, Donovan A, Delaby C, et al. Comparative studies of duodenal and macrophage ferroportin proteins. Am J Physiol Gastrointest Liver Physiol. 2006; 290: G156–63.
26. Vulpe CD, Kuo YM, Murphy TL, et al. Hephaestin, a ceruloplasmin homologue implicated in intestinal iron transport, is defective in the sla mouse. Nat Genet. 1999; 21: 195–99.
27. Nemeth E, Tuttle MS, Powelson J, et al. Hepcidin regulates cellular iron efflux by binding to ferroportin and inducing its internalization. Science. 2004; 306: 2090–93.
28. Rivera S, Nemeth E, Gabayan V, et al. Synthetic hepcidin causes rapid dose-dependent hypoferremia and is concentrated in ferroportin-containing organs. Blood. 2005; 106: 2196–99.
29. Ganz T. Hepcidin and iron regulation, 10 years later. Blood. 2011; 117: 4425–33.
30. Detivaud L, Nemeth E, Boudjema K, et al. Hepcidin levels in humans are correlated with hepatic iron stores, hemoglobin levels, and hepatic function. Blood. 2005; 106: 746–48.
31. Cazzola M, Finch C. Iron balance in thalassemia. Prog Clin Biol Res. 1989; 309: 93–100.
32. Pak M, Lopez MA, Gabayan V, et al. Suppression of hepcidin during anemia requires erythropoietic activity. Blood. 2006; 108: 3730–35.

Chapter 2

Hemostasis

INTRODUCTION

Hemostasis is the term used to describe the arrest of bleeding. Hemostasis represents an intricate, highly balanced interaction between blood vessels, platelets, plasma coagulation factors and fibrinolytic proteins in the formation and dissolution of blood clots. Under normal conditions, these balanced physiological processes maintain blood in a free-flowing state. Hemostasis may be categorized into primary (platelet plug formation), secondary (formation of a stabilized fibrin clot through the coagulation cascade), and tertiary (formation of plasmin for breakdown of fibrin via fibrinolysis) concurrent processes.

1. PRIMARY HEMOSTASIS

Primary hemostasis is essentially a function of platelets. In many medical textbooks and even in research articles, the terms *thrombocytes* and *platelets* are considered synonymous. Technically they are not so. Thrombocytes (clot cells) are cells found in non-mammalian vertebrates. Thrombocytes are mononucleated cells functionally similar to platelets. In contrast, platelets are non-nucleated cytoplasmic fragments of bone marrow megakaryocytes. Of course, the terms thrombopoiesis and thrombopoietin refer to platelets.

In clinical usage, thrombocytosis or thrombocytopenia refers to increased or decreased platelet count in the peripheral blood.

In 1882, Giulio Bizzozero was the first to describe the third morphological element of blood, totally unrelated to white and red cells. Earlier, platelets were variously considered as artefacts or precursors of red blood cells. Bizzozero not only accurately described its shape and size in circulating blood but also confirmed its shape change *in vitro*. "These elements underwent rapid morphological change, emitting long protrusions (Fig. 2.1) and within 2–3 minutes, forming granular aggregates as large as 80–100 µm. Most importantly, he described their role in thrombosis and hemostasis. "The thrombus material is constituted by few leukocytes plunged in large masses of platelets; this thrombotic material may be of a great value in stopping hemorrhages by closing discontinuities in a vessel wall. This is the first clear statement in the history of medicine of the physiological role of platelets in hemostasis. After the work of Bizzozero and the discovery of megakaryocytes by Wright, few advances in the knowledge of platelet biochemistry and pathophysiology occurred until the year 1960.[1]

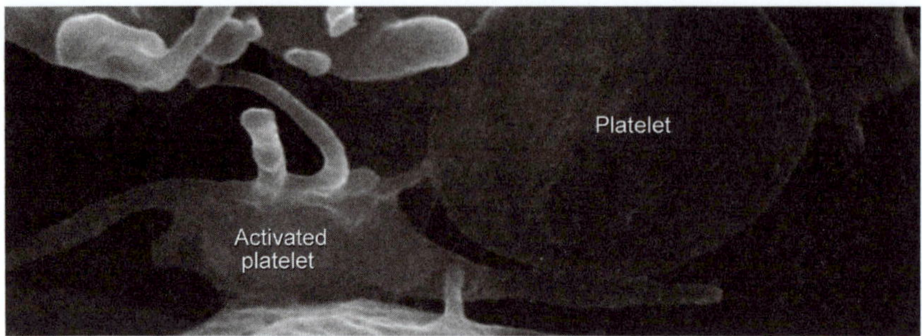

Fig. 2.1: The platelets

STRUCTURE OF PLATELETS

Circulating, resting platelets are biconvex. In a blood smear, platelets appear circular to irregular, lavender, and granular (Fig.2.2). They possess granular cytoplasm with no nucleus and their diameter, when seen in a Wright stained peripheral blood film, averages 2.5 μm.

Platelet Plasma Membrane

The platelet plasma membrane is a standard bilayer composed of proteins and lipids. Anchored within the membrane is a glycoprotein layer, which *is much thicker* (20–30 nm), than the analogous surface layer of leukocytes or erythrocytes. The plasma membrane invades the platelet interior, producing its unique surface-connected canalicular system (SCCS). The SCCS twists sponge-like throughout the platelet. It serves as the pathway for transport of substances into the cells and as conduits for the discharge of α-granule products secreted during the platelet release reaction.

Dense Tubular System

Parallel and closely aligned to the SCCS is the dense tubular system (DTS). The platelet dense tubular system, an internal smooth endoplasmic reticulum membrane system, occupies a pivotal position in the initiation and modulation of platelet activation. The best available evidence suggests platelet prostaglandins and thromboxane synthesis and an internal calcium store critical to platelet activation are both found in the platelet dense tubular system. Studies of the structural, physiologic and chemical pro-

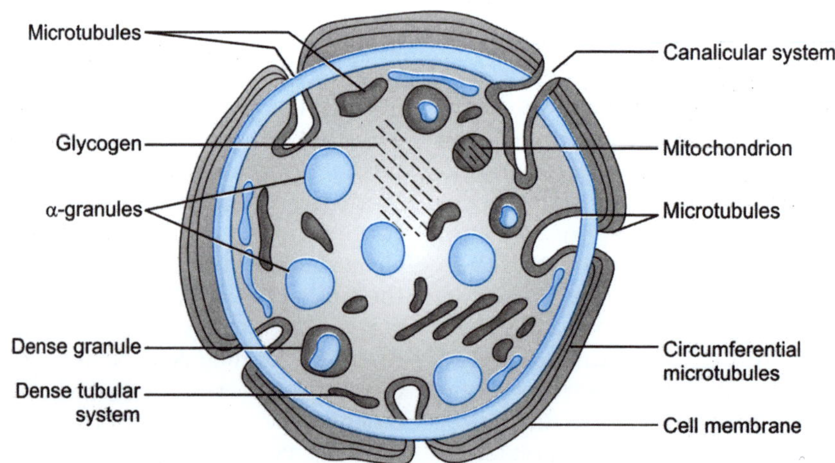

Fig. 2.2: Structure of a platelet

perties of thromboxane A_2 support the concept that this product of platelet prostaglandin synthesis acts to carry the calcium from the dense tubular system into the cytoplasm where the calcium is released to initiate contraction of the platelet contractile protein.

Microtubules

A circumferential bundle of microtubules maintains the platelet's discoid shape. The circumferential microtubules parallel the plane of the outer surface of the platelet and reside just within, although not touching, the plasma membrane. On cross section, microtubules are cylindrical, with a diameter of 25 nm. They also reassemble in long parallel bundles during platelet shape change to provide rigidity to pseudopods. In the narrow area between the microtubules and the membrane lies a thick meshwork of microfilaments composed of actin. Actin is contractile in platelets (as in muscle) and anchors the plasma membrane glycoproteins and proteoglycans.

Platelet Granules

There are 50 to 80 **α-granules** in each platelet. Unlike the nearly opaque dense granules, α-granules stain medium gray in osmium-dye transmission electron microscopy preparations. α-granules contain clotting mediators such as factor V, factor VIII, fibrinogen, fibronectin, platelet-derived growth factor, and chemotactic agents. As the platelet becomes activated, α-granule membranes fuse with the SCCS. Their contents flow to the nearby microenvironment, where they participate in platelet adhesion and aggregation and support plasma coagulation.

There are 2–7 **dense (δ) granules** per platelet which stain black (opaque) when treated with osmium in transmission electron microscopy. δ granules contain ADP, calcium, serotonin, which are platelet-activating mediators.

THROMBOPOIESIS

Platelets develop in the bone marrow from cells known as megakaryocytes. A mega-

karyocyte ("large nucleus cell") has a large lobulated nucleus.

During maturation, megakaryocyte progenitors undergo endomitosis, up to 128 times. Endomitosis is a process in which the cytoplasmic volume enlarges as the number of chromosomes multiplies without cellular division. The nucleus assumes a large lobulated appearance. Reaching the endothelium of the marrow sinusoids, the enlarged megakaryocyte forms cytoplasmic projections that protrude through endothelial pores into the lumen of the sinusoids. There, the projections presumably branch repeatedly forming formidable trees of proplatelets which segregate and release the platelets into the lumen of the sinusoids[3] (Fig. 2.3).

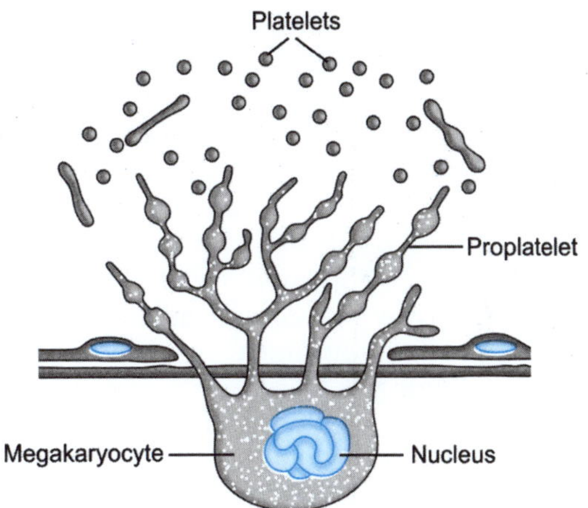

Fig. 2.3: Platelet formation by a megakaryocyte in the bone marrow (old theory)

In 1906, Wright introduced the initial concept given above. Almost a century later, studies on megakaryocytes producing platelets *in vitro* have revealed the details of platelet assembly and have led us back to the classical proplatelet theory of platelet release in which platelets fragment from the ends of megakaryocyte extensions. According to this theory, the entire megakaryocyte cytoplasm is converted into a mass of proplatelets, which are released from the cell. The nucleus is eventually extruded from the mass of proplatelets, and individual platelets are released from proplatelet ends[4] (Fig.2.4). In

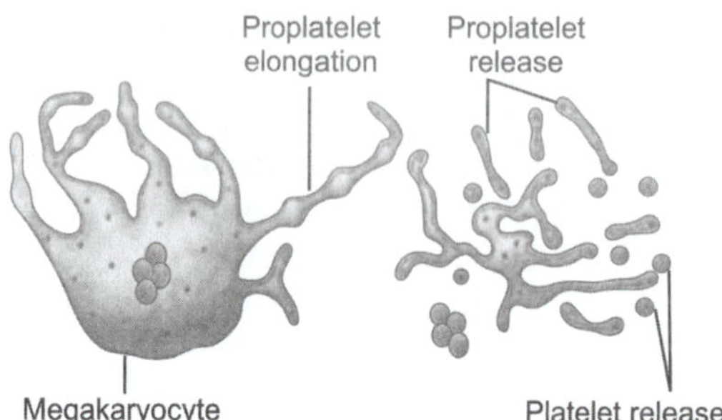

Fig. 2.4: Platelet formation in bone marrow (latest concept)

either scenario, each of these protoplatelet processes can give rise to 2000–5000 new platelets upon break-up.

Thrombopoietin

Thrombopoietin is a glycoprotein produced in the liver by both parenchymal cells and sinusoidal endothelial cells, as well as in the kidney by proximal convoluted tubule cells. Small amounts are also made by striated muscle and bone marrow stromal cells. In the liver, its production is augmented by interleukin 6 (IL-6). However, the liver and the bone marrow stromal cells are the primary sites of thrombopoietin production. Its negative feedback is different from that of most hormones in endocrinology. The effector regulates the hormone directly. Thrombopoietin is bound to the surface of platelets and destroyed, thereby reducing megakaryocyte exposure to the hormone. Therefore, the rising and dropping platelet concentrations regulate the thrombopoietin levels. Low platelets lead a higher degree of thrombopoietin exposure to the undifferentiated bone marrow cells, leading to differentiation into megakaryocytes and further maturation of these cells. On the other hand, high platelet concentrations lead to less availability of thrombopoietin to megakaryocytes.

ROLE OF PLATELETS IN PRIMARY HEMOSTASIS: PLATELET PLUG FORMATION

Primary hemostasis is brought about by three overlapping steps: Platelet adherence, platelet activation and platelet aggregation (Fig. 2.5).

Platelet Adherence

Normally, the endothelial cells express molecules that inhibit platelet adherence and activation while platelets circulate through the blood vessels. These molecules include nitric oxide, prostacyclin (PGI_2) and endothelial ADPase.

Endothelial cells are attached to the subendothelial collagen by an adhesive protein, von Willebrand factor (vWF) which these cells produce. vWF is also stored in the endothelial cells and secreted constitutively into the blood. When the endothelial layer is disrupted, damaged endothelial cells release vWF. Thus, subendothelial collagen and vWF anchor platelets to the subendothelium. vWF causes the platelets to change their shape showing adhesive filaments (extensions) that adhere to the subendothelial collagen on the endothelial wall.[6]

Platelet Activation

The intact endothelial lining inhibits platelet activation by producing nitric oxide,

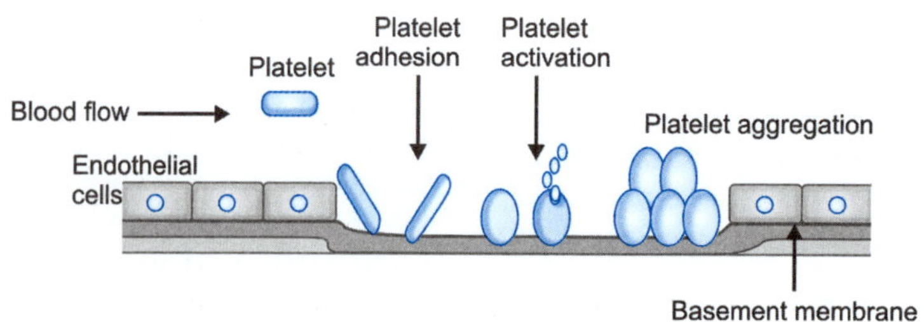

Fig. 2.5: Steps in formation of a platelet plug

endothelial ADPase, and PGI$_2$ (prostacyclin). Endothelial ADPase degrades the platelet activator ADP. After platelet adherence occurs, the subendothelial collagen binds to receptors on the platelet, which activates it. During platelet activation, the platelet releases a number of important cytokines and chemical mediators via degranulation. The released chemicals include ADP, vWF, thromboxane A$_2$, platelet-derived growth factor (PDGF), vascular endothelial growth factor (VEGF), serotonin, and coagulation factors. The extra ADP and vWF are especially important because they cause nearby platelets to adhere and activate, as well as release more ADP, vWF, and other chemicals. Platelet plug formation is considered a positive feedback process because ADP and vWF levels are successively increased as more and more platelets activate to form the plug. The other factors released during platelet activation perform other important functions. Thromboxane is an arachidonic acid derivative (similar to prostaglandins) that activates other platelets and maintains vasoconstriction. Serotonin is a short-lived inflammatory mediator with a vasoconstrictive effect that contributes to vascular changes associated with inflammation during an injury. The coagulation factors include factor V and VIII, which are involved in the coagulation cascade that converts fibrinogen into fibrin mesh after platelet plug formation.[6]

Mitochondrial hyperpolarization is a key event in initiating changes in morphology of platelets. Intraplatelet calcium concentration increases, stimulating the interplay between microtubule/actin filament complex.

Platelet Aggregation

The final step of platelet plug formation is aggregation of the platelets into a barrier-like plug. Receptors on the platelet bind to vWF and fibrinogen molecules, which hold the platelets together. Platelets may also bind to subendothelial vWF to anchor them to the damaged endothelium. The completed plug will cover the damaged components of the endothelium and will stop blood from flowing out of it, but if the wound is large enough, bleeding would not cease until the fibrin mesh from the coagulation cascade is produced, which strengthens the platelet plug. If the wound is minor, the platelet plug may be enough to stop the bleeding without the coagulation cascade.[7]

2. SECONDARY HEMOSTASIS (COAGULATION)

Although plasma proteins fibrinogen and prothrombin were discovered earlier, the first classic theory of coagulation was proposed by Paul Morawitz in 1905.[8] He convincingly described four "coagulation factors" in his scheme of coagulation (Fig. 2.6). According to him, in the presence of calcium and thromboplastin, prothrombin was believed to be converted to thrombin. In turn, the thrombin converted fibrinogen to fibrin, forming a fibrin clot. Later, fibrinogen, prothrombin, thromboplastin and calcium ions were labelled FI (clotting factor I), FII, FIII and FIV respectively, and thromboplastin was renamed tissue factor. Morawitz believed that all the ingredients of clotting were present in circulating blood and that the fact that such blood did not normally clot was due to the lack of a wettable surface in the blood vessels. This classic theory was accepted for the next forty years.

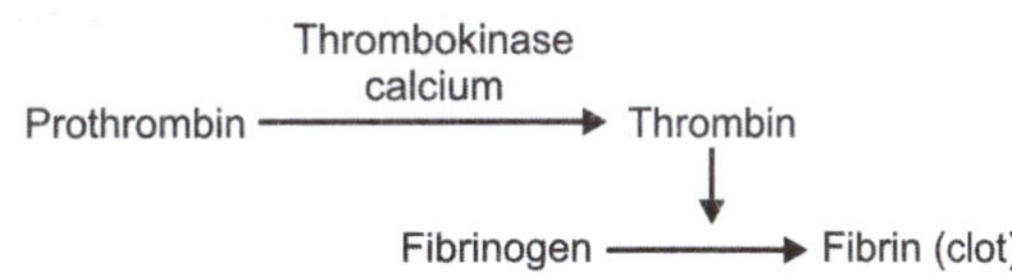

Fig. 2.6: Morawitz's theory of clotting of blood

Cascade Theory of Coagulation

In 1947, Owen[9] observed that the bleeding disorder in a young woman could not be explained by the 4-factor concept of coagulation proposed by Morawitz. He proposed that her plasma lacked "fifth" clotting factor (FV). By 1957, many additional coagulation factors were discovered: von Willebrand factor (vWF), FVII, FVIII, FIX, FX, and FXI.

The numeric system that was adopted assigned the number to the factor according to the sequence of discovery and not to the point of interaction in the cascade. However, what was not clear was how these multiple factors interacted to convert prothrombin to thrombin, resulting in the formation of a fibrin clot. In 1960s, 2 independent groups of biochemists Macfarlane[10] and Davie and Ratnoff.[11] Introduced a model of coagulation as a series of steps in which activation of one clotting factor led to the activation of another, culminating in a burst of thrombin generation. This model (Fig. 2.7) described each clotting factor as a proenzyme that could be converted to an active enzyme.

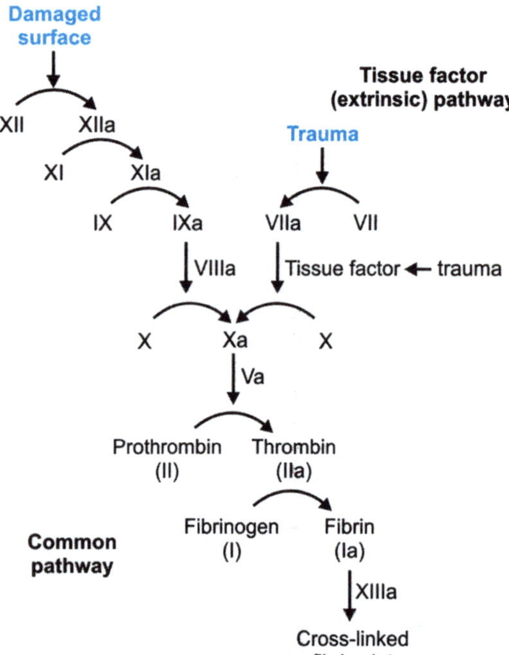

Fig. 2.7: Classic clotting cascade

The "cascade" or "waterfall" models suggested that the clotting sequences were divided into 2 pathways. Coagulation could be initiated via an "intrinsic pathway," so named because all the components were present in blood, or by an "extrinsic pathway," in which the subendothelial cell membrane protein, tissue factor (TF), was required in addition to circulating components. The initiation of either pathway resulted in activation of FX and the eventual generation of a fibrin clot through a common pathway.[12] The cascade or waterfall theory (Fig. 2.7) is now too well known to warrant detailed description.

These concepts represented a significant advance in the understanding of coagulation and served for many years as a useful model. However, by the end of 20th century, on the basis of many clinical and experimental observations, it began to be realized that the cascade/waterfall hypothesis does not fully and completely reflect the events of hemostasis *in vivo*.[13]

First, no explanation existed for the absence of a clinical bleeding tendency in deficiencies of FXII, prekallikrein, or high molecular weight kininogen, even though deficiencies in any one of these factors markedly prolong surface-activated coagulation assays for hemostasis *in vitro*. Second, no explanation existed for why FVIII or FIX deficiency caused clinically severe bleeding, even though the extrinsic pathway would be expected to bypass the need for FVIII and FIX.[14] These key observations led to the idea of revision of earlier models of coagulation. A vital observation was that a complex of FVIIa and TF activated not only FX but also FIX. More recent observations have led to the conclusion that activity of the FVIIa/TF complex is the major initiating event in hemostasis *in vivo*[15, 16], and the concept is known as cell-based theory of coagulation.

Cell-based Theory of Coagulation

The cell-based coagulation model of hemostasis enables a better understanding of the clinical problems seen in some coagulation disorders by emphasizing the central role of specific cell surfaces to control the hemostatic process. This model provides a potentially more accurate representation of the hemostatic process and facilitates the interpretation of coagulation tests and the pathophysiological mechanisms of clotting disorders such as hemophilia. Finally, the new cell-based coagulation theory can be considered a step forward in the understanding of major clinical events related to hemostasis. However, still, further investigations are being conducted in order to improve our understanding of the complex hemostatic mechanisms.[17]

The cascade theory was based on *in vitro* experiments in cell-free plasma. In the new theory, the cells, especially tissue factor-bearing fibroblasts or smooth muscle cells in the interstitial space and platelets occupy the central stage. *It may be reiterated the cell-based theory still utilizes almost all of the coagulation factors: FI to FXIII, except FXII. The difference is that this theory proposes that the clotting factors are not activated in the sequential manner nor in separate intrinsic and extrinsic streams as proposed by Macfarlane[10] and Davie and Ratnoff.[11]*

The concept of cell-based theory arose from the fact that except large molecular weight clotting factors FI and FVIII, many other clotting factors such as FII, FT (tissue factor III), FV, FVII, FIX and FX were found to exist in activated form in the extravascular interstitial fluid.[18] The cell-based theory consists of three overlapping stages (Fig. 2.8):

1. Initiation phase
2. Amplification phase
3. Propagation phase

1. Initiation Phase[19, 20]

The initiation phase is localized to the cells that express TF, which are found normally outside the vasculature. The FVIIa/TF complex activates small amounts of FIX and FX. FXa associated with its cofactor, FVa forms a prothrombinase complex on the surface of the TF-bearing cell. FV can be activated by FXa to produce the FVa required for prothrombinase assembly. Low-level activity of the TF pathway occurs at all times within the extravascular space. It is likely that FVII is bound to extravascular TF even in the absence of an injury and the extravascular FX and FIX can be activated as they pass through

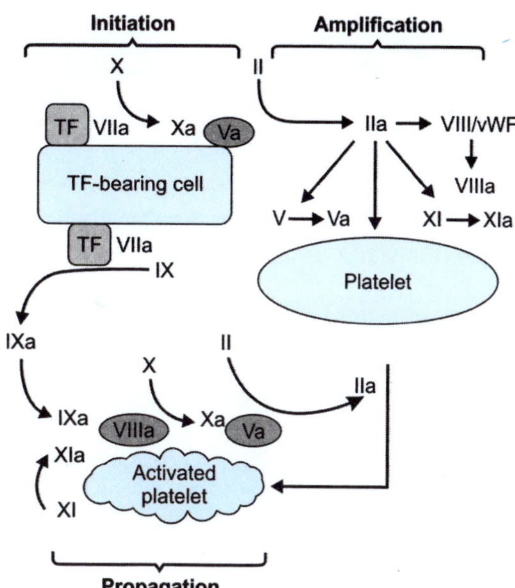

Fig. 2.8: The cell-based theory of coagulation

the tissues. The small amount of thrombin thus generated spontaneously is neutralized by antithrombin and tissue factor pathway inhibitor (TFPI) present in interstitial spaces. It has been observed that in the interstitial fluid, antithrombin and TFPI activity is present at a mean level of 40% of plasma activity.[11]

The coagulation process proceeds to the amplification phase only when damage to the vasculature allows platelets and FVIII (bound to vWF) to spill out into the extravascular tissues and to adhere to TF-bearing cells at the site of injury.

2. Amplification Phase[19, 20]

A small amount of thrombin generated on the TF-bearing cell has several important functions. A major function is the activation of platelets, exposing receptors and binding sites for activated clotting factors. As a result of this activation, the platelets release partially activated forms of FV onto their surfaces. Another function of the thrombin formed during the initiation phase is the activation of the cofactors FV and FVIII on the activated platelet surface. In this process, vWF is dissociated from the FVIII/vWF complex. The free vWF causes additional platelet adhesion and aggregation at the site of injury. In addition, small amounts of thrombin activate FXI and FXI on the platelet surface during this phase. By the end of the amplification phase, the stage is set for large-scale thrombin generation in the propagation phase.

Propagation Phase[19, 20]

As large numbers of platelets are recruited to the site of injury, the propagation phase of clot formation occurs on the surface of activated platelets. FIXa produced either from TF/FVIIa complex or from FXIa generated via thrombin action on FXI, in the presence of polyphosphate[21] will bind to FVIIIa on platelets in the presence of calcium ions to form tenase that is 50 times more efficient than TF/FVIIa complex in FX activation[22]), thus releasing large amounts of FXa. This FXa assembles with FVa and calcium to form prothrombinase on the activated platelet surface that will generate copious amounts of thrombin, which drive fibrin clot formation. Thus, TF/FVIIa appears to be the primary physiological trigger of *in vivo* coagulation.

Termination Phase[19, 20]

Once a fibrin platelet clot is formed over an area of injury, the clotting process must be limited to avoid thrombotic occlusion in surrounding normal areas of the vasculature. Unless controlled, clotting could propagate throughout the entire vascular tree. Three types of natural anticoagulants regulate clotting: Antithrombin III (ATIII) inhibits the activity of thrombin and other serine proteases, such as FIXa, FXa, FXIa.

Antithrombin III activity is markedly increased by binding with heparin. Protein C is a vitamin K-dependent plasma glycoprotein that, when activated by contact with thrombin, functions as an anticoagulant by inactivating FVa and FVIIIa. Protein C activity is enhanced by another vitamin K-dependent inhibitory cofactor, protein S. In addition, TF pathway inhibitor (TFPI), a protein secreted by endothelium, complexes to FXa and to TF/FVIIa, inactivating them to rapidly limit coagulation.

Even though the coagulation cascade model and common coagulation tests do not reflect the complexity of *in vivo* hemostasis, the coagulation tests are still valid. The coagulation tests have sensitivity to detect deficiencies of one or more coagulation factors and are therefore efficient to define changes in the coagulation factors of patients with a tendency to bleed. Though, the new cell-based hemostasis model incorporates the active participation of cellular structures to direct and control the process, the coagulation tests are still based on the cascade theory. According to Monroe and Hoffman[16] although the PT and aPTT do not reflect the role of inhibitors and do not necessarily identify the clinical risk of bleeding, they are not without value.

References

1. Gazzaniga V, Ottini L. The discovery of platelets and their function. Vesalius. 2001; VII: 22–26.
2. Fritsma GA. Platelet structure and function. Clin Lab Sci. 2015; 28: 125–35.
3. Dorn DC, Dorn A. Stem cell autotomy and niche interaction in different systems. World J Stem Cells. 2015; 7: 922–44.
4. Patel SR, Hartwig JH, Italiono JE (Jr). The biogenesis of platelets from megakaryocyte proplatelets. J Clin Invest. 2005; 115: 3348–54
5. Kaushansky K. Lineage-specific hematopoietic growth factors. N Engl J Med. 2006; 354: 2034–45.
6. Hemostasis | Boundless Anatomy and Physiology: Lumen Learninghttps://courses.lumen-learning.com/boundless-ap/chapter/hemostasis
7. Primary hemostasis | eClinpath. *www.eclinpath.com/hemostasis/physiology/primary-hemostasis/*
8. Morawitz P. The chemistry of blood coagulation [Die Chemie der Blutgerinnung] (R. Hartmann & P. F. Guenter, Trans.). Springfield, IL: Charles C Thomas. 1958.
9. Owren, PA. The coagulation of blood: Investigations on a new clotting factor. Acta Medica Scandinavica. 1947; 194: 521–49.
10. Macfarlane, R.G. An enzyme cascade in the blood clotting mechanism, and its function as a biological amplifier. Nature. 1964; 202: 498–99.
11. Davie EW, Ratnoff OD. Waterfall sequence for intrinsic blood clotting. Science. 1964; 145, 1310–12.

12. Luchtman-Jones L, Broze J. The current status of coagulation. Ann of Med. 1995; 27: 47–52.

13. Hoffman M, Monroe D. A cell-based model of hemostasis. Thrombois and Haemostasis. 2001; 85: 958–65.

14. Hoffbrand VA, Catovsky D, Tuddenham, E. (Eds). Postgraduate haematology (5th ed.). Malden MA: Blackwell. 2005.

15. Hoffman M, Monroe D. Rethinking the coagulation cascade. Current Hematology Reports. 2005: 4: 391–96.

16. Monroe DM, Hoffman M. The coagulation cascade in cirrhosis. Clin Liver Dis. 2009; 13: 1–9.

17. Ferreira CN, Sousa, LM, Carvalho MG. A cell-based model of coagulation and its implications. Rev. Bras. Hematol. Hemoter. 2010; 25 no. 25 Sao Paulo.

18. Le-Dzung T, Borg P, Thomas W, et al. Hemostatic factors in rabbit limb lymph: relationship to mechanisms regulating extravascular coagulation. Am J Physiol. 1998; 274 (Heart Circ. Physiol. 43): H769–776.

19. Dougald M, Monroe D, Hoffman M. What does it take to make the perfect clot? Arterioscler Thromb Vasc Biol 2006; 26: 41–48.

20. Riddel Jr JP, Aouizerat BE, Miaskowski C, et al. Theories of blood coagulation. J Pediatr Oncol Nurs. 2007; 24: 123–31.

21. Choi SH, Smith SA, Morrissey JH. Polyphosphate is a cofactor for the activation of factor XI by thrombin. Blood. 2011; 118: 6963–70.

22. Baugh RJ, Broze GJ, Jr, Krishnaswamy S. Regulation of extrinsic pathway factor Xa formation by tissue factor pathway inhibitor. J Biol Chem. 1998; 273: 4378–86.

Chapter 3

The Spleen

INTRODUCTION

Part of the mystery of the spleen is that its function is not clear from its anatomy, tucked behind the stomach high in the left upper quadrant of the abdomen. Spleen lies very close to the stomach and pancreas but, it is not part of the gastrointestinal tract, and has nothing to do with digestion. Antiquity considered it a gland, but it has no secretions that circulate through the body. Nicolas Massa, an Italian scientist, was the first to describe the morphology of spleen in 1536, as a soft and spongy structure. Unfortunately, the 16th century scientist, Andreas Vesalius, who described various structures of Anatomy in great detail, described spleen just in a few words: "An organ located behind the final left ribs." Vesalius excised the spleens of many animals without adverse effects, proving his contention that the spleen was not essential to life. The credit for the describing histological details of spleen goes to a 17th century researcher, Marcelo Malpighi, who described the morphology as well as histology of spleen, including its capsule, red pulp and white pulp (later named as malpighian corpuscles).[1]

As early as 1891, the role of spleen in the protection of the organism was demonstrated by Bardach who injected solution containing anthrax organisms into the abdomen of 25 splenectomized and 325 non-splenec-

tomized dogs. Of 25, 19 splenectomized dogs died, whereas only 5 of non-splenectomized dogs died.[1] However, the protective function of spleen came to be accepted by the medical fraternity as late as 1950s.

Barber surgeons of the 17th century removed spleens that had been eviscerated in battle, as they believed the organ was not necessary for life. This concept persisted into 20th century, since no bad effects were seen after removal of spleen for various reasons such as trauma, malignancy, etc. With the development of medical sciences of hematology and immunology in 1950, it came to be realized that spleen serves an important function in protection of the body against infections.[1]

In 1911, Kocher—an authority in the field of splenectomy—stated: "Injuries of the spleen demand excision of the gland. No evil effects follow its removal, while the danger of hemorrhage is effectually stopped." Therefore, from then onwards, everybody believed in splenectomy for every indication, whether, necessary or not[2], even though in 1919, researchers Dudley Morris and Frederick Bullock injected rats that had undergone splenectomy with plague bacillus. Animals that had lost their spleens had a mortality rate of 81%, much higher than the 39% in controls with a spleen. They warned surgeons that the spleen might be important in human immune defenses: "It is not improbable that the human body deprived of its spleen shows a similar increased susceptibility to infection".[3] However, their observation was completely ignored and the idea put forward by Kocher prevailed till 1950s.

In 1952, a piece of news alarmed everyone: King and Shumacker[4] reported fulminating sepsis among five infants below six months of age who were splenectomized for congenital hematological disorders. They emphasized the fact that the condition of "asplenia", resulting from surgical removal, causes a higher susceptibility to bacterial sepsis, from fulminating course to elevated mortality. This report led to re-thinking about the usefulness of human spleen in protection against infections.

STRUCTURE OF SPLEEN

Human spleen is enclosed in a capsule composed of dense connective tissue elastic fibers. Unlike in many animal species, capsule of human spleen contains very little smooth muscle fibers. This fact has been cited to question the ability of human spleen to act as a reservoir of red blood cells. Highly ramified system of trabeculae arising from the capsule penetrates and supports the pulp. However, in many mammals, the splenic capsule is rich in smooth muscle fibers which are innervated by sympathetic nerve fibers. In response to stimuli such as severe exercise or blood loss, the spleen contracts and pours significant amount of red cells into systemic circulation.

Splenic Parenchyma

The splenic parenchyma consists of white pulp and red pulp.

White Pulp

The white pulp is subdivided into the peri-arteriolar (or periarterial) lymphoid sheaths (PALS), the follicles, and the marginal zones. It is composed of lymphocytes, macrophages, dendritic cells, plasma cells, arterioles, and capillaries in a reticular framework (Fig. 3.1).

PALS

As the central arterioles enter the red pulp, they are surrounded by the PALS which are composed of lymphocytes and concentric layers of reticular fibers and flattened reticular cells. The PALS are divided into the inner PALS and the outer PALS. The inner PALS may stain slightly more intensely than the outer PALS due to its cellular composition of predominantly small lymphocytes. The

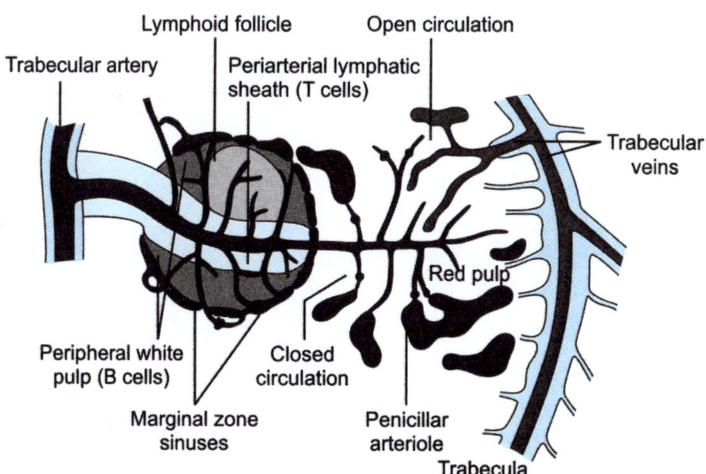

Fig. 3.1: The spleen

cells of the inner PALS are largely helper T cells, though smaller numbers of cytotoxic T cells may also be present, as well as inter-digitating dendritic cells, and migrating B cells. The outer PALS is populated by small and medium lymphocytes (both B and T cells), macrophages, and, upon antigenic stimulation, plasma cells. It is an important site of lymphocyte traffic where the formation of plasma cells occurs.[5]

Follicles

The follicles are continuous with the PALS and are typically found at bifurcation sites of the central arterioles. They are composed primarily of B cells with fewer follicular dendritic cells and helper T cell but typically do not contain cytotoxic T cells. The follicles have larger lymphocytes at the follicular center which is surrounded by a mantle zone or corona composed of small to medium lymphocytes. On antigenic stimulation, the follicles may contain germinal centers.[5]

Marginal Zone

The marginal zone is a unique region of the spleen situated at the interface of the red pulp with the PALS and follicles. Considered by many to be a separate compartment rather than part of the white pulp, it is designed to screen the systemic circulation for antigens and pathogens and plays an important role in antigen processing. A band of macrophages, the marginal zone metallophilic macro-phages, and the marginal sinus separate the marginal zone from the PALS and follicles.[5] The marginal zone blends into the red pulp. The marginal zone macrophages are impor-tant in clearance of microorganisms and viruses.[6]

Red Pulp

Three-fourths of the volume of human spleen consist of red pulp. Red pulp is comprised of four vascular structures in sequence: (i) Slender non-anastomosing arterial vessels (penicillar arterioles), (ii) splenic cords (cords of Billroth), the venous sinuses, and (iv) pulp vein. All these structures are supported by a reticular meshwork provided by fibroblasts (Fig. 3.1).

Cords of Billroth are found in the red pulp of the spleen between the sinusoids. The cords are comprised of fibrils and connective tissue cells with a large population of monocytes and macrophages. Erythrocytes pass through the cords of Billroth before entering the sinusoids. The passage into the sinusoids may be seen as a bottleneck, where erythrocytes need to be flexible in order to pass through. In disorders of erythrocyte shape and/or flexibility, such as hereditary spherocytosis, erythrocytes fail to pass through and get phagocytosed, causing extravascular hemo-lysis.

In human spleen, splenic sinuses (also known as splenic sinusoids) comprise approximately 1/3rd of the volume of red pulp. They consist of long anastomosing vascular channels which drain into pulp veins. These sinuses have unique endothelial cells which are arranged longitudinally, like narrow wooden planks of a barrel which run parallel to the long axis of the sinus. Tight junctions are present at regular intervals along thetheir lateral border (Fig. 3.2).[7]

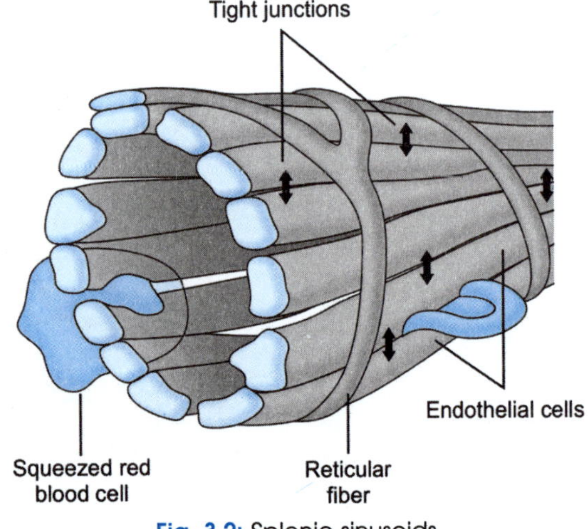

Fig. 3.2: Splenic sinusoids

SPLENIC CIRCULATION

Blood enters the spleen via splenic artery at the hilus. The branches of splenic artery penetrate deep into the organ before dividing into trabecular arteries. Splenic artery branches into trabecular arteries, which travel into the spleen within the fibrous tissue trabeculae. Trabecular arteries give off branches called *central arteries* which leave the trabecular sheath and enter the splenic parenchyma covered by periarterial lymphatic sheaths (PALS). Lymph nodules also surround the central arteries (white pulp). Each central arteriole gives rise to 15–30 radiating penicillar *arterioles*, which come out of lymphatic sheath and enter the red pulp. In the red pulp, some arterioles *open into pulp sinusoids* which drain into pulp veins (**closed circulation**). Some penicillar arterioles *open into splenic cords*. Blood percolates through the splenic cords into splenic sinusoids. Splenic sinusoids drain into pulp veins (**open circulation**). Small portions of the blood in the red pulp flows into the closed circulation. Most of the blood flows through the open circulation through the red pulp cords.[8] According to a recent report by Steiniger, et al[9], the human splenic red pulp harbors an entirely open circulatory system. Pulp veins drain into trabecular veins. Trabecular veins drain into splenic vein at the hilus.

FUNCTIONS OF SPLEEN

1. Blood Storage Function

In 1927, Barcroft and Stephens[10] noticed that in experimental dogs and cats, spleen contracts during exercise. Later studies confirmed that the spleen plays a role as reservoir of red blood cells in various animals such as dog, cat, sheep and horse. Spleen was found to contract in response to exercise, cold, hypoxia, bleeding, hypotension and circulating catecholamines. In severe exercise, splenic blood volume decreased to 54%, of the baseline value and hematocrit increased from 44 to 48%. The authors called it autotransfusion from the spleen. Later, many workers failed to replicate these results in humans. Hence, for many decades, the role of human spleen as a reservoir of RBCs was considered doubtful.

With the discovery of radioactive blood pool imaging technique, Sandler, et al[11] noted that after injection of radioactive red blood cells, splenic radioactivity decreased by 49% after exercise in 10 human subjects. Subsequently, Froelich et al[12], and Laub et al[13] also recorded similar results in human exercise. They concluded that human spleen plays a certain role in hemoconcentration recorded after exercise. However, other factors such as transfer of fluids from vascular to extravascular compartments also contribute to hemoconcentration observed after exercise.

The mechanism of decrease in splenic volume in humans is uncertain because smooth muscle fibers are very sparse in human splenic capsule. However, use of highly specific smooth muscle antibodies, Pinkus et al[14] found extensive network of smooth muscle fibers in the red pulp of human spleen.

The spleen is a *reservoir of platelets* also. It holds in reserve about one-third of all the body's platelets. These splenic platelets change places freely and rapidly with platelets in the general circulation. After leaving the bone marrow, the platelet lives for only nine days; it spends one-third of its life in the spleen. The spleen may hold platelets in reserve for times of danger; an injection of epinephrine quickly releases sequestered platelets into the blood. Thrombocytosis almost always follows the removal of the spleen, indicating the role of *spleen in the removal of aged platelets*. The high postoperative platelet count may last for months and even years.[15]

2. Culling and Pitting Function

The culling function of the spleen refers to its ability cull and destroy aged or abnormal red cells. Such unique ability is facilitated by the unique open circulation in spleen discussed above. The slow transit through the open circulation pathway gives sufficient opportunity to macrophages of the splenic cords to select and destroy defective red cells. The materials removed in normal persons include senescent red blood cells, acanthocytes and particles removed by pitting (see below). In disease states, spleen removes spherocytes, sickled cells, antibody coated red cells, white blood cells, and microorganisms.[16]

Pitting function of spleen refers to its ability to remove particles from intact red cells without destroying them. Pitting occurs as red cells within the cords attempt to reenter the circulation through the splenic sinusoids. To do this, they must pass through slit-like fenestrations of the sinus endothelium. As this occurs, the deformable portion of the cell bends to negotiate the opening, whereas the inclusion, which is nondeformable, is unable to pass through the narrow passage; thus, it is left behind to be phagocytized by resident macrophages. This function consists of pinching off particles-containing portions of red cells that have difficulty in squeezing through the pores between the splenic cords and sinusoids. The pinched cell seals itself; though the red cell may show the bite (Figs 3.2 and 3.3) Undesirable intracellular elements removed by the spleen include circulating particulate matter, Heinz bodies (denatured hemoglobin), Howell-Jolly bodies (nuclear remnants), and Pappenheimer bodies (iron granules).[16]

Evidence that supports the role of the spleen in culling and pitting is found in the fact that, after splenectomy, asplenic and hyposplenic patients lose their ability to clear damaged red cells and intra-erythrocyte

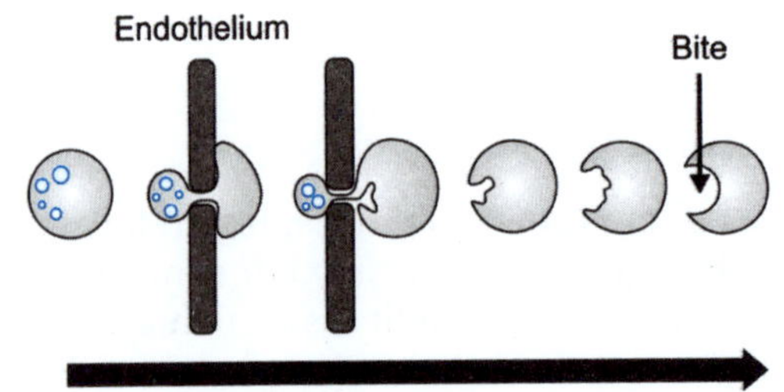

Fig. 3.3: Pitting function of spleen

inclusion bodies from the circulation. These patients display peripheral blood smears with an abnormal variety of erythrocytes, many with intracellular inclusions.

3. Immune Function

Because the spleen is composed of lymphocytic tissue, circulatory elements, and mononuclear phagocytic cell lines, it is ideally suited to play a coordinating role in the immune response. The organ's function in this respect includes its role in the nonspecific as well as the specific arms of the immune response, making it an important organ for immune homeostasis. The nonspecific immune functions of the spleen include the clearance of pathogens, the clearance of opsonized erythrocytes and platelets, and the production of complement. The ability of the spleen to clear blood-borne microorganisms and debris combined with its highly organized lymphoid compartment makes the spleen, the most important organ for antibacterial and antifungal immune reactivity.

The ability of the spleen to clear encapsulated bacteria is especially significant. Because these organisms have the ability to evade antibody and complement binding, their clearance depends on prolonged contact between pathogen and macrophage. It is well recognized that asplenic or hyposplenic patients are prone to a syndrome of fulminant septicemia, most often involving encapsulated bacteria.[17]

4. Iron Metabolism

Approximately half of total body iron in the adult human is present in hemoglobin in circulating red blood cells. Mature erythrocytes circulate until senescence or damaged and are then phagocytosed predominantly by the monocyte macrophages of the liver, bone marrow, and spleen. After red cell phagocytosis, hemoglobin is degraded, and the resultant heme moiety is cleaved to release free iron, which is then either stored within the macrophage or released into the circulation. This recycling of iron from senescent red cells supplies the bone marrow with ≈ 20 mg of iron per day for new red cell synthesis Liver and spleen macrophages are especially important for recycling of iron.

References

1. Petroianu A. Historical aspects of spleen and splenic surgeries in The Spleen Bentham e-book Pp. 3–19 (17). DOI: 10.2174/978160805273811-101010003 eISBN: 978-1-60805-273-8, 2011
2. Dianigi R, Boni L, Rausei S. History of splenectomy. Intern J Surg. Surgery. 2013; 11 Suppl 1: S42–43.
3. Morris DH, Bullock FD. The importance of the spleen in resistance to infection. Ann Surg. 1919; 70: 513–21.
4. King H, Shumacker HB (Jr). Splenic studies 1. Susceptibility to infection after splenectomy performed in infancy. Ann Surg. 1952; 136: 239–42.
5. Cesta F. Normal structure, function, and histology of the spleen. Toxicol Pathol. 2006; 34: 455–65.
6. Mebius RE, Kraal G. Structure and function of the spleen. Nat Rev Immunol. 2005; 5: 606–16.
7. Tewlbin F, Chamberlain JK, Weiss L. Microanatomy of mammalian spleen. In: The Complete spleen, Springer, 2014.
8. Kashimura M. Labyrinthine structure of arterial terminals in the human spleen, with special reference to "closed circulation". A scanning electron microscope study. Arch Histol Jpn. 1985; 48: 279–91.
9. Steiniger B, Bette M, Schwarzbach H. The open microcirculation in human spleens: A three-dimensional approach. J Histochem Cytochem. 2011; 59: 639–48.
10. Barcroft J, Stephens JG. Observations upon the size of the spleen. Am J Physiol. 1927; 64: 1–22.
11. Sandler MP, Kronenberg MW, Foremann FMB, et al. Dynamic fluctuations in blood and spleen radioactivity: Splenic contraction and relation to clinical radionuclide volume calculations. JACC. 1984; 3:1205–11.
12. Froelich JW, Strauss HW, Moore RH. Redistribution of visceral blood volume in upright exercise in healthy volunteers. J Nucl Med. 1988; 29: 1714–18.
13. Laub M, Hvid-Jacobsen K, Hovind P, et al. Spleen emptying and venous hematocrit in humans during exercise. J Appl Physiol. 1993; 74: 1024–26.
14. Pinkus GS, Warhol MJ, O'Connor EM, et al. Immunochemical localization of myosin in human spleen, lymph node and other lymphoid tissues. Am J Path. 1986; 123: 440–53.
15. Cuschieri A, Charles D. Disorders of the Spleen. Forbes. Boston, Blackwell Scientific, 1994.
16. Beck WS. Hematology, 5th Edi. MIT press, London, 1991
17. Weledji EP. Benefits and risks of splenectomy. Intern J Surgery. 2014; 12: 113–19.

Chapter 4

Regulation of Coronary Blood Flow

INTRODUCTION

Cardiovascular diseases, especially coronary heart disease (CHD), are epidemic in India. The Registrar General of India reported that CHD led to 26% of adult deaths in 2001–2003, which increased to 32% of adult deaths in 2010–2013.[1]

In view of the high prevalence of coronary artery disease all over the world, knowledge of regulation of coronary blood flow has attracted attention of research workers in cardiovascular physiology. Typically, angina pectoris is the first symptom of coronary heart disease. Interestingly, the first description of angina pectoris is found in the 1768 book, *"Some disorders of the breast"* by Heberden: "Those who are afflicted with it are seized, while they are walking, and more particularly when they walk soon after eating, with a painful and most disagreeable sensation in the breast, which seems as if it would take their life away, if it were to increase or to continue: The moment they stand still, all this uneasiness vanishes." Though the clinical description of angina pectoris was accurate, the author failed to correlate the symptoms with the heart disease, and presented it as a disorder of the breast.[2]

Regulation of coronary blood flow is mediated through multiple mechanisms including extravascular compressive forces, autoregulation of coronary blood flow, local metabolites, neural and hormonal influences and endothelial cells. Together, these mechanisms govern coronary flow and act to ensure an overall balance between myocardial oxygen delivery (supply) and metabolism (demand) in normal healthy individuals.

1. ROLE OF EXTRAVASCULAR COMPRESSIVE FORCES

i. **Phasic blood flow:** According to Ohm's law, flow = Δpressure/resistance), i.e. blood flow is directly proportional to the magnitude of the pressure gradient (arterial–venous) across the vascular tree and inversely proportional to the overall resistance of the vascular network. Vascular pressure itself is the difference between transmural expansive pressures forcing outward against the vascular wall and compressive tissue pressures counteracting that expansion. In the majority of organs, the compressive tissue force is constant and considered to make only a minimal yet consistent contribution to the pressure gradient. However, the periodicity of contractions in the heart and tremendous compressive forces of the myocardium result in a more complex pattern of flow in the coronary vascular system.[3.] In the coronary artery, the compressive forces of systole counteract the driving force for flow in the coronary circulation and, therefore, the majority of anterograde blood flow to the left ventricle occurs during diastole. Indeed, these myocardial compressive forces are so great during systole that the coronary circulation can undergo transient periods of retrograde blood flow (Fig. 4.1). This effect was first documented in the early 1930s.[4] The right ventricle generates much lower pressure during systole than left ventricle. Therefore, the right ventricular coronary blood flow is not significantly diminished during vascular contraction but rather shows a pattern of flow that is elevated during periods of high aortic pressure (i.e. systole) and diminished during periods of lower aortic pressure (i.e., diastole).[5] Taken together, these factors dictate that the left ventricle receives ~80% of its blood flow during the diastolic phase of the cardiac cycle.[6]

However, the phasic nature of left ventricular blood flow does not result in lower absolute flow values when compared to the right side of the heart. In a typical adult heart, baseline/resting left coronary blood flow typically ranges between ~0.5 and 1.0 ml/g/min. In contrast, right ventricular blood flow averages only ~0.3 to 0.6 ml/g/min. This phenomenon is explained by underlying differences in the rate of oxygen consumption between the left ventricle (~50–100 µl O_2/min/g) versus right ventricle (~30–50 µl O_2/min/g).[1]

ii. **Subendocardial *vs* subepicardial blood flow:** The nature of the physiologic interaction between myocardial contraction and coronary flow is further complicated. In a study, Rushmer and Thal[7], observed a disparity in compressive forces across the wall of the left ventricle. They reported that the degree of vascular compression is far greater in subendocardial layers, than in subepicardial region of left ventricular myocardium (Fig. 4.2). This disparity results in that subendocardial tissue receives a little if any flow during systole as the compressive forces are of sufficient magnitude to exceed the driving force of arterial pressure. Thus, the subendocardial myocardium receives flow practically only during diastole. By contrast, the lack of compressive forces at the subepicardial layers allow for perfusion throughout the entire cardiac cycle. This fact has an

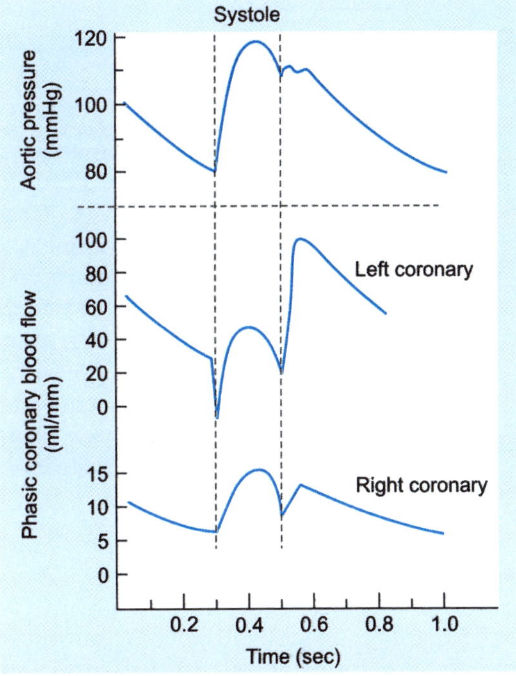

Fig. 4.1: Phasic coronary blood flow

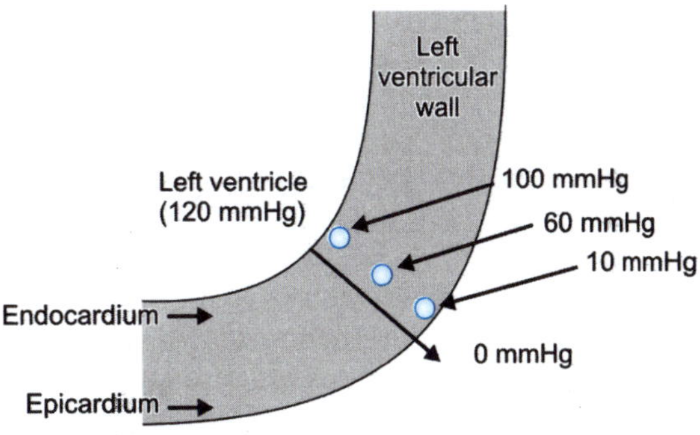

Fig. 4.2: Gradient of compressive forces across left ventricular wall

important clinical application. During myocardial ischemia, subendocardial region of the left ventricle is susceptible to maximum ischemic damage. According to Buja[8], ischemia first affects the *subendocardial* region, and tissue begins to die within 15–30 minutes of loss of blood supply. If the ischemia persists, myocardial infarcts evolve as a wavefront of necrosis, extending from subendocardium to subepicardium over a 3- to 4-hour period.

2. ROLE OF PERFUSION PRESSURE: AUTOREGULATION OF CORONARY BLOOD FLOW

Since, flow = Δpressure/resistance, any change in the perfusion pressure would be expected to change the blood flow rate proportionately. However, many circulations such as coronary, cerebral and renal, show the phenomenon of autoregulation. Autoregulation of blood flow is defined as *the intrinsic ability of an organ to maintain a constant blood flow*. Autoregulatory behavior is characterized as the ability of a vascular bed to maintain blood flow relatively constant across a wide range of perfusion pressures. In other words, this regulatory mechanism acts to preserve myocardial perfusion despite alterations in driving pressure. It may be stressed that the autoregulation of blood flow occurs only in a physiological range of perfusion pressure. Outside this range, lower or higher, the coronary blood flow becomes completely dependent on upstream driving pressure (responds passively to changes in pressure) (Fig. 4.3).[3]

At least two mechanisms seem to account for coronary blood flow autoregulation: (a) Myogenic theory, and (b) metabolic theory.

Myogenic Theory

An inherent property of the smooth muscle is that it contracts in response to stretch. Thus, if arterial pressure is suddenly increased, the arterioles are stretched and the vascular smooth muscle in their walls contracts in response to this stretch. The increased

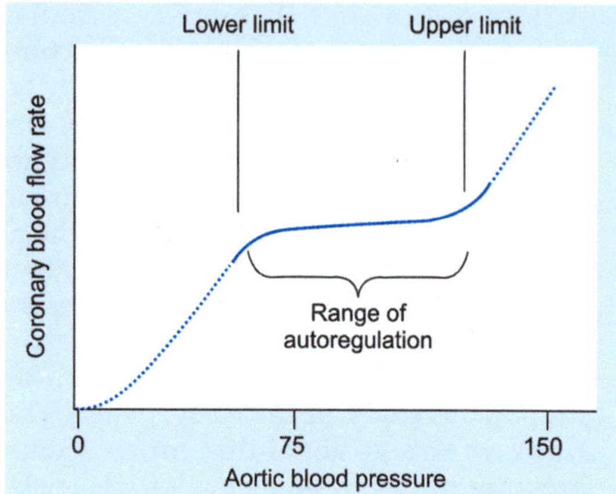

Fig. 4.3: Autoregulation of coronary blood flow

arteriolar resistance ensures that the flow does not increase significantly. If there is a sudden drop in pressure, there is a reduced stress placed on the arterial walls and therefore the vascular smooth muscle relaxes. The compensatory vasodilation ensures that blood flow to the tissue does not drop significantly despite changes in perfusion pressure (Fig. 4.4).[9]

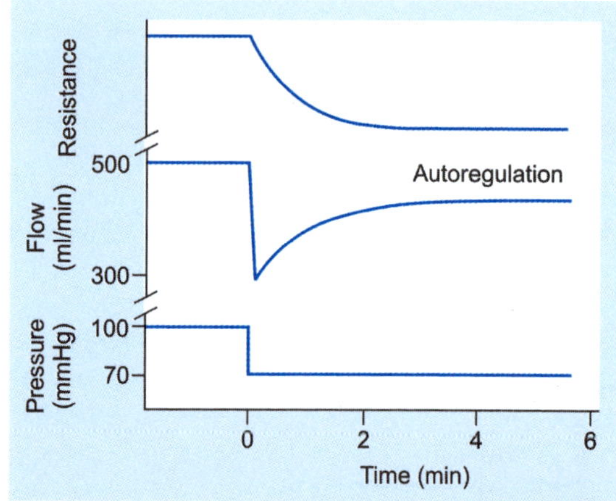

Fig. 4.4: Fundamental mechanism of autoregulation. A fall of perfusion pressure results in decreased vascular resistance

Although the coronary pressure-flow autoregulatory response is directionally consistent with a myogenic mechanism, definitive experiments to establish the contribution of myogenic responsiveness to pressure-flow autoregulation *in vivo* are confounded by numerous other pathways (e.g. metabolic and endothelial) with the potential to influence coronary microvascular resistance in response to changes in perfusion pressure.[3] While a few studies indicate that a myogenic mechanism is sufficient to achieve coronary pressure-flow autoregulation[10], there is also evidence to support a role for a metabolic as well as an endothelial-dependent component of this response. In particular, support for a metabolic mechanism comes from the observation that coronary autoregulation adjusts to the level of myocardial metabolism.[11]

Metabolic Theory

Actively metabolizing tissues produce certain vasodilatory metabolites which can regulate the blood flow to the tissue itself. The concentration of these metabolites is directly proportional to the level of metabolic activity performed by that particular tissue. Increased concentration of metabolites results in vasodilation of the arterioles, which results in decreased resistance to the blood flow. This increased blood flow is in coherence with the increased oxygen demands of the tissue. In contrast, the arterioles constrict if the concentrations of these metabolites decrease.[9] This mechanism is discussed in detail below.

3. METABOLIC CONTROL OF CORONARY BLOOD FLOW

Metabolic activity of the heart is the most important physiological mechanism regulating coronary vascular resistance. This mechanism ensures adequate O_2 delivery, as blood flow must increase in proportion to metabolic demand (e.g. increases in myocardial O_2 consumption during exercise) (Fig. 4.5). This parallel relationship between O_2 supply and demand is particularly important for the heart. Myocardium differs from other organs in two important respects. First, the heart has very little capacity for anaerobic metabolism. In fact, the heart normally consumes lactate produced by other tissues. Second, the heart consumes over 75%

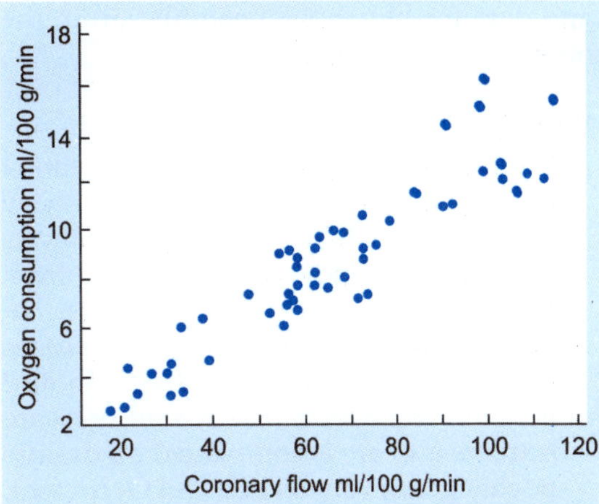

Fig. 4.5: Linear relationship between myocardial oxygen consumption and coronary blood flow

of the O_2 delivered to it at rest, and thus no significant O_2 reserve exists. Arterial blood typically has a pO_2 of 100 mmHg. Mixed venous blood, representing a sample from the entire body, normally has a pO_2 of 40 mmHg. In contrast, coronary venous blood has a pO_2 of approximately 18 mmHg under resting conditions. Therefore, increased oxygen demand of the heart during exercise can be met with by increased coronary blood flow rather than by greater oxygen extraction. During exercise, sympathetic-mediated vasodilation accounts for only about 25% of coronary vasodilation. However, the remaining 75% of exercise-induced coronary hyperemia is, at present, largely unexplained.[12, 13]

About one hundred years ago, Markwalder and Starling[14] opined that "the most potent agent in causing dilatation of the coronary vessels is non-volatile metabolite(s) produced by the heart muscle. By this means a local mechanism is available by which the heart muscle will increase the circulation through itself whenever increased demands are made on its functional capacity." The exact nature of the metabolite and the effector mechanisms activated remain unsettled; adenosine, pO_2, pCO_2 and pH have been considered possible vasodilators produced in response to increased myocardial activity. Recently, however, attention has been focused on H_2O_2 as a candidate for metabolic vasodilation in the coronary circulation.

Metabolic Messengers

i. **Adenosine:** Adenosine is a metabolite of cardiac myocytes and a potent coronary vasodilator.[14] The adenosine hypothesis proposes that adenosine is released from myocytes as myocardial pO_2 falls. No doubt, adenosine production increases during periods of severe hypoxia and ischemia[15]; however, these conditions are extreme and are accompanied by dramatic changes in venous pO_2 and O_2 content. During exercise, the changes in coronary venous pO_2 or O_2 content, reflective of tissue pO_2, are minimal.[16] Tune et al[17] reported that adenosine levels in the myocardial interstitium did not reach vasoactive levels during exercise. Thus, although adenosine may contribute substantially to the regulation of coronary blood flow during conditions of limited flow or extreme cardiac demand, it is doubtful that adenosine plays a key role in minute-to-minute matching of coronary flow to cardiac demand.[12] This conclusion is also supported by observations that adenosine receptor blockade does not impair myocardial contractile performance or causes metabolic evidence of ischemia.[18, 19]

ii. **Hypoxia, hypercapnia/H⁺:** Oxygen[20] and carbon dioxide[21] have both been considered as possible metabolic regulators of coronary blood flow. It has been suggested that changes in coronary blood flow that occur during alterations in myocardial metabolism and coronary perfusion pressure are correlated with changes in coronary venous pO_2. An increase in blood flow during hypercapnia has also been reported.[21, 22]

Case and colleagues[23], studied the role of carbon dioxide in regulating coronary blood flow and observed that coronary vascular resistance correlated well with coronary venous carbon dioxide concentrations. Despite its attractiveness as a regulator of coronary resistance vessel tone, carbon dioxide is not a likely mediator of the exercise-induced coronary vasodilation, as coronary venous carbon dioxide tension or pH remain essentially unchanged during exercise.[22]

Broten and Feigl[23] investigated the possibility that O_2 and CO_2 act synergistically in regulating coronary blood flow. They suggested that 20 to 30% of the coronary dilation observed over the autoregulatory range was due to an O_2/CO_2 sensitive mechanism. While these data suggest that there is a mechanism that responds to changes in O_2 and/or CO_2 to produce vasodilation in the coronary circulation, a direct effect of either O_2 or CO_2 on the coronary vasculature *in vivo* has yet to be demonstrated. Myers et al[24] have demonstrated that hypoxia causes dilation of isolated coronary microvessels through an endothelium-dependent mechanism. The vessels were isolated from the myocardium, indicating that the dilation was due to a direct effect of oxygen tension on the vascular endothelium and not a secondary effect of a metabolic vasodilator. However, the effect of hypoxia was not immediate (the dilation occurred over a 30-minute period) which does not support a role for oxygen in mediating rapid autoregulatory responses. Rather, the rapid vasoactive responses associated with changes in O_2 are likely mediated by production of some metabolite(s).

iii. **H_2O_2:** Over the past decades, the identification of a primary *metabolic feedback signal* that regulates coronary blood flow has proved elusive. Subsequently, the notion of *feed-forward regulation* of coronary blood flow has emerged. In a feedback system, the error signal for metabolic dilation is the metabolite produced during periods when oxygen demand exceeds delivery. In such a feedback system, once blood flow increases the error signal is eliminated when the production of metabolites returns to baseline.

In contrast, in a feed-forward signaling system, no error signal is generated. Rather, a metabolite is produced that is directly linked to oxygen utilization and, therefore, its level is regulated by oxygen utilization and not by a mismatch between demand and delivery. Vasodilatory signaling through reactive oxygen species, particularly hydrogen peroxide, has been proposed to function as feed-forward metabolic link that couples myocardial demand directly to coronary blood flow and oxygen delivery.[26]

Mitochondria are a major source of intracellular reactive oxygen species (ROS) generation due to leakage of electrons from the electron transport chain (ETC). The sequential reduction of molecular oxygen leads to formation of biological forms of ROS comprising of the superoxide anion (O_2^-), and hydroxyl radical (HO^-). Once generated, the superoxide is readily converted to more stable H_2O_2 by the enzyme manganese superoxide dismutase (MnSOD) (Fig. 4.6).

A distinction can be made into two groups of ROS: Free radicals and H_2O_2. Free radicals have an extremely high chemical reactivity due to the unpaired free electron [i.e. superoxide anion (O_2^-) and hydroxyl radical (OH^-)]. Other ROS like H_2O_2 is not considered a free radical since it lacks the free unpaired electron and thus have oxidizing rather than reactive effects.

Reports of the coronary vasodilatory actions of hydrogen peroxide (H_2O_2) began to emerge when H_2O_2 was identified as an endothelium-derived hyperpolarizing factor in human coronary arterioles. It has been demonstrated that mitochondrial production of superoxide (O^-) during periods of increased myocardial demand leads to increased release of H_2O_2 from myocytes and subsequent vasodilation of coronary arterioles.[25] Saitoh et al[26] also performed *in vivo* experiments demonstrating that production of H_2O_2 occurs in proportion to metabolism and that the vasodilatory action of metabolically produced H_2O_2 functions to couple coronary

blood flow to myocardial oxygen consumption (Fig. 4.6).

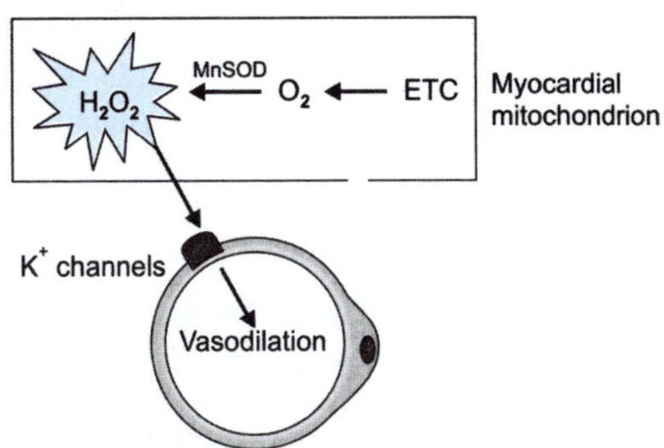

Fig. 4.6: Role of H_2O_2 in regulation of coronary blood flow (*see* text)

4. AUTONOMIC CONTROL OF CORONARY BLOOD FLOW

Innervation of the coronary vasculature by both the parasympathetic and sympathetic nervous systems has been documented in both animals and humans. Both sympathetic and parasympathetic nerve fibers are located within the vascular wall of large arteries as well as resistance arteries and arterioles with more dense innervation being present in resistance arteries and arterioles.[27] However, assessment of the direct influence of parasympathetic and sympathetic stimulation on coronary vascular resistance is complicated by changes in myocardial metabolism, which has a major role in determining coronary blood flow. A variety of techniques has been employed in studies to evaluate parasympathetic or sympathetic control of the coronary vasculature in the absence of autonomic influences on the myocardium and subsequent changes in heart rate, myocardial contractility, and perfusion pressure.

Sympathetic Control

Sympathetic control of the coronary circulation is exerted through both α-adrenergic and β-adrenergic effects. However, the direct effects of sympathetic stimulation of coronary vascular receptors are often masked by the metabolic vasodilation which occurs in response to sympathetic stimulation of the heart. Data obtained during conditions in which metabolic effects of sympathetic stimulation are controlled suggest that both α-adrenergic and β-adrenergic regulation of coronary vascular resistance may be important during physiological and pathological conditions.[13]

Alpha-adrenergic Regulation

The primary effect of α-adrenergic activation of coronary vessels appears to be a vasoconstriction of the vascular smooth muscle. However, a clear demonstration of coronary α-adrenergic receptor-mediated vasoconstriction has been difficult in the presence of competing metabolic influences secondary to augmented myocardial oxygen consumption. Under basal resting conditions, cardiac sympathetic activity is minimal so that α-adrenergic blockade has a negligible effect on coronary flow in awake resting dogs.[28] In contrast, exercise produces a greater increase in coronary blood flow after systemic α-adrenergic blockade than during control conditions.[29] *In short, α-adrenergic mechanisms act to blunt the increase in coronary flow that occurs during exercise.[30]*

Beta-adrenergic Regulation

It is difficult to separate from metabolic alterations in coronary vasodilation that result from β-adrenergic inotropic and chronotropic effects on the myocardium. In awake resting dogs[31], nonselective β-adrenergic blockade with propranolol decreased myocardial oxygen consumption, and this resulted in a parallel reduction of coronary blood flow; myocardial oxygen extraction was unchanged, suggesting that β-adrenergic control of the coronary circulation is minimal under resting conditions. However, during graded treadmill exercise, propranolol decreased coronary flow more than expected from the decrease in myocardial oxygen consumption, resulting in an increase in myocardial oxygen extraction and a decrease in coronary venous oxygen tension in dogs.[32] Similarly, in normal human subjects, β-adrenergic blockade with propranolol decreased myocardial blood flow during bicycle exercise out of proportion to the reduction of myocardial oxygen consumption, necessitating an increase in myocardial oxygen extraction.[33] The findings imply that β-adrenergic control of the coronary resistance vessels is minimal under resting conditions, *but β-adrenergic activation contributes to coronary vasodilation during exercise in a feed-forward manner.*

Parasympathetic Control

The coronary resistance vessels are richly innervated by the parasympathetic division of the autonomic nervous system. In dogs, pretreated with propranolol and paced to maintain a constant heart rate, stimulation of the vagosympathetic trunk produces coronary vasodilation independent of the cardiac effects of vagal stimulation. The coronary vasodilation produced by vagal stimulation was blocked by atropine.[34] To study the vasodilator influence of vagal activity on coronary blood flow during exercise, Gwirtz and Stone[35] administered atropine into a coronary artery of dogs during submaximal exercise. Atropine had no effect on heart rate or coronary blood flow, indicating that *parasympathetic effects on both the myocardium and coronary bed were negligible even at this level of exercise.*

In conclusion, although coronary blood flow is strongly responsive to local myocardial metabolic requirements, the autonomic nervous system provides a modulating influence that can alter the coupling between coronary flow and myocardial metabolism. During exercise, adrenergic activation exerts paradoxical effects that both oppose (alpha) and reinforce (beta) the increase in coronary flow that occurs in response to the increase in cardiac work. The net effect of sympathetic stimulation is β-adrenergic feed-forward vasodilation, which has been proposed to account for as much as 25% of the exercise hyperemia. However, cardiac denervation or pharmacological inhibition of autonomic control does not result in myocardial ischemia during exercise, implying that other vasodilator mechanisms act to compensate and mediate exercise hyperemia when autonomic control is blocked. These findings are consistent with the concept that autonomic control serves to optimize the matching of coronary blood flow to myocardial metabolic needs but is not essential for exercise hyperemia.[35]

5. VASCULAR ENDOTHELIUM

The vascular endothelium is the final common pathway regulating vasomotor tone. It modulates the contractile activity of the underlying smooth muscle through synthesis and secretion of vasoactive substances in response to blood flow, circulating hormones and chemical substances. Vasodilators are endothelium-derived relaxing factor, nitric oxide, prostacyclin and bradykinin. Vasoconstrictors include endothelin and thromboxane A_2. The net response depends on the balance between the two opposing groups.

References

1. Gupta R, Mohan I, Narula J. Trends in coronary heart disease epidemiology in India. Annals of Global Health. 2016; 82: 330–15.
2. Leach, A. History of Angina, *Res Medica* 1967, Special Issue, pp. 9–10.
3. Goodwill AG, Dick GM, Kiel AM, et al. Regulation of coronary blood flow: Effect of extravascular compressive forces: Phasic coronary blood flow.Compr Physiol. 2018; 7: 321–82.
4. Anrep GV, Saalfeld V. The effect of the cardiac contraction upon the coronary flow. J Physiol. 1933; 79: 317–31.
5. Gregg DE. Phasic blood flow and its determinants in the right coronary artery. Am J Physiol. 1937; 119: 580–88.
6. Feigl EO. Coronary physiology. Physiol Rev. 1983; 63: 1–205.
7. Rushmer RF, Thal N. The mechanics of ventricular contraction; a cinefluorographic study. Circulation. 1951; 4: 219–28.
8. Buja LM. "Myocardial ischemia and reperfusion injury". Cardiovascular Pathology. 2005;14: 170–75.
9. https://www.drbeen.com/blog/cvs-physiology-lecture-9-study-notes-autoregulation-blood-flow/
10. Cornelissen AJ, Dankelman J, VanBavel E, et al. Balance between myogenic, flow-dependent, and metabolic flow control in coronary arterial tree: A model study. Am J Physiol Heart Circ Physiol. 2002; 282: H22–24.
11. Mosher P, Ross J Jr, Mcfate PA, Shaw RF. Control of coronary blood flow by an autoregulatory mechanism. Circ Res. 1964; 14: 250–59.
12. Gorman MW, Tune DJ, Richmond KN, et al. Quantitative analysis of feed-forward sympathetic coronary vasodilation in exercising dogs. Journal of Applied Physiology. 2000; 89: 1903–11.
13. Muller-Dep JM. The coronary microcirculation in health and disease. ISRN Physiology 3013, Article ID 238979, 24 pages, http://dx.doi.org/10.1155/2013/23897
14. Markwalder J, Starling EH. A note on some factors which determine the blood flow through the coronary circulation. Journal of Physiology. 1913; 47: 275–285.
15. Berne RM. "The role of adenosine in the regulation of coronary blood flow," Circul Res. 1980; 47: 807–13.
16. Berne RM. Cardiac nucleotides in hypoxia: possible role in regulation of coronary blood flow." Amer J Physiol. 1963; 204: 317–22.
17. Heiss HW, Barmeyer J, Wink K. "Studies on the regulation of myocardial blood flow in man. I: training effects on blood flow and metabolism of the healthy heart at rest and during standardized heavy exercise." Basic Research in Cardiology. 1976; 71: 658–75.
18. Tune JD, Richmond N, Gorman MW, et al. "Adenosine is not responsible for local metabolic control of coronary blood flow in dogs during exercise." Am J Physiol. 2000: 278: H74–84.
19. Yada T, Richmond KN, Bibber R. Role of adenosine in local metabolic coronary vasodilation. Am J Heart Circ Physiol. 1999; 276: H1425–H1433.
20. Berne RM, Blackmon JR, Gardner TH. "Hypoxemia and coronary blood flow," J Clin Invest. 1957; 36: 1101–06.
21. Feinberg H, Gerola A, Katz LN. Effect of changes in blood CO_2 level on coronary flow and myocardial O_2 consumption. Am J Physiol. 1960; 199: 349–54.
22. Case RB, Felix A, Wachter M, et al. Relative effect of CO_2 on canine coronary vascular resistance. Circ Res. 1978; 42: 410–18.
23. Broten TP, Feigl EO."Role of myocardial oxygen and carbon dioxide in coronary autoregulation." Amer J of Physiol. 1992; 262: H1231–37.
24. Myers PR, Muller JM, Tanner MA. Effects of oxygen tension on endothelium dependent responses in canine coronary microvessels. Cardiovascular Research. 1991; 25; 885–94.

25. MiuraH, Bosnjak JJ, Ning G, et al. Role for hydrogen peroxide in flow-induced dilation of human coronary arterioles. Circulation Research. 2003; 92: e31–40.

26. Saitoh SI, Zhang C, Tuneetal JD. Hydrogen peroxide: a feed-forward dilator that couples myocardial metabolism to coronary blood flow, a rteriosclerosis, thrombosis, and vascular Biology. 2006; 26: 2614–21.

27. Hirsch EF, Borghard-Erdle AM. The innervation of the human heart. I. The coronary arteries and the myocardium.Arch Path. 1961; 71: 384–407.

28. Hodgson JM, Cohen MD, Sentpetery S, et al Effects of regional alpha- and beta-blockade on resting and hyperemic coronary blood flow in conscious, unstressed humans. Circulation. 1989; 79: 797– 809.

29. Gwirtz PA, Stone HL. Coronary blood flow and myocardial oxygen consumption after alpha-adrenergic blockade during submaximal exercise. J Pharmacol Exp Ther. 1981; 217: 92–98.

30. Duncker DJ, Bachie RJ. Regulation of coronary blood flow during exercise. Physiol Rev. 2008; 88: 1009–86.

31. Heyndrickx GR, Pannier JL, Muylaert P, et al. Alteration in myocardial oxygen balance during exercise after beta-adrenergic blockade in dogs. J Appl Physiol. 1980; 49: 28–33.

32. Gorman MW, Tune JD, Richmond KN, et al. Feed forward sympathetic coronary vasodilation in exercising dogs. J Appl Physiol. 2000; 89: 1892–1902.

33. Jorgensen CR, Wang K, Wang Y, et al. Effect of propranolol on myocardial oxygen consumption and its hemodynamic correlates during upright exercise. Circulation. 1973; 48: 1173–1182.

34. Broten TP, Miyashiro JK, Moncada S, et al. Role of endothelium-derived relaxing factor in parasympathetic coronary vasodilation. Am J Physiol Heart Circ Physiol. 1992; 262: H1579–84.

35. Gwirtz PA, Stone HL. Coronary blood flow changes following activation of adrenergic receptors in the conscious dog. Am J Physiol Heart Circ Physiol. 1982; 243: H13–19.

Mechanisms of Increased Cardiac Output and Skeletal Muscle Hyperemia during Exercise

INTRODUCTION

In physiological terms, any activity involving muscular contraction is an exercise. So, it includes not only athletic sports but also routine daily activities, like walking, running to catch a bus, or lifting a heavy bag of groceries, etc. Exercise not only involves neuromuscular coordination but also produces a considerable disturbance of internal environment. As a result, almost all the body systems undergo readjustments. Study of exercise physiology gives an excellent scope to discuss the interplay of different organ systems in the maintenance of constant internal environment.

A Synopsis of Cardiovascular Adjustments during Exercise

During exercise, oxygen consumption of the skeletal muscle increases 20-folds as compared to the resting values. Numerous cardio-respiratory adjustments occur in the body to meet with increased demand of oxygen and nutrients in the exercising muscles (myocardium, respiratory muscles and contracting muscles) and to dissipate heat via cutaneous vasodilation. To sustain the increased metabolic demand of these tissues, increased oxygen and nutrient delivery are accomplished by (i) increased cardiac output, (ii) increased blood flow to the active tissues, (iii) greater microvascular surface area available for exchange in the active tissues, and (iv) greater oxygen extraction from the blood in the active tissues.

The changes include large increases in heart rate and cardiac contractility to increase cardiac output, increased rate and depth of respiration which requires enhanced blood flow to respiratory muscles, vasodilation and increased blood flow in the contracting skeletal muscles, and vasoconstriction in the renal, splanchnic, and inactive skeletal muscle vascular beds (Fig. 5.1). These changes result in the regional redistribution of the cardiac output that allows diversion of greater percentage of cardiac output to the metabolically active tissues. Moreover, increased peripheral resistance in metabolically inactive tissues compensates for markedly decreased peripheral resistance in the metabolically active tissues and helps the blood pressure to be maintained or even increased during exercise. These alterations are coordinated by the sympathetic nervous system, which directs increased sympathetic outflow to the heart, resulting in increased cardiac output, and evokes selective vasoconstriction in peripheral organs (e.g. kidneys, small and large intestines, and non-exercising skeletal muscles). Changes in cutaneous blood flow during

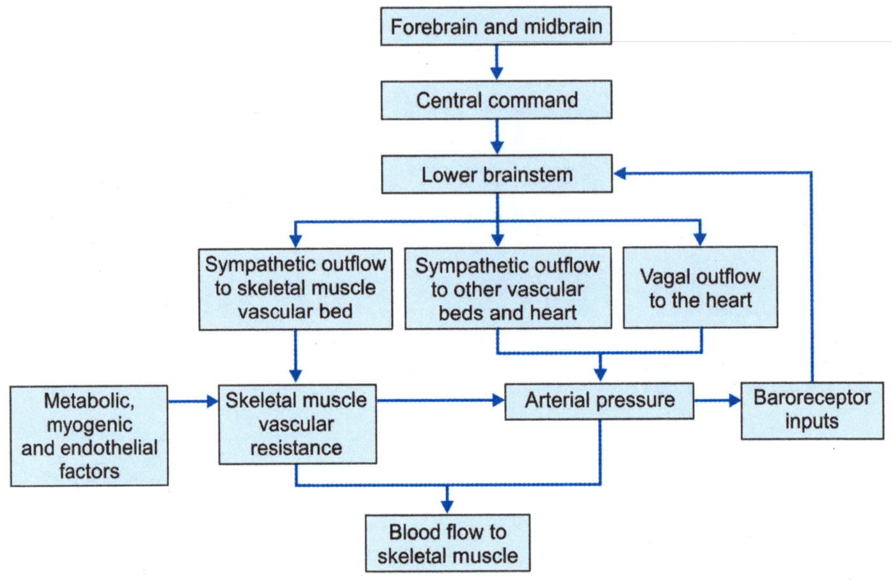

Fig. 5.1: Cardiovascular regulation during exercise

exercise depends on ambient temperature and the rise in body core temperature induced by increased metabolism with exercise, which increases skin blood flow to dissipate heat.[1,2] Mechanisms of increased cardiac output, increased skeletal muscle blood flow, and changes in splanchnic blood flow shall now be discussed.

I. CARDIAC OUTPUT DURING EXERCISE

An increase in heart rate and force of cardiac contraction leading to increased cardiac output is hallmark of cardiovascular changes during exercise. Exercise induces parasympathetic withdrawal and sympathetic activation, which are a function of exercise intensity and the muscle mass recruited. Parasympathetic withdrawal aims at increasing heart rate (HR), while the engagement of the sympathetic activity aims at increasing HR, at enhancing myocardial contractility to increase stroke volume (SV). In this autonomic modulation at least two neural mechanisms are involved in normal physiological response: One is a central mechanism (central command), while the other is peripheral (exercise pressor reflex).

a. Central Command

The increase in heart rate at the onset of exercise was documented as early as 1913 by the pioneering exercise physiologists, August Krogh and Johnas Lindhard[3] (Fig. 5.2). Based on this rapid response, they concluded "that the mechanism which shall produce the abrupt changes must be a nervous mechanism" and "we think that the evidence is in favor of an irradiation of impulses from the motor cortex" to the cardiovascular center. Later, the irradiation of impulses from the motor cortex to medullary cardiorespiratory centers came to be known as "central command."

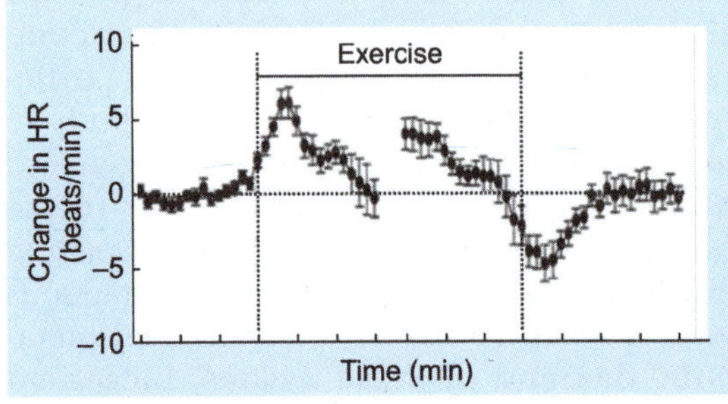

Fig. 5.2: Abrupt changes in the heart rate at the start and the end of exercise

In the central command mechanism, the activation of regions of the brain responsible for motor unit recruitment also activates the cardiovascular control areas located in the medulla. It is thought that central command establishes a basal level of sympathetic activity to the cardiovascular apparatus closely linked to the intensity of the effort.[4]

Nonetheless, an attempt to directly identify the response in sympathetic nerve activity during exercise has been conducted. Ninomiya et al[5] recorded for the first time cardiac and renal sympathetic nerve activity (CSNA and RSNA) during voluntary body movement in freely moving conscious cats and found abrupt increases in the sympathetic outflows prior to the onset of body movement. Tsuchimochi et al[6] observed an exercise–intensity-dependent increase in CSNA at the onset of treadmill exercise in conscious cats. Taken together, central command does not have uniform influence on the sympathoadrenal system at the onset of exercise, but selectively augments sympathetic outflow to specific organs such as heart, kidneys, and skin.

In exercising muscles, it is difficult to separate the central command component from peripheral component. The problem was solved by first observing cardiovascular response to graded loads and observing the response after partial neuromuscular blockade by tubocurarine. Increasing work load in normal subjects resulted proportional increase in CV response. However, in individuals with partial neuromuscular block, even mild load produced large CV response, indicating that the greater effort to lift the load resulted in greater central command (Fig. 5.3). In this experimental condition, the chemical milieu of the contracting muscle is not altered and hence peripheral signals cannot be responsible for cardiovascular response.[7,8]

b. Exercise Pressor Reflex

Alam and Smirk, in 1937[9] demonstrated the role of peripheral mechanism(s) in cardiovascular changes during exercise. They

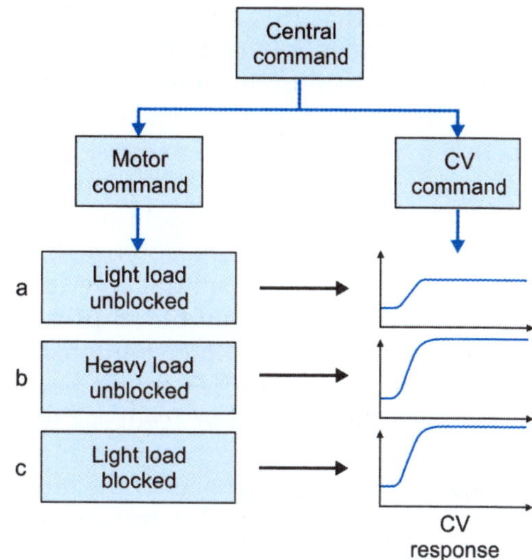

Fig. 5.3: Graded cardiovascular response to increasing workload (a and b). In (c), cardiovascular response is as great with light load in partially blocked person as with heavy load in unblocked individuals

recorded blood pressure in human volunteers during and after the end of exercise. They observed increased blood pressure during exercise which rapidly declined after the end of exercise. The experiment was repeated but circulation to the limb was arrested at the end of exercise. In the second group, the blood pressure remained elevated for several minutes even after cessation of exercise. The phenomenon was called "metabolic pressure reflex". Since then a large bulk of evidence has demonstrated that metabolic reflex coming from skeletal muscle evokes cardiovascular adjustments during exercise.

Several substances have been demonstrated to be able to activate the metaboreflex, such as lactic acid, potassium, bradykinin, arachidonic acid products, ATP, deprotonated phosphate, and adenosine. Moreover, studies with ^{31}P nuclear magnetic resonance spectroscopy revealed that the metaboreflex can be activated by decrements in intramuscular pH. These findings are interpreted with the concept that the metaboreflex is activated whenever blood flow to contracting muscles is insufficient to warrant oxygen delivery and/or metabolites washout, thereby suggesting that this reflex corrects any possible mismatch between blood flow and metabolism in the muscle. However, there is evidence that in humans the metaboreflex can be active even during mild exercise, when there is sufficient O_2 delivery to the muscle. In this situation there is no evident mismatch between muscle flow and metabolism, thereby demonstrating the essential role of the metaboreflex in the normal blood pressure response even for light exercise intensities. Therefore, the metaboreflex might be responsible for a tonically active feedback to the cardiovascular control areas which induce cardiovascular changes whenever the muscle metabolism is activated by muscle contractions, even at mild intensities of effort.[10]

In 1971, Coote et al[11] demonstrated that the muscle pressor reflex could be elicited by ventral root stimulation. Then, McCloskey and Mitchell[12] showed the involvement of group III/IV afferents in this cardiovascular reflex. This component is known as the muscle "metaboreflex." It was later demonstrated that mechanical changes in muscles and tendons can also elicit cardiovascular responses. This component has been termed "mechanoreflex." These two reflexes of muscular origin together constitute the *exercise pressor reflex*. It is well established that these two reflexes have their afferent arm in groups III and IV nerve endings within the muscle, with type III nerve afferents mainly acting as mechanoreceptors and type IV as metaboreceptors. These receptors collect information concerning the mechanical and metabolic conditions of contracting muscles and send this piece of information to cardiovascular controlling centers located in the *medulla*, where the information is integrated and elaborated. Then, cardiovascular medullary centers organize the hemodynamic response to exercise taking into account the mechanical and metabolic status of the working muscle.[13]

II. EXERCISE HYPEREMIA

Exercise hyperemia refers to the increase in skeletal muscle blood flow that occurs during muscular activity. Because this increase in blood flow occurs in response to increased cell metabolism, exercise hyperemia is also referred to as active or functional hyperemia. The term, active or functional hyperemia also applicable to the increase in flow to any organ that experiences an increase in parenchymal cell metabolism.[2]

Initial Versus Later Vasoregulatory Mechanisms in Exercise Hyperemia

Blood flow increasing within one second of the onset of muscular activity is referred to as the *fast vasodilator response*. This is followed by a larger *vasodilation that is sustained* and dependent on the magnitude and duration of exercise. This second vasodilatory phase is thought to result from the production of vasodilator metabolites by the contracting muscle cells that can be modulated by other local vasoregulatory mechanisms. Since it takes some time for active muscles to produce metabolites, which then have to diffuse to nearby arterioles, followed by binding to their receptors and activation of the second messenger signaling mechanisms that cause relaxation of vascular smooth muscle, the fast vasodilator mechanism must result from other processes, which are not exactly clear. Withdrawal of sympathetic tone does not explain this initial vasodilator phase because sympathectomy or treatment with ganglionic blockers does not alter the time to onset of this fast response. Some workers have proposed that muscle pump may be responsible for this early increase in muscle blood flow during exercise.[2]

Mechanisms Responsible for Exercise Hyperemia

Mechanical Factors

Arterial inflow to active skeletal muscle decreases during contractions and increases when the muscle relaxes. In contrast, the venous outflow increases during rhythmic contractions but decreases during muscle relaxation. These mechanical effects of exercise are due to increased extravascular pressure during rhythmic contractions, which expel blood through the venous system and increase the arteriovenous pressure gradient. This muscle pump mechanism can contribute up to 60% of the hyperemic response, at least under some conditions, and provides an attractive explanation for the observation that going from supine to upright position produces a twofold increase in exercise hyperemia (which is difficult to ascribe to other known mechanisms that contribute to the increased flow). The muscle pump also provides an explanation for the fast vasodilator mechanism.[2]

Role of Sympathetic Innervation in Exercise Hyperemia

Skeletal vessels have extensive adrenergic sympathetic innervation and there is a tonic vasoconstrictor discharge. In many animals, cholinergic sympathetic innervation has also been reported. Possibly, exercise hyperemia may be at least partly be due to decreased vasomotor discharge or increased cholinergic sympathetic discharge:

i. **Activation of cholinergic sympathetic fibers:** Bülbring and Burn[14], in 1935, reported for the first time the existence of sympathetic cholinergic nerve to skeletal muscle contributing to exercise vasodilatation in the cat and dog. It is histologically verified that sympathetic cholinergic nerves innervate blood vessels of skeletal muscle in several species such as the cat, dog, sheep, etc. but there is no histological evidence for the existence of sympathetic cholinergic nerves to skeletal muscle in the monkey and man.[15] Even in the dog possessing the sympathetic cholinergic system, exercise hyperemia of hindlimb contracting muscle is not significantly influenced by surgical sympathectomy or ganglionic or muscarinic blockade. Taken together, it has been currently accepted that the cholinergic sympathetic nervous system is not responsible for exercise hyperemia.

ii. **Withdrawal of sympathetic vasoconstrictor activity or sympatholysis in exercise hyperemia:** During high intensity exercise the adrenergic sympathetic nerve activity is markedly increased to elevate cardiac output and peripheral resistance. It contributes to the control of blood pressure during exercise. It also helps to redistribute the blood flow from inactive muscles as well as splanchnic area to active tissues. In contrast, there is intense vasodilation in the actively contracting muscles. Is this a result of withdrawal of sympathetic activity in the skeletal vessels? Theoretically yes, but there is no known mechanism by which sympathetic discharge to the skeletal vessels can be increased in inactive muscles as well as decreased in active muscles. Nowadays, it is believed that, during exercise, there is increased sympathetic discharge to all the skeletal vessels. In active muscles, effect of increased vasomotor discharge is counteracted by local vasodilator metabolites. This phenomenon is known as functional sympatholysis.[16]

Metabolic Mechanisms in Exercise Hyperemia

Although it is clear that the increased interstitial concentration of vasodilator metabolites derived from active muscle cells plays a major role in producing exercise hyperemia, it does not appear that one single metabolite accounts for metabolic vasodilation of arteriolar vascular smooth muscle. Alterations in the type of contraction (isometric *vs* isotonic) and the mass of muscle involved in the contractile activity may also produce differential release of metabolites.

While a large number of metabolites have been suggested to mediate exercise hyperemia, the most compelling evidence suggests that adenosine, potassium ions, and osmolarity changes may be quite important. In addition, decreases in arteriolar pO_2 (hypoxemia) that may occur in exercise owing to enhanced diffusion of the gas across the walls of these vessels may also contribute to active hyperemia. The notion that release of ATP from contracting myocytes and by erythrocytes in response to hypoxia and mechanical deformation of red cells as they pass through microvessels of contracting skeletal muscle is also gaining support.[16]

Criteria for Consideration of a Metabolite as a Vasodilator[17]

In 1983, Shepherd[17] laid down the following criteria:

1. The substance should be present in the skeletal muscle.
2. The substance should have access to skeletal resistance vessels.
3. The concentration of the substance should be sufficient to cause vasodilation and concentration should increase in proportion to the increase in muscle activity.
4. Exogenous administration of the substance should produce a prolonged and sustained vasodilation without producing any sensation (pain) to the subject.
5. Pharmacological or physiological maneuvers that alter the blood flow responses to exercise should have similar effects on the vasodilator responses to any putative substance given exogenously.

In line with these criteria, the following metabolites have been considered as chemical mediators responsible for exercise hyperemia:

a. **K^+:** Any muscle action potential leads to release of potassium ions from skeletal muscle cells which, under conditions of high action potential frequency cannot be compensated for by the re-uptake by the sodium potassium ATPase pump or the washout by the bloodstream. As a result, interstitial potassium concentration increases. This is reflected by an enhanced K^+-efflux from working skeletal muscle which occurs early after initiation of exercise.

    Thus K^+ seem to fulfill some of the criteria set by Shepherd.[17] First, K^+ is released during skeletal muscle repolarization. Second, K^+ is released from the contracting muscles in proportion to its metabolic activity. During exercise interstitial K^+ concentrations as high as 10 mM have been reported. But, in dogs, K^+ depletion did not disrupt the relationship between oxygen consumption and blood flow during exercise. Moreover, exogenous administration of K^+ is *very painful* and does not produce sustained vasodilation. In view of these observations, it is unlikely that K^+ plays an obligatory role in exercise hyperemia.[13]

b. **Lactate or H^+:** Early studies on vasodilatory metabolites focused on lactate.[18] With the onset of heavy exercise, there is an increase in interstitial and venous concentrations of lactate. Particularly in the transition from rest to heavy exercise there is a net production of lactic acid by muscle. The increase in lactate formation is not necessarily due to a lack of cellular oxygen. Indeed it has been shown in human muscle under exercise that the lactate efflux is unrelated to cellular pO_2 but increases with increasing oxygen consumption. Though lactate has been shown to act principally as a dilator of vascular smooth muscle, a study in humans demonstrated that interstitial lactate concentration as measured with microdialysis did not correlate with the changes in muscle blood flow at 30–60% of maximal work capacity, thus challenging its role as a major mediator of exercise hyperemia under submaximal exercise.[19]

c. **Adenosine:** In exercising or hypoxic skeletal muscle an enhanced breakdown of adenosine triphosphate (ATP) occurs during the resynthesis of which, adenosine is formed. Part of it can probably be released from skeletal muscle and reach the cells of the arteriolar walls. Exogenous application of adenosine induces strong dilatation.[20] Furthermore, in humans, a strong correlation between leg blood flow during exercise and interstitial adenosine concentrations as measured by microanalysis has been reported.[21] Adenosine receptor antagonist theophylline reduced exercise hyperemia in humans but not in dogs. These discrepancies suggest that adenosine need not necessarily play an essential role in exercise hyperemia, but rather it may be an important mediator under experimental conditions that involve skeletal muscle hypoxia.

To conclude, muscle-derived interstitial potassium, lactate and adenosine concentrations increase during exercise and all can reach by diffusion into the vascular smooth muscle of adjacent arterioles as well as the pericytes located in the capillary areas. This makes a contribution to exercise hyperemia likely and plausible but it is difficult to quantitate their respective contributions.[20]

d. **Endothelial derived vasodilators:** The main vasodilators released by vascular endothelium are NO and prostaglandins. In 1969, prostaglandins were proposed to be involved in muscle blood flow regulation based on the findings that infusion of prostaglandins into the brachial artery increased blood flow. A role for prostaglandins in exercise hyperemia was later supported by the findings that both plasma and interstitial prostacyclin and prostaglandin E_2 concentrations were increased during muscle contractions in the forearm and leg, respectively. However, infusion of cyclooxygenase (COX) inhibitors to inhibit the formation of prostaglandins, has shown no effect on blood flow at rest or during exercise in the human forearm.

In the late 1980s, Vallance and coworkers blocked NO synthase (NOS) by infusion of NG-monomethyl-L-arginine (L-NMMA) and showed a 50% reduction in resting forearm blood flow. The importance of NO for resting blood flow, blood flow in recovery from exercise, and blood flow during passive movement has since been widely confirmed. During exercise, however, inhibition of NO formation has been shown not to reduce blood flow. Thus, neither NO nor prostanoids appear to be obligatory for exercise hyperemia (Fig. 5.4). However, experiments in which the synthesis of NO and prostanoids have been inhibited simultaneously have demonstrated a clear reduction in blood flow

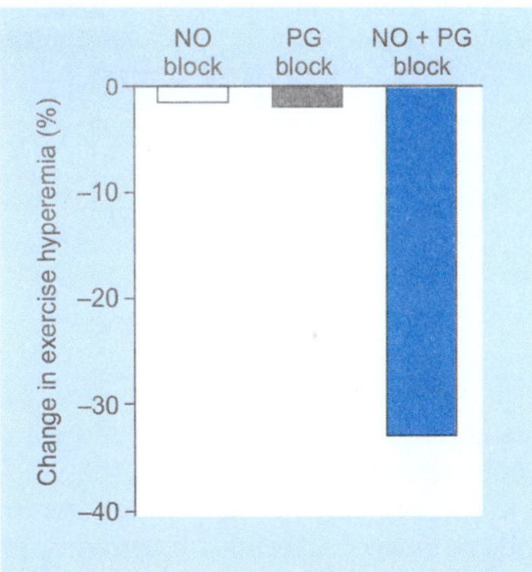

Fig. 5.4: Effect of nitric oxide (NO) and prostaglandins blockade on skeletal muscle hyperemia during exercise

during exercise. The observation that single inhibition of a system has no effect on exercise hyperemia, whereas combined inhibition markedly lowers blood flow suggests that there is a compensatory formation of the other vasodilator so that adequate blood flow is achieved. Direct interactions between the vasodilator systems may explain this redundancy.[22]

Conducted Vasodilation

Arterial circulation in skeletal muscle is typically comprised of feed arteries and branch arterioles including first-, second- and third order terminal arterioles. While distal branch arterioles are embedded in the skeletal muscle, proximal feed arteries are not directly involved in vasomotor regulation involved in exercise hyperemia. However, vasodilation in feed arteries and arterial network outside the skeletal muscle has to coordinate with skeletal vessel hyperemia essential for supplying greater amount of blood to the exercising muscles. The mechanisms responsible for intramuscular vasodilation have been discussed in detail above. So, the mechanism of vasodilation in extramuscular arteries need to be elucidated. In this context, a concept of conducted vasodilation is currently popular.

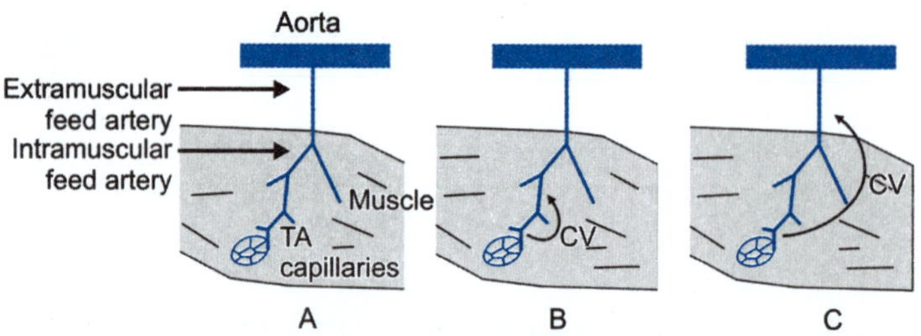

Fig. 5.5A to C: Conducted vasodilation (CV). TA: Terminal arteriole

In this concept, dilatation originating in the microcirculation embedded in the skeletal muscle ascends through gap junctions in the smooth muscle of arterioles to proximal feed arteries (Fig. 5.5). The resulting dilation of feed arteries allows increased blood flow to the muscles.[23]

SPLANCHNIC BLOOD FLOW DURING EXERCISE

At rest, the splanchnic organs receive 20% of the cardiac output, but only consume 10–20% of the available oxygen. To a certain extent, blood may be safely redistributed from the splanchnic organs to the working muscles and skin. The splanchnic vascular bed may therefore act as a reservoir of blood for the increased needs of skeletal muscle during exercise. Sympathetic nervous system (SNS) activity is massively increased in response to exercise. This causes increased splanchnic vascular resistance that may decrease splanchnic blood flow (SBF), despite the massive rise in cardiac output associated with physical exercise. In fact, SBF can decrease by 80% during maximum intensity exercise.[24] The importance of the SNS activity was demonstrated in a study in healthy individuals and patients with spinal cord injury at two levels (high and low). The effects of exercise with an arm-crank test on the portal vein flow and femoral artery, were measured. In normal persons and those with high level spinal cord lesions, a 30% reduction in portal vein flow was observed, but in high spinal cord injury (sympathetic denervation), the portal vein flow remained unchanged.[25]

The effect of exercise on the GI blood flow is dependent on various factors including exercise duration, environmental temperature and prandial state.[26] The effect of exercise duration on SBF was studied by Rehrer et al.[24] A gradual reduction in portal vein flow from 20% after 10 min to 80% after 1 hour of exercise was observed. The reduction in SBF during exercise is more pronounced in high environmental temperatures. Kenney et al[27] compared the effects of exercise in 22°C and 36°C. Exercising in 36°C resulted in an additional 17% decrease in SBF compared with 22°C at the same exercise intensity.

Thus, there is a redistribution of blood flow during exercise and diversion of blood flow from the splanchnic viscera to the working muscles. The mechanism of this redistribution is still unclear. One mechanism would be that the splanchnic vasoconstriction simply reflects generalized increased sympathetic tone during exercise. In working muscles the increase in sympathetic drive could be opposed by the effect of local vasodilator metabolites.

In view of these reports of decreased splanchnic blood flow during exercise, it would be prudent that the exercise is better avoided during the immediate postprandial period. This would be particularly important when the meal is a large one or environmental temperature is high.[28]

References

1. Joyner MJ, Casey DP. Regulation of increased blood flow (hyperemia) to muscles during exercise: A hierarchy of competing physiological needs. Physiol Rev. 2015; 95: 549–601.

2. Korthuis RJ. Skeletal Muscle Circulation. San Rafael (CA): Morgan & Claypool Life Sciences; 2011. Chapter 4, exercise hyperemia and regulation of tissue oxygenation during muscular activity. Available from: https://www.ncbi.nlm.nih.gov/books/NBK57139

3. Krogh A,Lindhard J. The regulation of respiration and circulation during the initial stages of muscular work. J Physiol. 1913; 47,112–36.

4. Matsukawa K, Liang N, Ishii K. Central command: Feed forward control of the system during exercise. Phys Fitness Sports Med. 2012; 1: 573–77.

5. Ninomiya I, Matsukawa K, Nishiura N. Central and baroreflex control of sympathetic nerve activity to the heart and kidney in a daily life of the cat. Clin Exp Hypertens. 1988; 10 Suppl 1: 19–31

6. Tsuchimochi H, Matsukawa K, Komine H, Murata J. Direct measurement of cardiac sympathetic efferent nerve activity during dynamic exercise. Am J Physiol Heart Circ Physiol. 2002; 283: H1896–906.

7. Eldridge, FL, Millhorn, DE, Kiley, JP, et al. Stimulation by central command of locomotion, respiration and circulation during exercise. Respir Physiol 1985; 59: 313–27.

8. Turner DL.Cardiovascular and respiratory control mechanisms during exercise: An integrated view. J Exp Biol. 1991; 160: 309–40.

9. Alam M, Smirk FH. Observations in man upon a blood pressure raising reflex arising from the voluntary muscles. J Physiol. 1937; 89: 372–83

10. Crisafulli A, Marongiu E, Ogoh S. Cardiovascular reflexes activity and their interaction during exercise. Biomed Res Int. 2015; 2015: 394183. Published online 2015 Oc18. doi: 10.1155/2015/394183

11. Coote JH, Hilton SM, Perez-Gonzalez JF. The reflex nature of the pressor response to muscular exercise. J Physiol. 1971; 215: 789–804.

12. McCloskey DI, Mitchell JH. Reflex cardiovascular and respiratory responses originating in exercising muscle. J Physiol. 1972; 224: 173–186

13. Adreani CM, Hill JM, Kaufman MP. Responses of groups III and IV muscle afferents to dynamic exercise. Jo Appl Physiol. 1997; 82: 1811–17.

14. Bülbring E, Burn, JH. The sympathetic dilator fibres in the muscles of the cat and dog. J. Physiol (Lond.). 1935; 83, 483–501

15. Bolme P, Fuxe K. Adrenergic and cholinergic nerve terminals in skeletal muscle vessels. Acta Physiol Scand. 1970; 78: 52–59.

16. Hong KS, Kim K. Skeletal muscle contraction-induced vasodilation in the microcirculation. Journal of Exercise Rehabilitation 2017; 13: 502–7.

17. Shepherd JT. Circulation to skeletal muscle. In: *Handbook of Physiology. The Cardiovascular System. Peripheral Circulation and Organ Blood Flow.* Bethesda, MD: Am Physiol Soc. 1983; 3: 319–370.

18. Hilton SM. Experiments on the post-contraction hyperaemia of skeletal muscle. J Physiol 1953; 120: 230–45.

19. Sarelius I, Pohl U. Control of muscle blood flow during exercise: local factors and integrative mechanisms. Acta Physiol (Oxf). 2010; 199: 349–65.

20. Lott ME, Hogeman CS, Vickery L, et al. Effects of dynamic exercise on mean blood velocity and muscle interstitial metabolite responses in humans. Am J Physiol Heart Circ Physiol. 2001; 281: H1734–41.

21. Hellsten Y, Maclean D, Radegran G, et al. Adenosine concentrations in the interstitium of resting and contracting human skeletal muscle. Circulation. 1998; 98: 6–8.

22. Hellsten Y, Nyberg M, Jensen LG, et al. Vasodilator interactions in skeletal muscle blood flow regulation. J Physiol. 2012; 590: 6297–305.

23. Bagher P, Segal SS. Regulation of blood flow in the microcirculation: Role of conducted vasodilation. Acta Physiol (Oxf). 2011; 202: 271–284.

24. Rehrer NJ, Smets A, Reynaert H, Goes E, De MK. Effect of exercise on portal vein blood flow in man. Med Sci Sports Exerc. 2001; 33: 1533–7.

25. Thijssen DH, Steendijk S, Hopman MT. Blood redistribution during exercise in subjects with spinal cord injury and controls. Med Sci Sports Exerc. 2009; 41: 1249–54.

26. Steege RWF, Kolkman JJ. Review article: the pathophysiology and management of gastro-intestinal symptoms during physical exercise, and the role of splanchnic blood flow. Aliment Pharmacol Ther. 2012; 35:516–28.

27. Kenney WL, Ho CW. Age alters regional distribution of blood flow during moderate-intensity exercise. J Appl Physiol. 1995; 79: 1112–19.

28. Qamar MI, Read AE. Effects of exercise on mesenteric blood flow in man. Gut. 1987; 28: 583–87.

Chapter 6

Blood Pressure Regulation during Exercise

INTRODUCTION

The fundamental principles of regulation of arterial blood pressure are well established. For example, arterial blood pressure is positively related to cardiac output and total peripheral vascular resistance; with the latter being strongly influenced by sympathetically-mediated vasoconstriction. The benefits of maintaining blood pressure at a 'normal' level are also widely appreciated. A decrease in blood pressure may decrease blood flow and oxygen delivery to peripheral tissues and potentially compromise cellular and organ system function. Conversely, an increase in blood pressure may cause devastating end-organ damage.[1] From these considerations, one would imagine that the blood pressure is very tightly regulated around a 'set point' via adaptive changes in cardiac output and peripheral resistance. Blood pressure changes during exercise, a physiological activity, may be studied to test this hypothesis. As the exercise continues, mean blood pressure increases linearly with increase in workload, whereas systolic blood pressure increases markedly, and diastolic pressure falls mildly. Systolic blood pressure reaches peak values of between 200 and 240 millimeters of mercury in normotensive persons. Because mean arterial pressure is equal to cardiac output times total peripheral resistance, the observed increase in mean arterial pressure results from an increase in cardiac output that outweighs a concomitant decrease in total peripheral resistance. Thus, increase in mean arterial pressure is a normal and desirable response, the result of a resetting of the arterial baroreflex to a higher pressure. Without such a resetting, the body would experience severe arterial hypotension during intense muscular activity.[2,3] Regulation of blood pressure, especially resetting of baroreflex and post-exercise hypotension have been discussed in this chapter.

MECHANISM(S) OF INCREASED BP DURING EXERCISE

i. **Central command:** In 1913, Krogh and Lindhard[4] observed an abrupt increase in pulmonary ventilation and pulse rate on start of exercise. The latent period of less than 1 second could not be brought about by any chemicals arising from working muscles. They concluded that abrupt increase was due to "irradiation of motor impulses" from the motor cortex to the respiratory center and cardiac centers. When the intensity of exercise increases, the cortical irradiation also increases proportionately. The cortical irradiation was afterwards named "central command" by Goodwin et al[5] in the 1970s. Experiments intentionally exaggerating the central command were performed using a neuromuscular blockade[6] (Fig. 5.3) or exercise in patients with muscular weakness[7], and the results supported cortical irradiation. Central command has been defined as a *feed forward mechanism* involving parallel activation of motor, respiratory and cardiovascular centers during exercise. Central command is thought to be crucial for rapid cardiovascular regulation operating in parallel with, or even preceding, the start of exercise. The role of central command in cardiovascular changes during exercise has been already discussed in Chapter 5.

ii. **Role of exercise pressure reflexes:** Alam and Smirk[8,9] demonstrated that dynamic calf exercise evoked an increase in blood pressure, which was maintained by circulatory arrest after cessation of exercise (which would abolish the effect of central command). Then, once blood flow was restored, blood pressure fell. These experiments demonstrated that metabolic signals originating in exercising skeletal muscle can also mediate cardiovascular adjustments to exercise. This response was called "metaboreflex."Several substances have been demonstrated to be able to activate the metaboreflex, such as lactic acid, potassium, bradykinin, arachidonic acid products, ATP, deprotonated phosphate, and adenosine. The role of exercise pressor reflex in increasing cardiac output during exercise has been discussed earlier in Chapter 5.

Arterial Baroreflex Function during Exercise

Arterial baroreceptors play a central role in the regulation of blood pressure by their action on heart rate, stroke volume as well as vascular resistance (Fig. 6.1). Any increase in blood pressure results in reflex decrease in cardiac output (bradycardia and decreased force of cardiac contraction), as well as peripheral resistance, thereby restoring the blood pressure (COXPR) to normal. The effect of exercise on the arterial baroreflex function

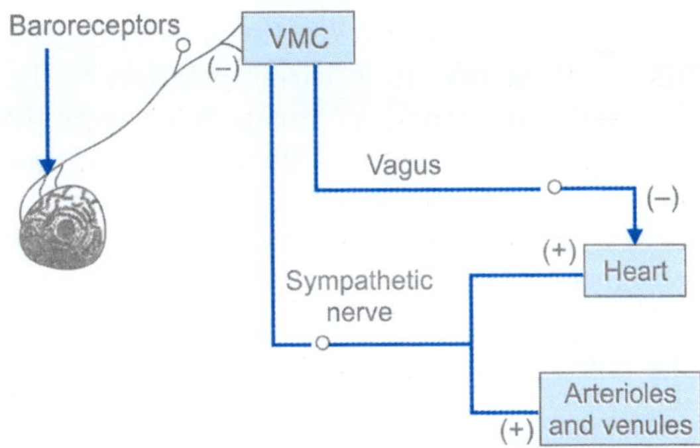

Fig. 6.1: Regulation of blood pressure

has been investigated during the last five decades. Earlier reports suggested that arterial baroreceptors became non-functional during exercise. This conclusion was based on the observation that during exercise both blood pressure and heart rate increased progressively with the increase of workload, whereas baroreceptor response involves a decreased in heart rate when blood pressure increases. Therefore, early research suggested that the arterial baroreflex was "switched off" as it was "physiologically unnecessary" for the cardiovascular adjustments to exercise or alternatively that the sensitivity of the reflex was significantly decreased during exercise to increase both HR and arterial blood pressure[10,11] Indeed, Iellamo et al[12] reported that the sensitivity of the arterial baroreflex is gradually attenuated from rest to heavy dynamic exercise. However, Potts et al[13] were the first to report in human studies, that full baroreflex stimulus-response curve was well preserved during increasing exercise workload. They suggested that the *carotid baroreflex is reset* during dynamic exercise and it functionally operates around the exercise-induced increase in arterial blood pressure. Ogoh et al[14] demonstrated that the operating point of the carotid baroreflex was progressively shifted upward and rightward (Fig. 6.2). The shift of the stimulus-response curve to the higher arterial blood pressure and HR allows the baroreflex to operate at the prevailing arterial blood pressure during exercise as effectively as operating at rest, and it also preserves the *reflex gain*. Further information arose from additional studies showing that this resetting occurs in direct relation to the intensity of effort, without a change in sensitivity.[15, 16] In view of these observations, nowadays, exercise-induced "resetting" of the arterial baroreflex function has been well accepted.

Why is Baroreflex Resetting Important?

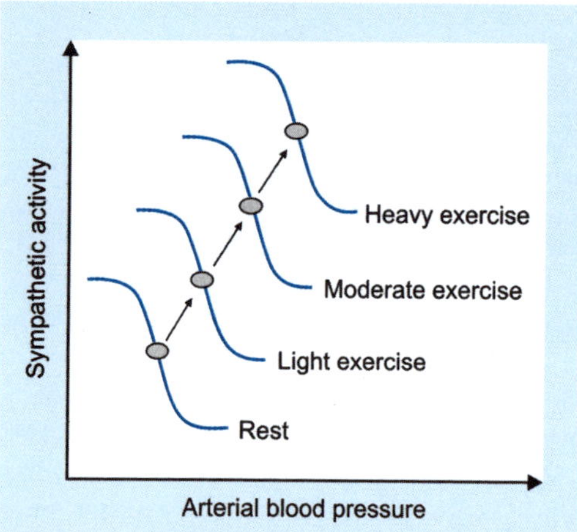

Fig. 6.2: Resetting of arterial baroreceptors during exercise

The "resetting"of the arterial baroreflex is essential to evoke and maintain an effective autonomic nervous system modulation and an adequate cardiovascular adjustment to exercise. Otherwise, uncontrolled increase in cardiac output and peripheral resistance may increase the blood pressure to dangerous high levels. Sheriff et al[17] provided direct evidence of this antagonistic effect of the baroreflex by comparing the muscle metaboreflex pressor response in dogs before and after arterial baroreceptor denervation. *They showed that the pressor response was increased by about 200% after baroreceptor denervation, as compared to the same dogs with intact baroreceptors.*

Similar findings have been reported in humans in whom carotid baroreceptors had been therapeutically denervated. In these subjects, the arterial blood pressure response to exercise was higher than in normal individuals[18, 19] These findings provide proof that the baroreflex acts to finely balance the oppo-

sing effects of sympathetic vasoconstriction and metabolic vasodilation, and it also acts to partly restrain the arterial blood pressure response to exercise by buffering activation of the increase in sympathetic activity due to the central command and the exercise pressor reflex.[15]

Cardiopulmonary Baroreflex

Along with central command and exercise pressor reflex, cardiopulmonary baroreflex can also modulate arterial baroreflex during exercise. Cardiopulmonary baroreceptors are low pressure baroreceptors that monitor blood volume. Several studies have shown the interaction between carotid and cardiopulmonary baroreflexes. Alteration in cardiopulmonary baroreceptor load during dynamic exercise affects not only the prevailing exercise-induced arterial blood pressure but also the resetting of the arterial baroreflex.[20] Ogoh et al[21] increased central blood volume (cardiopulmonary baroreceptor load) by increasing pedal frequency to enhance the muscle pump at the same amount of central command. Under these conditions, the magnitude of exercise-induced increases in arterial blood pressure was reduced and carotid baroreflex reset leftward and downward during dynamic exercise. Moreover, Volianitis et al[22] reported that when leg cycling was added to arm-cranking exercise, arterial blood pressure was reduced below that of arm exercise alone and resulted in relocation of the operating point of the carotid baroreflex-MAP curve to the lower arterial blood pressure despite greater activation of central command and the exercise pressor reflex. Therefore, these results indicate that the cardiopulmonary baroreflex plays an important role in baroreflex resetting during exercise and it operates together with central command and the exercise pressor reflex.

POST-EXERCISE HYPOTENSION

Post-exercise hypotension (PEH) is a phenomenon of a prolonged decrease in resting blood pressure for many minutes or even hours following acute exercise (Fig. 6.3). Although PEH was first documented by Hill in 1897 following a 400-yard dash, it was only after Fitzgerald's 1981 report of the effect of jogging, on his own labile hypertension, that the scientific community began to systematically examine this phenomenon.[23]

Magnitude of the Blood Pressure Decline

Numerous studies have confirmed the existence of post-exercise hypotension in normotensive and hypertensive individuals. In young healthy normotensive individuals, moderate-intensity dynamic exercise for 30–60 min produces post-exercise blood pressure reductions on the order of 5–10 mmHg in the supine position that last several hours. Typically, post-exercise hypotension is more severe in hypertensive individuals, reaching 20 mmHg and lasting up to 12 hours. The magnitude of pressure reductions is exaggerated in the seated or standing positions.[24]

Mechanisms of Post-exercise Hypotension

The expression of post-exercise hypotension results from the integration of a variety of obligatory and situational components. Obligatory components include: (1) Re-resetting of the baroreflex (which generally results in sympathoinhibition of sympathetic nerves to muscle vascular beds) and (2) a sustained histaminergic vasodilation of the previously exercising skeletal muscle vascular beds. Situational components include fluid deficit, hot environment and orthostasis.[24]

In comparison with pre-exercise level, post-exercise hypotension is characterized by a persistent drop in systemic vascular resistance. With the termination of exercise, cardiac output declines from high exercising values more rapidly than systemic vascular resistance recovers. This imbalance in the two determinants of arterial pressure results in a hypotension that is maintained for hours. Several other aspects of the post-exercise hemodynamic state merit comments.[24]

Post-exercise drop in cardiac output is multifactorial. Most important reason, of course, is the termination of central command. Moreover, during exercise, the rhythmically contracting skeletal muscles in the leg reduce the degree of venous pooling by squeezing veins, in effect, pumping blood back to the heart. This "muscle pump" activity is absent during passive recovery from exercise. Third, the increase in venous pooling, in conjunction with the loss of plasma volume associated with exercise, leads to a reduction in central venous pressure (~ 2 mmHg supine) and cardiac filling pressure (preload). Thus, it is generally accepted that basically post-exercise hypotension is due to a persistent drop in systemic vascular resistance in the face of post-exercise decrease in cardiac output.[24]

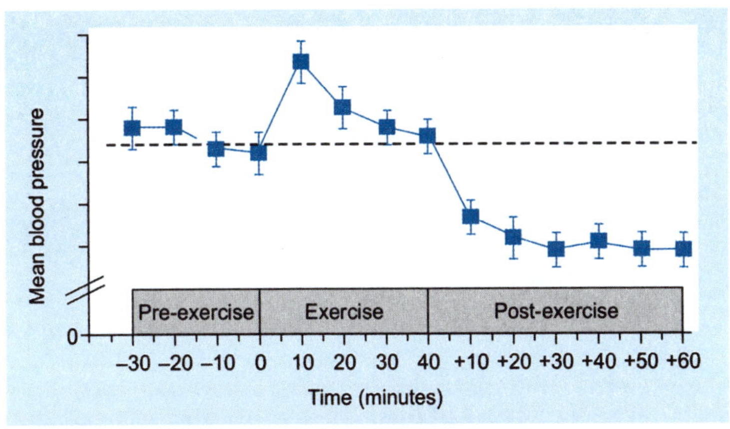

Fig. 6.3: Post-exercise hypotension

Resetting of Baroreflex

In an experiment involving moderately intense exercise of leg muscles only, both forearm and calf vascular resistance decreased in parallel with systemic vascular resistance by ~ 30%. Thus, the vasodilation that underlies post exercise hypotension is not restricted to the sites of active skeletal muscles (i.e., the legs) but involves inactive regions as well (e.g. the arms).[22] Thus, some generalized (neural/humoral) mechanism seems to be involved.

Multiple studies have documented that the amount of activity in the sympathetic nerve fibers that control vasoconstriction in the leg is inhibited during post-exercise hypotension in humans.Under resting conditions, muscle sympathetic nerve activity is under strong regulation by the arterial baroreflexes. During exercise, the baroreflex is "reset" to a higher operating point and sympathetic activity is increased. After exercise, these reflexes are reset to lower pressure (Fig. 6.4), so that sympathetic outflow from the central nervous system is lower than pre-exercise levels. The neural mechanisms involved in baroreflex resetting during exercise and post-exercise hypotension are unknown.[25]

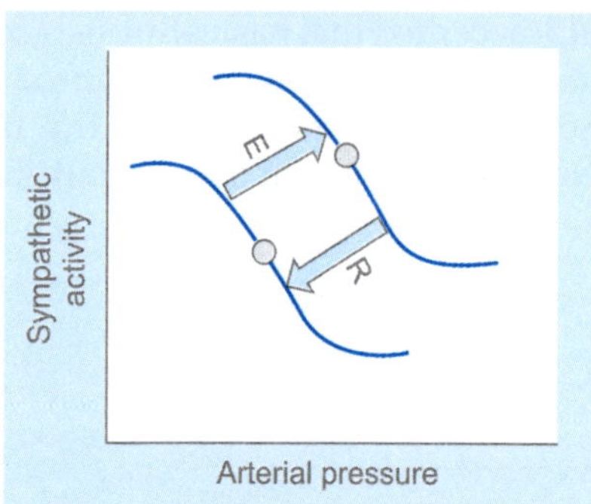

Fig. 6. 4: Baroreflex during exercise (E) and recovery (R)

Local Vasodilatory Mechanism(s)

The work of Halliwill and colleagues, using combined high-dose histamine H_1 and H_2 receptor antagonists, showed that histamine has an obligatory role as a primary mediator of sustained post-exercise vasodilation.[26, 27]

Intensity of post-exercise vasodilation following 60 minutes of moderate-cycling is inhibited by 80% when histamine receptors are blocked (Fig. 6.5). The remaining 20% that is not affected by histamine blockade is presumed to be the result of resetting of the baroreflex and sympathetic withdrawal. It is now clear that histamine, formed and released within active skeletal muscle tissue, and subsequent activation of histamine H_1 and H_2 receptors, mediates sustained post-exercise vasodilation (Fig. 6.6). To date, the exercise-related trigger for histamine release remains elusive.[28]

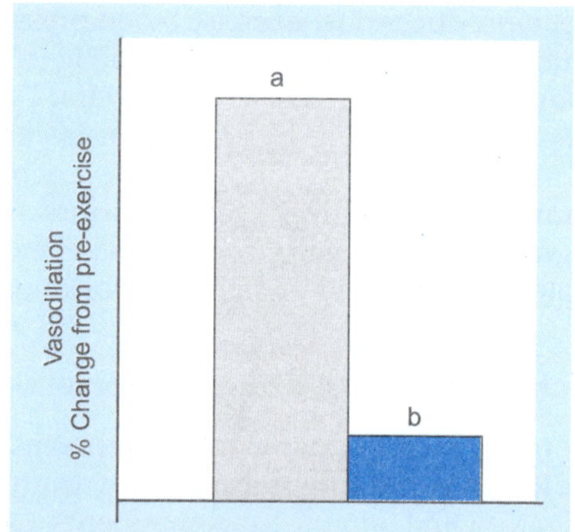

Fig. 6.5: Effect of histamine receptor blockade on post-exercise vasodilation (b) as compared to control (a)

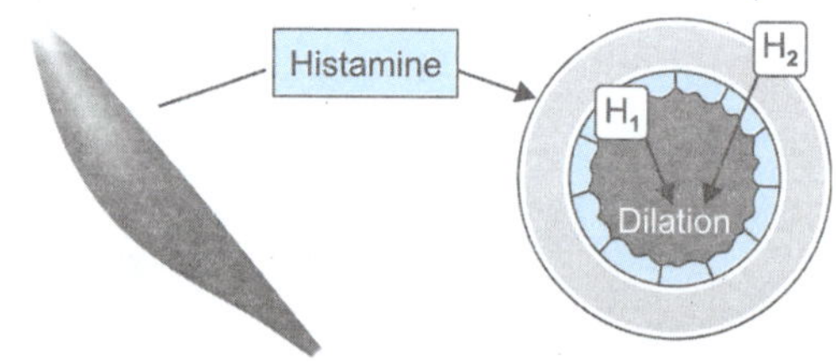

Fig. 6.6: Role of histamine in post-exercise vasodilation

Histamine is well recognized for its potent vasodilator effects, but is difficult to measure when released into blood or produced locally, because it is metabolized quickly or taken up by basophils in the blood stream. Studies that readily demonstrated vasodilatation which was blocked by histamine receptor antagonists have not shown changes in circulating histamine in either plasma or whole blood during recovery from exercise. This observation suggests that histamine is released locally without significant spillover to the circulation. This is corroborated by recent work demonstrating that when exercise is performed unilaterally, post-exercise histaminergic vasodilatation is restricted to the previously exercised leg.[29]

Several possible mechanisms may increase intramuscular histamine during recovery from exercise. Mast cells located within the connective tissue layer surrounding skeletal muscle fascicles and those also found near blood vessels may degranulate, releasing histamine locally.[29]

Role of Environmental Factors in Post-exercise Hypotension

Many of the environmental factors that have an impact on post-exercise hypotension can be predicted by understanding the fundamental hemodynamic changes. It is expected that exercise in a hot environment will exacerbate post-exercise hypotension in two ways: A greater loss of plasma volume due to sweating and a greater drop in vascular resistance due to vasodilation of the cutaneous circulation. In contrast, fluid replacement may lessen post-exercise hypotension. Active recovery, by maintaining action of the muscle pump, is likely reduce post-exercise hypotension compared with passive recovery, but this has not been studied systematically. Seated upright or standing during recovery from exercise exaggerates the magnitude of venous pooling. This could further reduce central venous pressure, causing cardiac output to fall, and result in reduced orthostatic tolerance.[24]

Orthostatic Hypotension and Syncope

It can be expected that post-exercise hypotension is likely to predispose to orthostatic hypotension and syncope. The ability to defend against orthostatic stress is compromised by enhanced venous pooling, reduced venous return, and reduced sympathetic outflow. Furthermore, the baroreflexes are reset to defend pressure at a lower level. Thus, orthostatic tolerance is often reduced in healthy individuals during the first hours after exercise, and syncopal episodes are not uncommon immediately after exercise. Studies have not identified what factors might influence the incidence of orthostatic intolerance after exercise or determined the extent to which post-exercise hypotension in the supine position would predict orthostatic intolerance after exercise.[24]

References

1. Chapleau MW. Baroreceptor Reflexes. In: Robertson D, Biaggioni I, Burnstock G, Low P, Paton J, editors. Primer on the Autonomic Nervous System. Academic Press; 2012: p. 161–165.
2. Rowell LB. Human cardiovascular control. New York: Oxford University Press, 1993.
3. Raven PB, Chapleau MW. Blood pressure regulation XI: Overview and future research directions. Eur J Appl Physiol. 2014; 114: 579–86.
4. Krogh A, Lindhard J. The regulation of respiration and circulation during the initial stages of muscular work. J Physiol. 1913; 47: 112–36.
5. Goodwin GM, McCloskey DI, Mitchell JH. Cardiovascular and respiratory responses to changes in central command during isometric exercise at constant muscle tension. J Physiol. 1972; 226: 173–90.
6. Galbo H, Kaer M, Secher NH. Cardiovascular, ventilatory and catecholamine responses to maximal dynamic exercise in partially curarized man. J Physiol. 1987; 389: 557–68.
7. Innes JA, De Cort SC, Evans PJ, et al. Central command influences cardiorespiratory response to dynamic exercise in humans with unilateral weakness. J Physiol. 1992; 448: 551–63.
8. Alam M, Smirk FH. "Observations in man upon a blood pressure raising reflex arising from the voluntary muscles." J Physiol. 1937; 89: 372–83.
9. Alam M, Smirk FH. Observations in man on a pulse-accelerating reflex from the voluntary muscles of the legs. Upon a blood pressure raising reflex arising from the voluntary muscles. J Physiol. 1938; 92: 167–77.
10. Pickering TG, Gribbin B, Petersen ES, et al. "Comparison of the effects of exercise and posture on the baroreflex in man." Cardiovascular Research. 1971; 5: 582–586.

11. Cunningham DJ, Petersen ES, R. Peto R, et al. "Comparison of the effect of different types of exercise on the baroreflex regulation of heart rate." Acta Physiol Scandin. 1972; 86: 444–55.

12. Iellamo F. "Baroreflex control of heart rate during exercise: a topic of perennial conflict. J of Appl Physiol. 2001; 90: 1184–85.

13. Potts JT, Shi XR, Raven PB. "Carotid baroreflex responsiveness during dynamic exercise in humans." AmJ Physiol—Heart and Circulatory Physiology. 1993; 265: H1928–38.

14. Ogoh S, Fisher JP, Dawson EA, et al. Autonomic nervous system influence on arterial baroreflex control of heart rate during exercise in humans. J Physiol. 2005; 566: 599–611.

15. Papelier Y, Escourrou P, Gauthier JP, et al. "Carotid baroreflex control of blood pressure and heart rate in men during dynamic exercise." J Appl Physiol. 1994; 77: 502–506.

16. Norton KH, Boushel R, Strange S, et al. Resetting of the carotid arterial baroreflex during dynamic exercise in humans. J Appl Physiol. 1999; 87: 332–38.

17. Sheriff DD, O'Leary DS, Scher AM, et al. Baroreflex attenuates pressor response to graded muscle ischemia in exercising dogs. Am J Physiol Heart Circ Physiol. 1990; 258: H305–10.

18. Walgenbach SC, Donald DE, Inhibition by carotid baroreflex of exercise-induced increases in arterial pressure. Circ Res. 1983; 52: 253–62.

19. Timmers HJML, Wieling W, Karemaker JM. "Cardiovascular responses to stress after carotid baroreceptor denervation in humans." Ann the New York Acad Sciences. 2004; 1018: 515–19.

20. Ogoh S, Brothers RM, Barnes Q, et al. Effects of changes in central blood volume on carotid-vasomotor baroreflex sensitivity at rest and during exercise. J Appl Physiol. 2006; 101: 68–75.

21. Ogoh S, Fisher JP, Fadel PJ, et al. Increases in central blood volume modulate carotid baroreflex resetting during dynamic exercise in humans. J Physiol. 2007; 581: 405–18.

22. Volianitis S, Yoshiga CC, Vogelsang T, et al. Arterial blood pressure and carotid baroreflex function during arm and combined arm and leg exercise in humans. Acta Physiol Scand. 2004; 181: 289–95.

23. MacDonald JR. Potential causes mechanisms, and implications of post- exercise hypotension. J Human Hyperten. 2002: 16: 225–236.

24. Halliwill JR. Mechanisms and clinical implications of post-exercise hypotension in umans. Exercise and Sport Sciences Reviews: 2001; 29: 65–70.

25. Halliwill JR, Taylor JA, Eckberg DL. Impaired sympathetic vascular regulation in humans after acute dynamic exercise. J Physiol (Lond.) 495: 279–88.

26. Lockwood JM, Wilkins BW, Halliwill JR. H1 receptor-mediated vasodilatation contributes to postexercise hypotension. J Physiol. 2005; 563: 633–642.

27. McCord JL, Beasley JM, Halliwill JR. H_2-receptor-mediated vasodilation contributes to post-exercise hypotension. J Appl Physiol. 1985; 100: 67–75.

28. Romero SA, Minson CT, Halliwill JR. The cardiovascular system after exercise. J Appl Physiol. 1985; 122: 925–32.

29. Halliwill JR, Buck TM, Lacewell AN, et al. Post-exercise hypotension and sustained post-exercise vasodilatation: What happens after we exercise? Exp Physiol. 2013; 98: 7–18.

Chapter 7

Exercise Hyperpnea

INTRODUCTION

Ventilation increases rapidly from the first breath at the onset of dynamic exercise, and this ventilation increase is tightly coupled with a change in workload or metabolic rate. This increase in ventilation is called "exercise hyperpnea." The mechanisms that control it have been a matter of debate in the field of exercise physiology for over a century. The first report on the exercise hyperpnea was documented early in the last century by Douglas et al in 1912.[1] Douglas exercised on a treadmill breathing into a rubber bag carried on his back, now known a 'Douglas bag'. After such exercise at different work intensities, the volume of air and the O_2 and CO_2 concentration in the air collected in the bag were measured. Douglas reported increase in pulmonary ventilation, accompanied by increased oxygen consumption and CO_2 production; all proportionate to the severity of exercise.

Claude Bernard's concept of maintenance of homeostasis of "milieu intérieur" assumes that living organisms tend to maintain constant the composition of their internal environment. According to this concept, the increase in O_2 extracted by the muscles from the blood must be replenished, and the CO_2 and H^+ produced by the muscles must be eliminated during exercise. If alveolar ventilation and alveolar-capillary diffusion increase in proportion to the increase in metabolic rate during exercise, these requirements will be met. The mechanisms that regulate breathing during exercise have fascinated physiologists for well over a century.[2]

MECHANISMS OF EXERCISE HYPERPNEA

Two possible mechanisms are: (A) inputs from the peripheral and/or central chemoreceptors, and (B) neural inputs from the higher center and/or peripheral receptors.[3, 4]

A. Chemoreceptor Mechanisms

i. **Peripheral (venous) chemoreceptors:** Exercising muscles have increased O_2 consumption and CO_2 production, leading to decreased pO_2 and oxygen content and increased pCO_2 and CO_2 content of the venous blood (Fig. 7.1). In a typical negative feedback control system, this deviation could be detected by venous chemoreceptors and impulses sent to the medullary respiratory centers for corrective measures in the form of increased pulmonary ventilation. But, for over a century, researchers of exercise physiology have looked for such venous chemoreceptors, without any success.

Mammals do not possess any known receptive mechanisms capable of directly

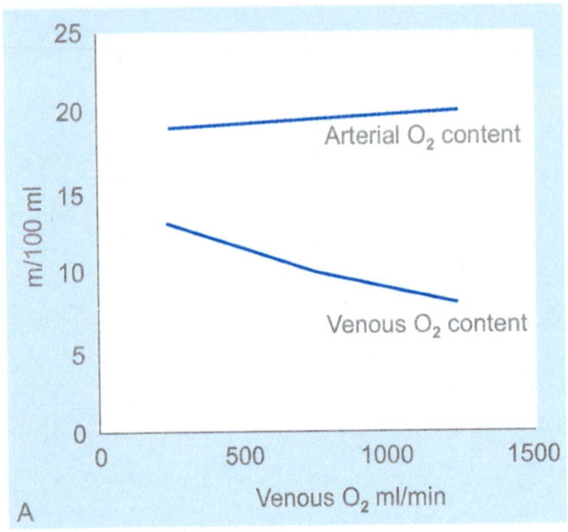

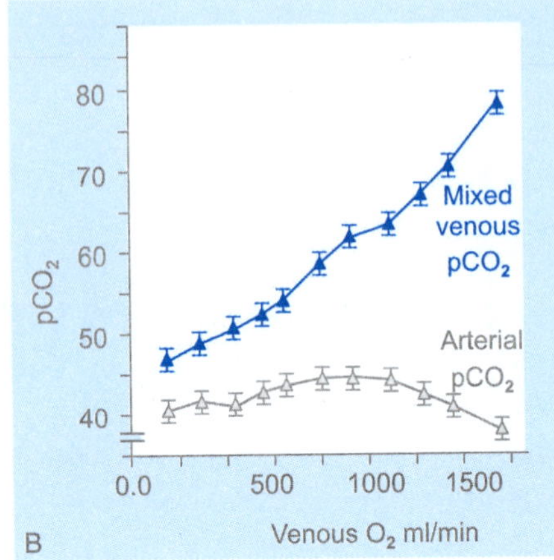

Fig. 7.1: Arterial and venous blood oxygen content (A) and pCO$_2$ (B) in relation to severity of exercise

sensing the rate at which CO$_2$ and O$_2$ are exchanged in the peripheral tissues or levels of these gases in venous blood. Hence, one of the most puzzling challenges in physiology is to how respiratory neurons adjust their output to variables which cannot be directly monitored or do not change during exercise. Particularly perplexing is that there is no known mechanism to account for an over twentyfold increase in breathing representing the largest non-volitional drive to breathe.

ii. **Central (arterial) chemoreceptors:** Arterial pCO$_2$ is a powerful stimulator for the chemoreceptors. Between rest and maximum exercise in man, metabolic rate or the rate of oxygen consumption, increases by up to ~ 23-fold, and breathing increases by up to ~ 35-fold to match or exceed metabolic rate. If breathing matches metabolic rate, the systemic arterial blood gases, i.e. the partial pressures of oxygen (pO$_2$) and of carbon dioxide (pCO$_2$) should remain constant and the stability of pCO$_2$ is the simplest indication of how well matching occurs. Since they do match during exercise, but elevated arterial pCO$_2$ has never been recorded even during severe exercise. So, it is concluded that the arterial chemoreceptors do not provide the primary drive for exercise hyperpnea, but rather act as a check preventing excessive ventilation, a sort of a

fine tuning of ventilation during exercise. Further evidence for the absence of any role of arterial chemoreceptors in exercise hyperpnea was provided by observations that carotid chemoreceptor ablation in man does not prevent breathing matching metabolic rate either at rest or during moderate exercise.[5]

B. Neurogenic Mechanisms

a. Central Command

A major hallmark of the exercise hyperpnea is that the hyperpnea begins simultaneously with the onset of exercise (Fig. 7. 2), which was documented as early as 1913 by the pioneering exercise physiologists, August Krogh and Johnas Lindhard.[6] Based on this rapid response, they concluded "that the mechanism which shall produce the abrupt-changes must be a nervous mechanism" and

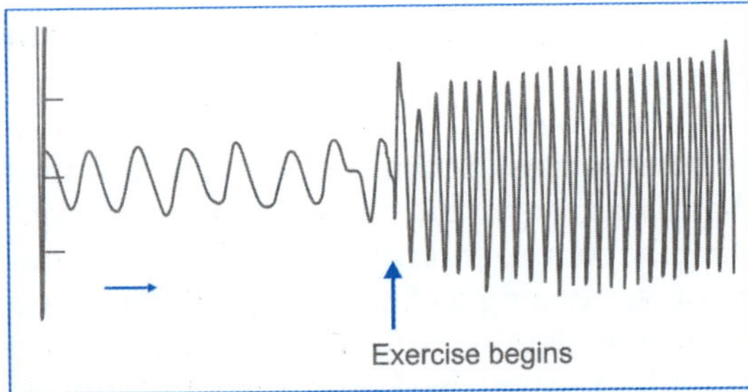

Fig. 7.2: Abrupt increase in ventilation at the start of exercise

we think that the evidence is in favor of an irradiation of impulses from the motor cortex" to the respiratory center. Later, the irradiation of impulses from the motor cortex to medullary cardiorespiratory centers came to be known as "central command".

Possible Role of Hypothalamus (Hypothalamic Locomotor Region—HTLR) in Central Command

The results obtained from early studies, which were concerned primarily with determining the site of origin of the central command, demonstrated that respiratory changes occurred during spontaneous locomotion in decorticate cats.[7] The ventilatory responses preceded the start of locomotion, and were found to disappear after ablation of the hypothalamus. These findings suggested that the respiratory and locomotor responses were not dependent upon the conscious sensory or motor areas of the cortex. More recent evidence provides additional support to the hypothesis that the central command mechanism originates from the hypothalamus.

Eldridge et al[8] have demonstrated that the electrical stimulation of the paraventricular region of the hypothalamus elicits parallel ventilatory and locomotor responses in decorticate cats. In addition, the activation of this region by an injection of picrotoxin elicited the identical response to electrical stimulation. Picrotoxin is particularly useful as a stimulating agent because it only activates cell bodies and not passing axons like electrical stimulation. Although the results from these experiments suggest that the central command mechanism may originate from the hypothalamus, the evidence is still circumstantial, since the feedback receptors located in the contracting muscles were not ablated and could therefore have provided the stimulus to ventilation. In order to remedy this weakness, Eldridge et al[9] studied the response to electrical or chemical stimulation of the hypothalamus in the anesthetized and paralyzed cat. Ventilatory and locomotor responses to the stimulation were measured by recording from the phrenic and biceps femoris nerves, respectively. Their findings were similar to those observed during spontaneous or evoked locomotion, and provided a convincing argument for the hypothalamic origin of the central command stimulus to ventilation during exercise. The central command hypothesis was further supported by results obtained from experiments which utilized the technique of retrograde transport of horseradish peroxidase to show the existence of efferent connections from the paraventricular region of the hypothalamus to the spinal cord, and to the regions of the medulla associated with respiration like the ventrolateral nucleus tractus solitarius and the ventral surface of the medulla.[10] Therefore, experimental evidence obtained to date suggests that a central command mechanism, which can operate independently of the motor cortex and feedback from the exercising muscles, may have a prominent role in eliciting increases in ventilation during exercise.[11]

Possible Role of Motor Cortex in Central Command

Despite the findings implicating the hypothalamus, it is also possible that volitional cortical mechanisms, via connections directly from the motor cortex to respiratory motoneurons in the spinal cord, may also have a role in the simultaneous activation of ventilation and locomotion during exercise.[11] This hypothesis is supported by the findings of Lipski et al[12], which demonstrated that cortical stimulation in the cat was capable of eliciting short latency excitatory responses in intercostal and phrenic motoneurons. In addition, transcranial magnetic stimulation of the motor cortex in man produces a detectable change in diaphragmatic electromyographic activity if the activity of this muscle is facilitated by voluntary activation. Moreover, during exercise there was an increase in facilitation which was equivalent to that seen during voluntary inspiration, and this finding also suggests that the cerebral cortex may contribute significantly to the ventilatory drive during exercise.[13]

b. Peripheral Neurogenic Mechanisms

The ventilatory responses to voluntary and passive exercise were studied by Morikawa et al in 20 healthy subjects. In the normal subjects, minute ventilation increased abruptly from the first breath after the onset of the two types of exercise; the increase in ventilation was greater in active (voluntary) exercise than in passive leg movements (Fig. 7.3). These responses to passive leg exercise were also studied in 23 patients with spinal cord transection at the level of T5–12. In the patients with spinal cord transection, there was no change in minute ventilation with passive exercise (Fig. 7.4).[14] In case of patients with spinal cord transection, obviously, the afferent input from passively moving legs could not reach the medullary respiratory centers. These results suggest that ventilatory responses at the onset of mild exercise at least partly related to drives from the moving limbs.[14]

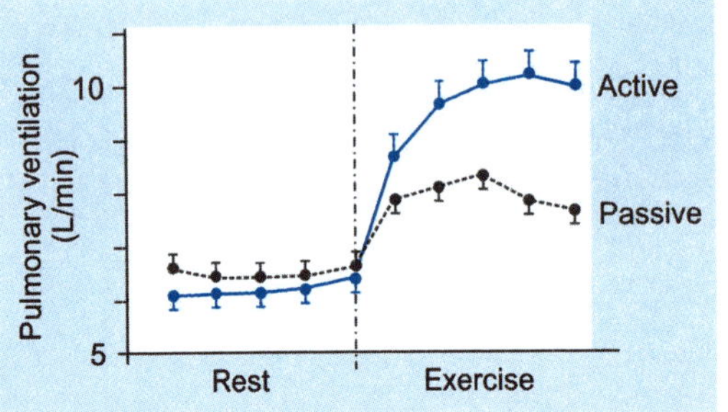

Fig. 7.3: Effect of voluntary and passive leg movements on pulmonary ventilation in humans, after Morikawa et al[14]

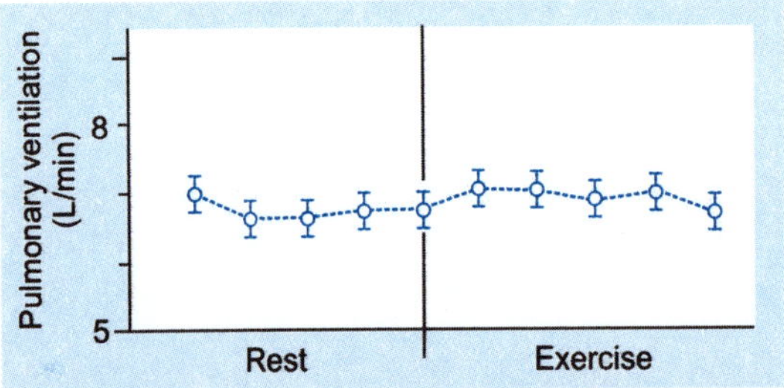

Fig. 7.4: Effect of passive leg movements in paraplegic patients on pulmonary ventilation, after Morikawa et al[14]

Mechanical information concerning muscle contraction is sensed by the mechanoreceptors in the muscles, tendons, or joints and reflexively stimulates the respiratory center in the medulla oblongata through afferent sensory nerve fibers, thus producing increased ventilation. This mechanism has been called "afferent neurogenic drive," or "peripheral neural reflex," Even, passive limb movement produces a rapid increase in ventilation at the initial stage of movement, indicating the involvement of afferent neurogenic drives from moving limbs as one of the causes of exercise hyperpnea. This mechanism is strongly supported by experiments showing that increased ventilation induced by electrical stimulation or manual passive limb movement disappeared in partial spinal cord resected or anesthetized animals[15, 16], or human patients with a spinal cord transection as mentioned earlier.[14] Now, it is well established that the pathways from the peripheral receptors to the respiratory center go through group III and/or IV thin afferent fibers. The endings of group III fibers are mainly connected to mechanoreceptors. They respond quickly at the onset of muscle contraction (within 200 ms) and begin to fire from a lower level of contraction force. On the contrary, the endings of group IV fibers are located near blood and/or lymphatic vessels, and mainly connected to metabo- or chemo-sensitive receptors. They respond later between roughly 5–20 from the onset of contraction, and have a higher threshold. It is assumed that peripheral neurogenic drive is mainly sensed by mechanoreceptors and goes through group III afferents from the onset of dynamic exercise. In spite of contradictory results observed by some investigators, the general consensus has been that the peripheral neural reflex mechanism has a role in exercise hyperpnea.[17]

Redundancy in Multiple Regulatory Systems

It is accepted that central command originated from the motor cortex or HTLR, and/or afferent drives mainly from mechanoreceptors in the muscles, transmitted through group III afferents, are the main causes of exercise hyperpnea. Whether central command or peripheral neurogenic drive plays a predominant role in exercise hyperpnea is still a matter of debate. It should be reasonable to consider that both are involved in exercise hyperpnea though a single mechanism should be able to elicit nearly the whole ventilatory response during exercise. It can be likened to a backup system that allows the system to function in the event that one of them is impaired. Waldrop et al[18] observed that simultaneous activation of both HTLR and muscle afferents in cats evoked increases in ventilation that were less than the sum of the increase evoked by separate activation. They also demonstrated that additional HTLR stimulation during muscular contraction caused a large increase in ventilation while additional muscular contraction during HTLR stimulation showed a small increase, indicating the predominance of hypothalamic central command. Many researchers now support the idea that a redundant multiple regulation system is the most reasonable way to explain exercise hyperpnea induced by two neurogenic mechanisms.[17]

References

1. Douglas CG, Haldane JS, Henderson Y, et al. The physiological effects of low atmospheric pressures, as observed on Pike's Peak, Colorado.(Preliminary Communication.) Proc Roy SocLond. 1912; 85: 65–67.
2. Forster HV. Recent advances in understanding mechanisms regulating breathing during exercise. *J Physiol.* 2014; 592: 429–31.
3. Kao FF, Cunningham DJ, Lloyd BB. An experimental study of the pathways involved in exercise hyperpnea employing cross circulation techniques. In *The Regulation of Human Respiration*, 461–502. F.A. Davis Company, Philadelphia, PA, USA, 1963.
4. Weissman ML, Whipp BJ, Huntsman DJ, et al. Role of neural afferents from working limbs in exercise hyperpnea. J Appl Physiol. 1980; 49: 239–48.
5. Parkes MJ. Evaluating the importance of the carotid chemoreceptors incontrolling breathing during exercise in man. Hindawi Publishing Corporation Bio Med Research International. 2013, Article ID 893506, 18 pageshttp//dx.doi.org/10.1155/2013/893506
6. Krogh A, Lindhard J. The regulation of respiration and circulation during the initial stages of muscular work. J Physiol. 1913; 47: 112–36.
7. Ransom SW, Magoun HW. Respiratory and pupillary reactions induced by electrical stimulation of the hypothalamus. Arch Neurol Psychiatry. 1933; 29: 1179–93.
8. Eldridge FL, Millhorn DE,Walldrop TJ. Exercise hyperpnea and locomotion: parallel activation from hypothalamus. Science. 1981; 211: 844–46.
9. Eldridge FL, Millhorn DE, Kiley JP. Stimulation by central command of respiration locomotion and circulation during exercise, Respir Physiol. 1985; 59: 313–37.
10. Holstege G. Some anatomical observations on the projections to the brainstem and spinal cord. J Comp Neurol. 1987; 260: 98–126.
11. Mateika JH, Duffin J. A review of the control of breathing. Eur J Appl Physiol. 1997; 71: 1–27.
12. Lipski JA, Martin-Body RL. Morphological properties of respiratory motor neurons in cats as revealed by intercellular injection of horse-radish peroxidase. J comp Neurol. 1987; 260: 423–34.
13. Mier A, Murphy K, Shea S, et al. Diaphragmatic EMG in response to cortical magnetic stimulation during sleep. Am Rev Respir Dis. 1990; 141: A 379–385.
14. Morikawa T, Ono Y, Sasaki K, et al. Afferent and cardiodynamic drives in the early phase of exercise hyperpnea in humans. 1 Nov, 1989 https://doi.org/10.1152/jappl.1989.67.5.2006
15. Pan LG, Forster HV, Wurster RD, et al. Effect of partial spinal cord ablation on exercise hyperpnea on ponies. J Appl Physiol. 1990; 69: 1821–127.
16. Hida W, Shindoh C, Kikuchi Y, et al. Ventilatory response to phasic contraction and passive movement in graded anesthesia. J Appl Physiol. 1986; 61: 91–97.
17. Ishida K, Miyamura M. Neural regulation of respiration during exercise: Beyond the conventional central command and afferent feedback mechanisms. J Phys Fitness Sports Med. 2012; 1: 235–45.
18. Waldrop TG, Mullins DC, Millhorn DE. Control of respiration by the hypothalamus and by feedback from contracting muscles in cats. Respir Physiol. 1986; 64: 317–28.

Enteric Nervous System

INTRODUCTION

The story of enteric nervous system began more than 150 years ago with the discovery of neurons in the wall of the intestine by Auerbach and Meissner.[1] Cajal, a neuro-anatomist, using silver staining, demonstrated different types of neurons in the gut.[2] The functional significance of these neurons was described for the first time by WM Bayliss and EH Starling in 1899.[3] They isolated a loop of dog intestine, increased the pressure within the loop, and observed a muscular movement, which caused forward movement of the contents. They called it "the law of the intestine", which is now known as the peristaltic reflex. Next, they cut all neural communications between the gut and the CNS. Since the reflex activity was unaffected, they introduced the term "the local nervous system". Trendelenburg confirmed these findings by demonstrating that the peristaltic reflex could be elicited *in vitro* in the isolated gut of a guinea pig, without participation of the brain, spinal cord, dorsal root, or cranial ganglia. "This finding is unique; no other peripheral organ had such a highly developed intrinsic neural apparatus. If you cut the connection linking the bladder or the skeletal muscles to the CNS, and all motor activity ceases. In spite of cutting CNS connection to the gut, function persists".[4]

ENTERIC NEURON PLEXUSES

The enteric nervous system (ENS) was originally thought to be part of the autonomic component of the peripheral nervous system, and the neurons in the gut wall were thought to be postganglionic parasympathetic neurons. The idea that the gut has a "brain of its own" arose when it was found that intestinal peristaltic contractions were coordinated reflexes involving the intramural nerves and that the majority of enteric neurons did not contact the parasympathetic axons of the central nervous system directly. Subsequent examination of the functional and chemical diversity of enteric neurons revealed that the enteric nervous system closely resembles the central nervous system. Enteric nervous system contains some 100 million neurons, approximately the number found in the spinal cord. The ENS is remarkably brain-like, both structurally and functionally. Its neuronal elements are not supported by collagen and Schwann cells, like those in the rest of the peripheral nervous system, but by glia that resemble the astrocytes of the CNS. These glia do not wrap individual axons in single membranous invaginations; rather, entire bundles of axons are fitted into the invaginations of enteric glia. The ENS is also vulnerable to what are generally thought of as brain lesions: Both the Lewy bodies associated with

Parkinson's disease and the amyloid plaques and neurofibrillary tangles identified with Alzheimer's disease have been found in the bowels of patients with these conditions.[5] The enteric nervous system may perhaps best be regarded as a displaced part of the central nervous system that retains communication with it through sympathetic and parasympathetic afferent and efferent neurons.

In the enteric nervous system, the nerve cell bodies are grouped into small ganglia that are connected by bundles of nerve processes forming two major plexuses, called the myenteric (or Auerbach's) plexus and the submucous (or Meissner's) plexus (Fig. 8.1). The density of neurons is high in the myenteric plexus, less in the submucous plexus.[6]

i. The **myenteric plexus** lies between the longitudinal and circular layers of muscle and extends the entire length of the gut. The myenteric plexus is also present in the striated muscle portion of the esophagus, where it innervates motor end plates with the inhibitory neurotransmitter nitric oxide. This innervation is unique to the esophagus.[7] Myenteric plexus primarily provides motor innervation to the two muscle layers, but there are numerous projections from the myenteric plexus to the submucosal ganglia and to enteric ganglia of the gallbladder and pancreas. There are also a substantial number of projections from the myenteric neurons to the sympathetic ganglia.

ii. The **submucous plexus**, located in the submucosa between the circular muscle layer and the muscularis mucosa, is best developed in the small intestine, where it plays an important part in secretory control. Besides innervating the glandular epithelium, neurons in the submucous plexus innervate the muscularis mucosa, intestinal endocrine cells, and submucosal blood vessels. A ganglionated plexus similar to the submucous plexus is found in the gallbladder, cystic duct, and common bile duct, as well as in the pancreas.[8]

The enteric ganglia consist of tightly packed nerve cell bodies, terminal bundles composed of nerve fibers, and glial cells and their processes. Glial cells are an integral component of the enteric nervous system, and they outnumber enteric neurons. Enteric glial cells resemble the astrocytes of the central nervous system, and their laminar extensions cover most of the surface of enteric neuronal cell bodies.

The enteric nervous system consists chiefly of two main morphological types of neurons. Type I neurons have many club-shaped processes and a single long, slender process, whereas type II neurons are multipolar and have many long, smooth processes.[9] The four primary targets of the enteric nerves of the gut

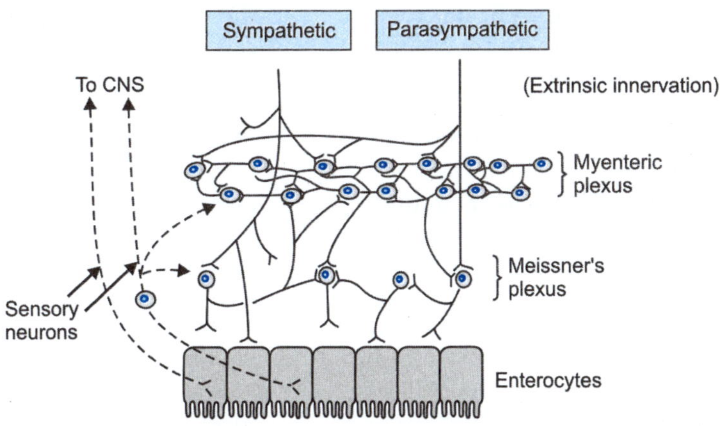

Fig. 8.1: Extrinsic and intrinsic innervation of the gut

are smooth muscle cells responsible for gastrointestinal motility; mucosal secretory cells; gastrointestinal endocrine cells; and the gastrointestinal microvasculature that maintains mucosal blood flow during intestinal secretion.

CHEMICAL NATURE OF ENTERIC NEURONS

The chemical mediators of the enteric nervous system were initially thought to be limited to neurotransmitters such as acetylcholine and serotonin. Subsequent research added purines to the list, such as ATP, amino acids such as γ-aminobutyric acid, and peptides such as vasoactive intestinal polypeptide. More recently, nitric oxide has emerged as a neurotransmitter in the enteric nervous system, as it has in the central nervous system. Overall, more than 20 candidate neurotransmitters have now been identified in enteric neurons, but, neurotransmitter functions have been clearly defined for only a few of these substances, including acetylcholine, substance P, vasoactive intestinal polypeptide, and nitric oxide. A wide variety of neurons that perform different functions may use the same neurotransmitter.[10]

CLASSIFICATION OF NEURONS OF ENS

The neurons that make up the enteric nervous system can be classified as intrinsic afferent neurons, interneurons, motor neurons, and secretomotor and vasomotor neurons (Fig. 8.2).

I. **The intrinsic primary afferent neurons (IPANs)** are located in both the myenteric and submucous plexuses. They constitute the sensory limb of all *intrinsic motor* and *secretomotor* activities of the intestine. *Morphologically, intrinsic afferent neurons are type II neurons.* They project circumferentially to interneurons in the surrounding myenteric and submucous plexuses. Electrophysiologically, the sensory neurons are characterized by prominent after-hyperpolarization and are therefore called AH neurons. After-hyperpolarization inhibits further excitation. They are all cholinergic and may

or may not contain substance P.[10] Three types of intrinsic afferent neurons have been described:

a. *Mucosal chemosensors:* Intestinal reflexes can be elicited by chemicals applied to the lumen. Effective stimuli included acid and alkaline solution and 5-HT.

b. *Mucosal mechanoreceptors:* Enteric reflexes are also elicited by mechanical stimuli, such as stroking, applied to the mucosa.

c. *Stretch responsive neurons:* These afferent neurons, involved in enteric motility reflexes, are evoked by distension.[11]

II. **Interneurons** are interposed between the primary afferent neurons and the motor or secretomotor neurons. Interneurons involved in motor reflexes are directed orally or aborally and are designated as ascending or descending, respectively. Motor neurons form multisynaptic pathways in the length of the gut that control the distances along the intestine for which peristaltic waves are propagated. Several subgroups of interneurons have been defined on the basis of their neurotransmitter content, but their various physiologic roles are unknown.

III. **Motor neurons**

a. *Longitudinal muscle motor neurons.* This relatively large class (25%) of small neurons with short projections to the longitudinal muscle (type I). They receive synaptic inputs from the enteric primary afferent neurons and from ascending and descending interneurons.

b. *Circular muscle motor neurons* are either excitatory or inhibitory. The *excitatory motor neurons project locally or orally* to the circular muscle, and their main neurotransmitters are acetylcholine and substance P. *The inhibitory motor neurons in the circular muscle project caudally* and contain vasoactive intestinal polypeptide and nitric oxide.[10]

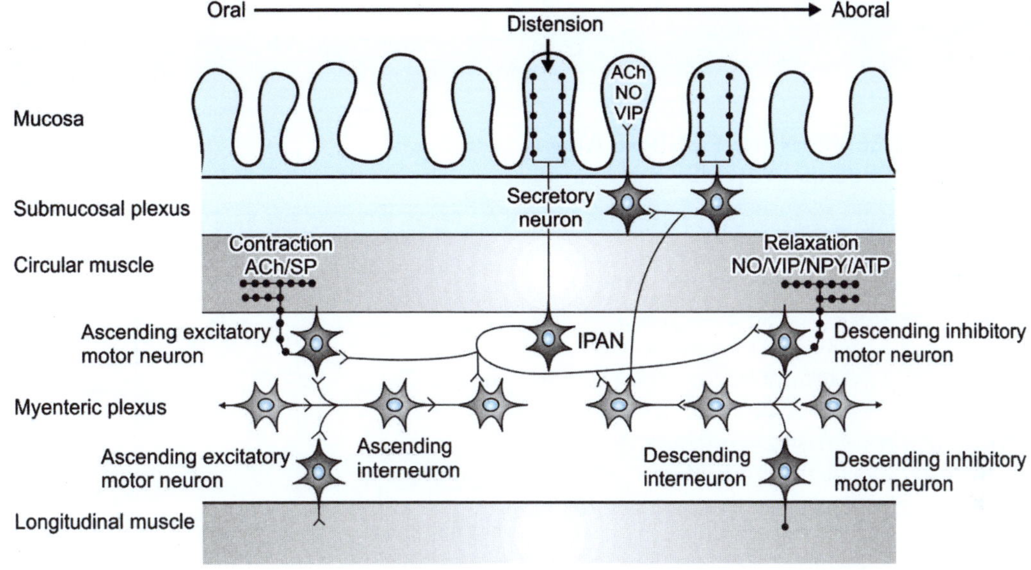

Fig. 8. 2: Types of neurons in enteric nervous system. IPAN: Intrinsic primary afferent neuron

IV. **Secretomotor and vasomotor neurons:** Intestinal secretions and blood flow are controlled by secretomotor and vasomotor neurons, respectively. Intrinsic afferent neurons, with processes in the mucosa, are activated by luminal chemical or mechanical stimuli on enterochromaffin cells of the mucosa causing release of 5-HT (Fig. 8.3). Intrinsic afferent neurons, with processes in the mucosa are stimulated by 5-HT. The intrinsic afferent neurons exert direct control over secretomotor and vasomotor neurons by releasing acetylcholine and vasoactive intestinal polypeptide. Most secretomotor neurons have their cell bodies located in the mucosal plexus, which is a network of fine nerve fiber bundles that lies beneath the mucosal epithelium. Some neurons project from the submucosa to the myenteric plexus, whereas others supply the muscularis mucosae. There are two main types of intestinal secretomotor neurons: Cholinergic and non-cholinergic. Acetylcholine released from the cholinergic neuron acts on muscarinic receptors on the mucosal epithelium. The non-cholinergic neurons appear to mediate most of the local reflex responses and utilize vasoactive intestinal polypeptide as a transmitter. Vasomotor (dilator) neurons also have their cell bodies in the submucous layer. The delicate balance of epithelial transport processes (mainly secretions) and blood flow is accomplished by the intrinsic secretomotor neurons through local reflex circuits. These reflexes are under extrinsic modulation, primarily via the sympathetics.[1, 11]

PERISTALTIC REFLEX CIRCUITRY

The accepted circuitry of the peristaltic reflex involves the following chain of events: (1) Mechanical or chemical stimuli in the gut lumen activate intrinsic primary afferent neurons (IPANs) residing in both plexuses; (2) IPANs activate interneurons that project in both oral (ascending) and aboral (descending) directions; (3) ascending interneurons activate excitatory motor neurons that cause smooth muscle contraction by releasing acetylcholine (ACh) and neuropeptides while descending interneurons produce relaxation below the point of stimulation by activating inhibitory motor neurons that release nitric oxide (NO), purines and other inhibitory molecules (Fig. 8.3).[12]

INTERACTIONS BETWEEN THE CENTRAL AND THE ENTERIC NERVOUS SYSTEMS

Although the enteric nervous system can function independent of the central nervous system, the latter has an important role in coordinating the diverse functions of the enteric nervous system. The enteric nervous system is well connected to the central autonomic neural network in the central nervous system through both motor and sensory pathways of the sympathetic and the parasympathetic nervous systems (Fig. 8.4).

The relative roles of the ENS and CNS differ considerably along the digestive tract. Movements of the striated muscle of esophagus are determined by neural pattern generators in the CNS. Likewise, the CNS has a major role in monitoring the state of the stomach and, in turn, controlling its contractile activity and acid secretion, through vago-vagal reflexes. In contrast, in the small intestine and colon, the ENS contains full reflex circuits, including sensory neurons, inter-neurons and several classes of motor neuron, through which muscle activity, transmucosal fluid fluxes, local blood flow and other functions are controlled. The CNS has control of defecation, via the defecation centers in the lumbosacral spinal cord. The importance of

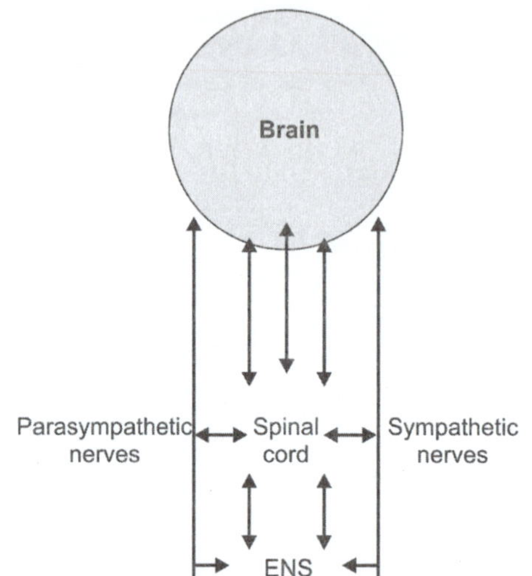

Fig. 8.4: Interaction between central nervous system and enteric nervous system

the ENS is emphasized by the life-threatening effects of some ENS neuropathies. By contrast, removal of vagal or sympathetic connections with the gastrointestinal tract has minor effects on GI function. Voluntary control of defecation is exerted through pelvic connections, but cutting these connections are not life-threatening and other functions are little affected.[13]

Motor Input from the Central Nervous System

The parasympathetic motor pathways consist of the vagus nerves that control the motor and secretomotor functions of the upper gastro-

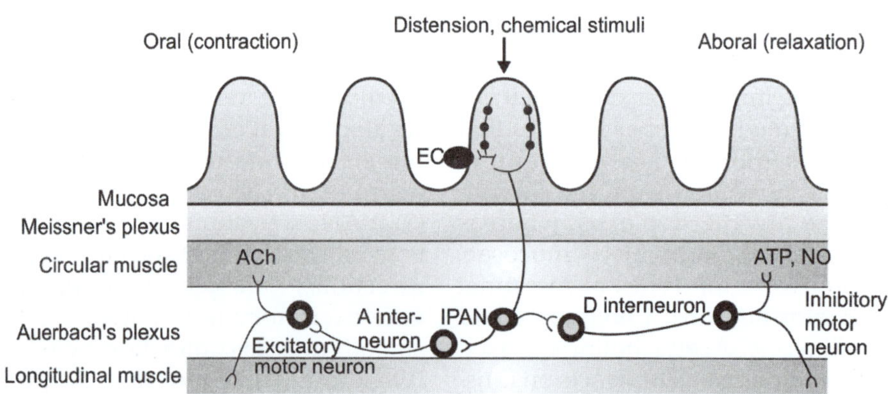

Fig. 8.3: Peristaltic reflex circuitry. EC: Enterochromaffin cell; IPAN: Intrinsic primary afferent neuron; D interneuron: descending interneuron; A interneuron: Ascending interneuron

intestinal tract and the sacral nerves that regulate the functions of the distal colon and rectum. The parasympathetic preganglionic neurons are all cholinergic and exert excitatory effects on enteric neurons through nicotinic and, in some regions, muscarinic receptors. These neurons provide rich connections to the myenteric neurons in the upper gastro-intestinal tract and the distal colon and anorectum. In the small bowel, however, vagal preganglionic neurons innervate only small clusters of select myenteric neurons[14] which may serve as command neurons or pattern generators. These differences in the intensity of innervation by parasympathetic pregang-lionic fibers reflect the fact that the central nervous system exerts more direct control in the most proximal (i.e. the esophagus and stomach) and the most distal (i.e. the rectosigmoid) parts of the gastrointestinal tract and less direct control on the functions of the small intestine and proximal colon. The sympathetic fibers entering the gut are adrenergic, postganglionic fibers with cell bodies in the prevertebral ganglia. They have at least four distinct targets in the gut: secretomotor neurons containing vasoactive intestinal polypeptide, presynaptic cho-linergic nerve endings, submucosal blood vessels, and gastrointestinal sphincters. Adrenergic nerve cell bodies are not found in the enteric plexuses.[10]

Sensory Output to the Central Nervous System

Neurons that carry sensory information to the central nervous system are termed primary afferent neurons (Fig. 8.1). They are carried in the vagal and splanchnic nerves. Vagal pri-mary afferent neurons have nerve cell bodies in the nodose ganglia. It has been estimated that 80% of the fibers in the vagal trunk are afferent fibers. The primary vagal afferent neurons in the smooth muscle layer are sen-sitive to mechanical distension of the gut; they have very low thresholds and convey infor-mation about physiologic motor activity in the gut. Certain vagal primary afferent neurons in the mucosa are sensitive to luminal concen-trations of glucose, amino acids, or long-chain fatty acids, whereas others respond to a wide variety of chemical and mechanical stimuli.[15]

Splanchnic primary afferent neurons have their endings in the gut wall and their cell bodies in the dorsal root ganglia. These afferent neurons are nociceptors and are involved in sensing pain in the gastrointes-tinal tract. They respond to high-intensity mechanical, thermal, and chemical stimuli that damage or threaten tissues.[16]

INTERSTITIAL CELLS OF CAJAL (ICC)

Interstitial cells of Cajal were initially identi-fied by Cajal in 1893[2], but their function remained a mystery for over a century. On the basis of staining characteristics with methy-lene blue and silver chromate, Cajal con-sidered them to be primitive neurons. Later on, ultrastructural studies suggested that ICC are either primitive muscle cells or fibroblast-like cells (Fig. 8.5).

Smooth muscles of alimentary tract reveal two types of potentials: Slow waves and functional spike potentials (Fig. 8.6). Basic electrical rhythm (BER) described also as slow waves, originates within longitudinal smooth muscle layer only. Initiators of BER are inter-stitial cells of Cajal. Smooth muscle cells do not have special ion mechanisms, necessary to generate slow waves. Cajal cells are charac-terized by ability to cyclic, spontaneous depolarization called slow waves. This is how they act as pacemaker in the alimentary tract. Slow waves migrate from these cells toward myocytes of longitudinal layer through gap junctions and nexuses, inducing electrotonic energy within internal circular layer. Most of the studies were carried out on cells of Cajal isolated from canine or mice intestine. Spontaneous and rhythmic electric activity was always observed in these cells, in comparison to smooth myocytes which at the same condition never showed spontaneous activity. Removal of ICC from a fragment of smooth muscle causes complete or almost complete lack of slow waves in the remaining part of smooth muscle.[17]

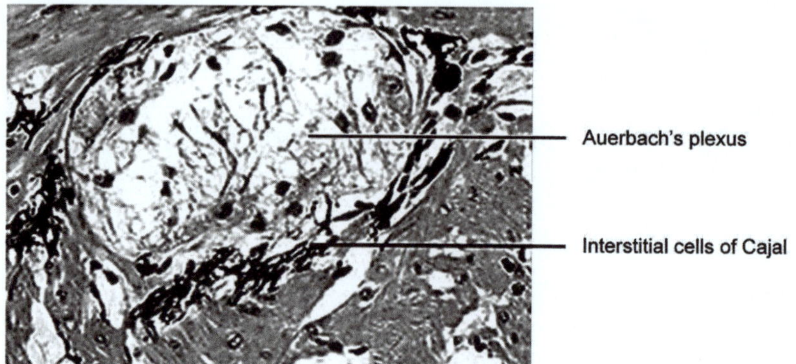

Fig. 8.5: Immunostain shows interstitial cells of Cajal surrounding the Auerbach's plexus

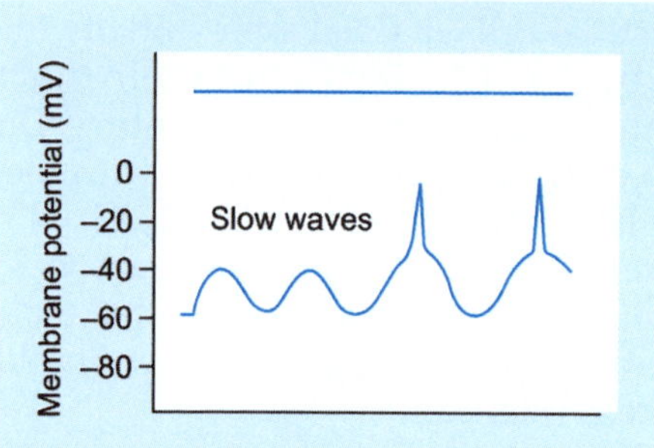

Fig. 8.6: Slow wave activity and spikes in smooth muscle of the intestine

FUNCTIONS OF THE ENTERIC NERVOUS SYSTEM

1. Control of Intestinal Motility

Whereas, esophageal and gastric motility is regulated by vagal control, intrinsic reflexes of the enteric nervous system are essential to the generation of the patterns of motility of the small and large intestines. The major muscle movements in the small intestine are: Mixing activity; propulsive reflexes that travel for only small distances; the migrating myo-electric complex; peristaltic rushes; and retro-pulsion associated with vomiting. The enteric nervous system is programmed to produce these different outcomes.

The smooth muscle sphincters restrict and regulate the passage of the luminal contents between regions. In general, reflexes that are initiated proximal to the sphincters relax the sphincter muscle and facilitate the passage of the contents, whereas reflexes that are initiated distally restrict retrograde passage of

contents into more proximal parts of the digestive tract.

The progress of the contents in an oral to anal direction is inhibited when (extrinsic) sympathetic nerve activity increases. To achieve this, transmission from enteric exci-tatory reflexes to the muscle is inhibited and the sphincters are contracted. The post-gang-lionic sympathetic neurons utilize nora-drenaline as the primary transmitter. Under resting conditions, the sympathetic pathways exert a little influence on motility. They come into action when protective reflexes are activated.[10]

2. Regulation of Fluid Exchange and Local Blood Flow

The enteric nervous system regulates the movement of water and electrolytes between the gut lumen and tissue fluid compartments. It does this by directing the activity of secre-tomotor neurons that innervate the mucosa in the small and large intestines and control its permeability to ions. Neurotransmitters of secretomotor neurons are vasoactive intes-tinal peptide (VIP) and acetylcholine. Secretion is integrated with vasodilatation. Most secretomotor neurons have cell bodies in submucosal ganglia.

Local blood flow to the mucosa is regulated through enteric vasodilator neurons so that the mucosal blood flow is appropriate to balance the nutritive needs of the mucosa and to accommodate the fluid exchange between the vasculature, interstitial fluid and gut lumen.

There are no intrinsic vasoconstrictor neurons. Overall blood flow to the gut is regulated from the CNS, via sympathetic vasoconstrictor neurons. In times of hypotension or hypovolemia, even they occur during digestion, the sympathetic can divert blood flow away from the gastrointestinal tract.[18]

3. Defense Reactions

Enteric neurons are involved in a number of defense reactions of the gut. Defense reactions include diarrhea to dilute and eliminate toxins, exaggerated colonic propulsive activity that occurs when there are pathogens in the gut, and vomiting.

Fluid secretion is provoked by noxious stimuli, particularly by the intraluminal presence of certain viruses, bacteria and bacterial toxins. This secretion is due in large part to the stimulation of enteric secretomotor reflexes. The physiological purpose is undoubtedly to rid the body of pathogens and their products. However, if the pathogens overwhelm the body's ability to cope, the loss of fluid (diarrhea) can become a serious threat to the organism.[18]

4. Entero-enteric Reflexes

Signals between different regions of the gut are carried both by hormones (such as cholecystokinin, gastrin and secretin) and by neural circuits. For example, when nutrients enter the small intestine, secretion of digestive enzymes from the pancreas occurs.

Entero-enteric reflexes regulate *motility* of one region of the intestine in relation to other. The entero-enteric reflex that is initiated by fat in the distal intestine and slows transit in the proximal small intestine is referred to as the "ileal brake". This response is mediated through enteric nervous system. Information is also exchanged between separate regions of the bowel along *extramural pathways* that bypass the circuits in the enteric nervous system. Distension of segments of intestine activates a reflex causing sympathetic inhibition of motility in another regions of the intestine. The reflex pathways between the intestine and prevertebral sympathetic ganglia are responsible for this type of entero-enteric reflexes. These reflexes, activated by overdistension in one region of the intestine, result in inhibited motility (paralytic ileus) in a distant region.[1,18]

References

1. Hansen BM. The enteric nervous system I: organization and classification. Pharmacol Toxicol. 2003; 92: 105–13.
2. Cajal SR. Sur les ganglions, et plexus nerveuxd'intestin. C R Soc Biol. 1893; 5: 217–23.
3. Bayliss WM, Starling EH. The movements and innervation of the small intestine. J Physiol Lond. 1899; 24: 99–143.
4. Trendelenburg, P. Physiologische und pharmakologische Verucheuber die Dunndarmperistaltik. Naunyn-Schmiedebergs Arch. Exp. Pathol. Pharmakol. 1917; 81: 55–129.
5. Gershon MD. The enteric nervous system: a second brain. Hosp Pract. 1999; 15; 34: 31–32.
6. Costa, M, Brookes SJ, Hennig G. Anatomy and physiology of the enteric nervous system. Gut. 2000, 47 (Suppl 4): iv15–19.
7. Worl J, Mayer B, Neuhuber WL. Nitrenergic innervation of rat esophagus: focus on motor end plates. J Auton Nerv Syst 1994: 49: 227–233.
8. Kirchgessner AL, Gershon MD. Innervation of the pancreasby neurons in the gut. J Neurosci. 1990; 10: 1626–42.
9. Furness JB, Costa M. The enteric nervous system. New York: Churchill Livingstone, 1987.
10. Goyal RK, Ikuo H. The enteric nervous system. N Engl J Med 1996; 334: 1106–15.
11. Furness JB. Types of neurons in the enteric nervous system. J Auton Ner Sys. 2000; 81: 87–96.
12. Grubišiæ V, Gulbransen BD.Enteric glia: the most alimentary of all glia. J Physiol. 2017; 595: 557–70.
13. Furness JB, Callaghan BP, Rivera LR, et al. The enteric nervous system and gastrointestinal innervation: integrated local and central control. AdvExp Med Biol. 2014; 817: 39–71.
14. Kirchgessner AL, Gershon MD. Identification of vagal efferent fibers and putative neurons in the enteric nervous system of the rat. J Comp Neurol. 1989; 285: 38–53.
15. Mei N. Intestinal chemosensitivity. Physiol Rev. 1985; 65: 211–37.
16. Sengupta JN, Gebhart, GF. Gastrointestinal afferent fibers and sensation. In: Johnson LR, ed. Physiology of gastroinrestinal tract. 3rd edn. New York: Raven Press, 1995.
17. Pasternak A, Szura M. Gills K, et al. Interstitial cells of Cajal—systematic review. Folia Morphol. 2016; 75: 281–86
18. Furness JB. Enteric nervous system. Schlorpedia. 2007; doi: 10.4249/scholarpedia. 4064.

Chapter 9

Gastrointestinal Hormones

INTRODUCTION

The publication of an article by Bayliss and Starling in 1902 is generally regarded as the birth of general endocrinology and endocrinology of gastrointestinal tract.[1] They discovered that infusion of hydrochloric acid into a denervated loop of the small intestine caused secretion of pancreatic juice in dogs. These findings led them to a conclusion that a chemical substance released by the acidified intestine and transported by the bloodstream to the pancreas triggered stimulation of pancreatic secretion. They also showed that an acid extract of the intestinal mucosa given intravenously stimulated pancreatic secretion. Those findings confirmed that the exocrine pancreas was stimulated by a chemical factor produced in the intestinal mucosa and released by luminal acid.[1] They called this substance secretin. The term "hormone" was first introduced by Starling in 1905 to describe a chemical substance that transmits information between cells.[2] Between 1900 and 1960, only three gastrointestinal hormones (secretin, gastrin and cholecystokinin) were discovered. In the next 40 years, the development of novel analytic methods that helped to isolate active substances from tissue extracts and determine their chemical composition and molecular structure led to discovery of as many as ten gastrointestinal hormones.

The function of the gastrointestinal system is regulated by gastrointestinal neurons, as well as hormones produced by enteroendocrine cells. The gut is innervated by two types of neurons: (i) Intrinsic neurons consisting of two plexuses located within the gut, namely the myenteric plexus (Auerbach's plexus) and the submucosal plexus (Meissner's plexus). (ii) The extrinsic innervation of the gastrointestinal tract is a part of the parasympathetic and sympathetic nervous systems.

ENTEROENDOCRINE CELLS

The enteroendocrine cells (EECs) within the gut do not form separate endocrine glands. Gastrointestinal EECs are restricted to the mucosa, predominately located within its deeper half, and comprise only a small minority (<1%) of the overall epithelial cell population. They are mostly lying isolated from one another interspersed by nonendocrine epithelial cells. The frequency of EECs is highest in the duodenum and falls steadily in the small intestine. The hormones are stored inside the endocrine cells in the form of secretory granules for immediate release by exocytosis. The enteroendocrine cells may be divided into two groups: Open-type cells and closed-type cells (Fig. 9.1). The open-type endocrine cells have apical cytoplasmic processes that reach the intestinal lumen and

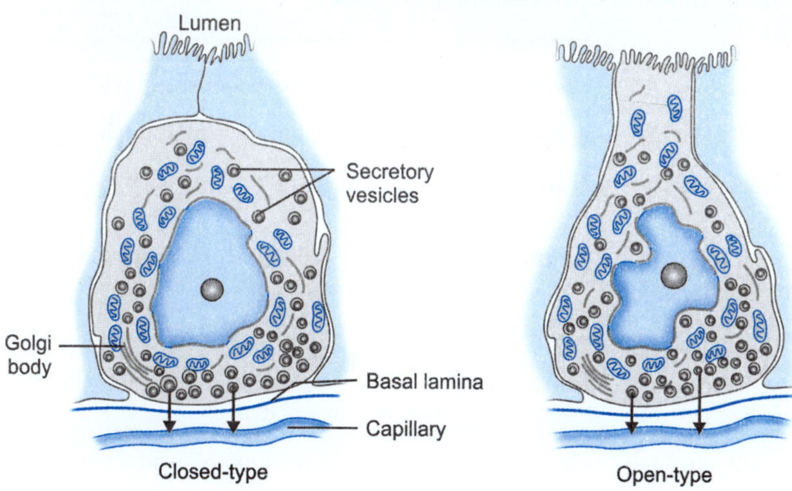

Fig. 9.1: Closed- and open-types of enteroendocrine cells

receive chemical information from intestinal contents. Examples of open-type of neuro-endocrine cells include G cells, S cells, I cells, etc. Enterochromaffin-like cells (ECL cells) are example of closed-type endocrine cells which do not make contact with the intestinal lumen and respond only to neural or hormonal signals.[3]

The open-type of EECs also acts as sensors of luminal contents, particularly nutrients. Those EECs 'open' to the lumen thereby function as transepithelial signal transduction conduits, with apical physiochemical signals resulting in basolateral exocytosis of bio-logical mediators (Fig. 9.2). These cells act either in a classical endocrine fashion or by a paracrine effect on proximate cells or afferent nerve terminals of enteric nervous system as well as extrinsic innervation of the gut. Their diffuse distribution is thus functionally appropriate. Non-nutrient chemical factors also regulate EEC activity. For example, secretin is secreted in response to duodenal acidification, and 5-HT secreting EECs have mechanosensory properties.[4,5]

Gastrointestinal hormones exert their effects by multiple pathways using paracrine, autocrine, neurocrine, as well as endocrine routes. Peptidyl hormones produced by the enteroendocrine cells of the gut can be

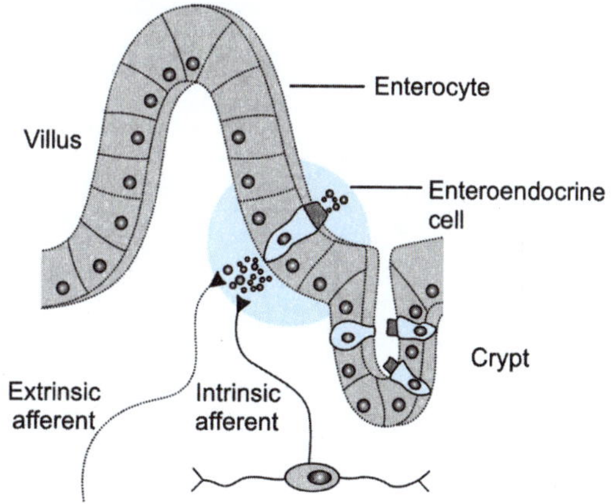

Fig. 9.2: The transepithelial transducer function of enteroendocrine cells

classified by similarity in structure and function into several main categories or hormone families described in Table 9.1.

TURNOVER OF GASTROINTESTINAL MUCOSAL CELLS

A single multipotent class of stem cells gives rise to the various differentiated cell types of the **gastric mucosa**. Proliferative cells are located in the isthmus region of the gland (the part of the gland immediately adjacent to the opening into gastric pit). Some cells migrate

Table 9.1: List of enteroendocrine cells, their secretions and functions[6]

Enteroendocrine cell	Secretion	Function
A cell (stomach)	Ghrelin	Appetite control, growth hormone release
D cell (stomach, small intestine)	Somatostatin	Inhibition of gastrin release
Enterochromaffin-like (ECL) cell (stomach)	Histamine	Gastric acid secretion by parietal cells
Enterochromaffin cell (small intestine, large intestine)	Serotonin	Facilitation of intestinal motility reflexes and secretion
G cell (stomach)	Gastrin	Stimulation of gastric acid secretion
I cell (proximal intestine)	Cholecystokinin	Bile secretion, exocrine pancreas secretion, and satiety
K cell (proximal small intestine)	Gastric inhibitory peptide	Stimulation of insulin release
L cell (distal small intestine, large intestine)	Glucagon-like peptide-1, peptide YY, and oxyntomodulin	Stimulation of carbohydrate uptake, slowing of intestinal transit, appetite regulation, insulin release
M cell (small intestine)	Motilin	Initiation of migrating myoelectric complex
N cell (small intestine)	Neurotensin	Control of smooth muscle contraction
P cell (stomach)	Leptin	Appetite regulation, reduction of food intake
S cell (proximal small intestine)	Secretin	Stimulation of exocrine pancreatic secretion

upwards, becoming mucus-secreting surface epithelial cells and surviving 2–3 days; other cells within the proliferative zone differentiate into parietal cells, chief cells or ECL cells and migrate towards the base of the gland, surviving 50–190 days (Fig. 9.3).[7]

In the **small intestine**, stem cells, intercalated with Paneth cells, are located at the crypt base (Fig. 9.4). These stem cells continuously generate rapidly proliferating transit-amplifying (TA) cells, which occupy the remainder of the crypt. TA cells differentiate into the various functional cells on the villi (enterocytes, tuft cells, goblet cells and enteroendocrine cells) to replace the epithelial cells being lost at the villus tip. Epithelial turnover occurs every 3–5 days. New Paneth cells are supplied from the TA cells every 3–6 weeks.[7]

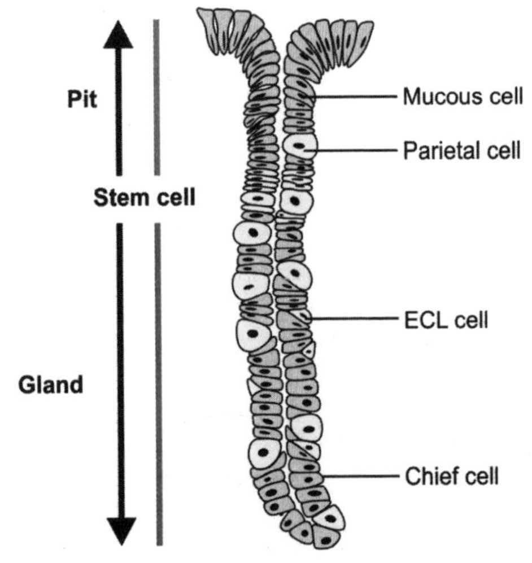

Fig. 9.3: Differentiation of gastric mucosal stem cells

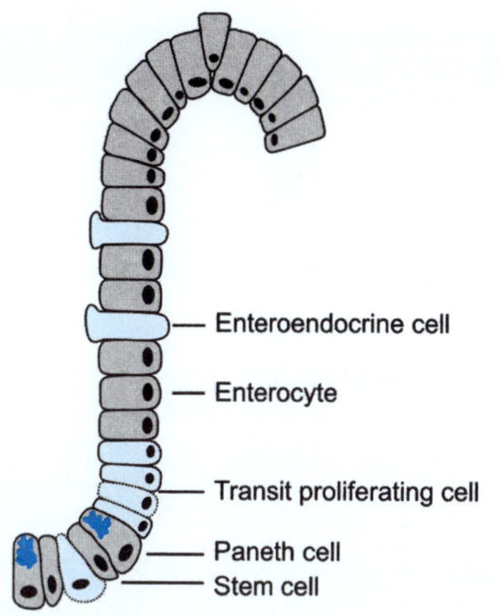

Fig. 9.4: Turnover of epithelial and enteroendocrine cells in the small intestine

GASTROINTESTINAL ENDOCRINES

I. Gastric Endocrines

a. Gastrin

Gastrin is produced and released by **G cells** in the stomach. G cells are open-type endocrine cells. Their apical processes project into the gastric lumen and thus the luminal content influences their secretory activity. G cells are found mainly in the pyloric mucosa of the stomach and a few in duodenal mucosa. Ninety percent of the gastrin is produced by G cells in the stomach and the remaining 10% is produced by G cells in the duodenum. Gastrin is present in three biologically active molecular forms: 34-amino acid peptide called the "big gastrin" (G-34), 17-amino acid peptide called the "little gastrin" (G-17) and 14-amino acid peptide called the "mini-gastrin" (G-14). The enzymatic processing of preprogastrin produces all of the above mentioned physiologically active forms of gastrin (Fig. 9.5). The fourteen C-terminal amino acids are identical in all three forms (G-14, G-17 and G-34). The active forms of gastrin, G-14, G-17 and G-34, are about equipotent in their ability to stimulate acid secretion. However, G-17 is the principal secretion of the antrum. Its plasma concentration rises during the digestive phase, and it is believed to be the main physiologic form of the hormone that is released in response to a meal. During the interdigestive phase, G-34 is the principal form found in the circulation.

b. Gastric Somatostatin

Vagal stimulation increases secretion of gastrin. The post-ganglionic fibers of the vagus nerve that innervate the G cells of the stomach release gastrin releasing peptide (GRP) (also known as bombesin), which stimulates the G cells to release gastrin. Simultaneously, vagal fibers also innervate and inhibit, *via* the enteric nervous system, D cells of pyloric antrum. **D cells** secrete somatostatin which acts in paracrine fashion to suppress gastrin release form G cells. Thus, vagal stimulation not only stimulates G cell secretion directly but also inhibits somatostatin secretion by D cells; both the actions complement each other to promote gastrin release. The secretion of somatostatin and the secretion of gastrin are closely related to pH of the gastric content. The secretion of somatostatin by D cells is directly stimulated by increased concentration of H^+ leading to inhibition of gastrin and gastric acid secretion.

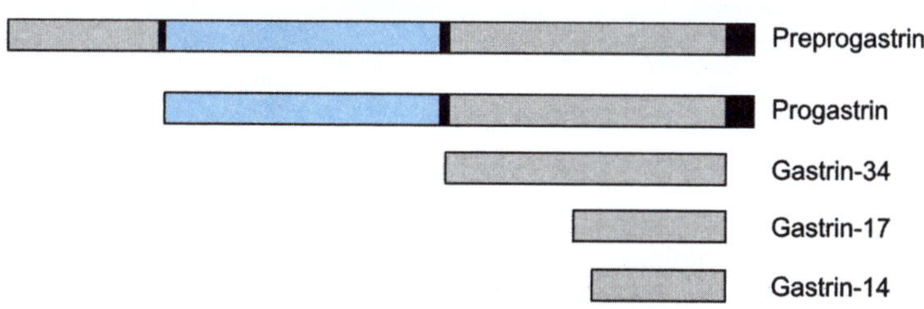

Fig. 9.5: Biogenesis of gastrins

Gastrin release is partially inhibited at an antral luminal pH of 3.5 and completely inhibited at a luminal pH of 2.0 (Fig. 9.6).

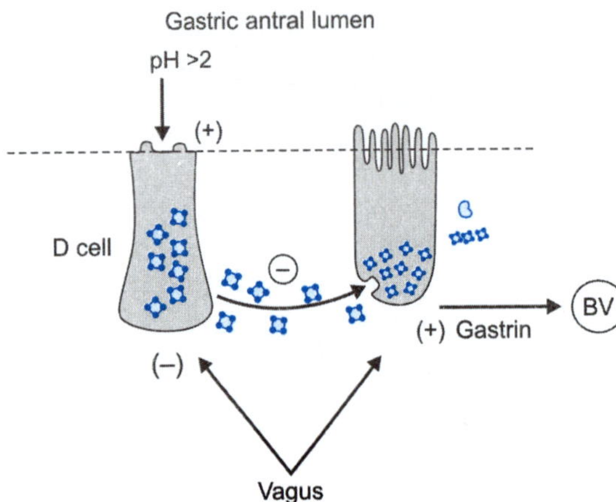

Fig. 9.6: Control of secretion of gastrin by vagus (BV: Blood vessels)

Ingestion of a meal is the strongest stimulant of the secretion of gastrin. Small peptides, aromatic amino acids (e.g. phenylalanine and tryptophan) act directly on the G cell to stimulate gastrin release. Elevated calcium in serum or in gastric lumen, alcoholic beverages, and coffee are also effective stimuli for gastrin secretion.[8] Distension of the gastric walls causes reflexive gastrin release. These reflexes operate through long, vagovagal and short, enteric nervous system pathways.

A primary physiological effect of gastrin is the stimulation of gastric acid secretion by the parietal cells of the gastric glands in the gastric mucosa. Gastrin stimulates HCl secretion mainly *via* binding to the enterochromaffin-like cells (ECL), and to a lesser extent, by direct action on parietal cells.[3] ELC cells release histamine in response to gastrin. Histamine, in turn, stimulates HCl secretion by binding to the H_2 receptor on nearby parietal cells (Fig. 9.7).[9]

c. Gastric Histamine

The endocrine systems of the antrum and corpus contribute to the gastric acid secretion in different ways. The corpus epithelium contains endocrine cells that regulate acid secretion through paracrine mechanisms, the most important being histamine-releasing **enterochromaffin-like (ECL)** cells and somatostatin-releasing D cells. The two main endocrine cell types of the corpus differ from antral endocrine cells in that they seem not to respond directly to luminal chemicals. Instead, they act as integrators of neuro-humoral signals. Thus, gastrin releases histamine from ECL cells, which in turn increases acid secretion by acting at parietal cell histamine-H_2 receptors (Fig. 9.7). Both parietal and ECL cells are inhibited by somatostatin released from corpus D cells in response to a variety of neurohumoral agents, e.g. noradrenaline, vasoactive intestinal polypeptide (VIP), calcitonin gene-related peptide (CGRP) and cholecystokinin (CCK).[10]

Besides increasing gastric secretion, gastrin inhibits gastric emptying, stimulates both

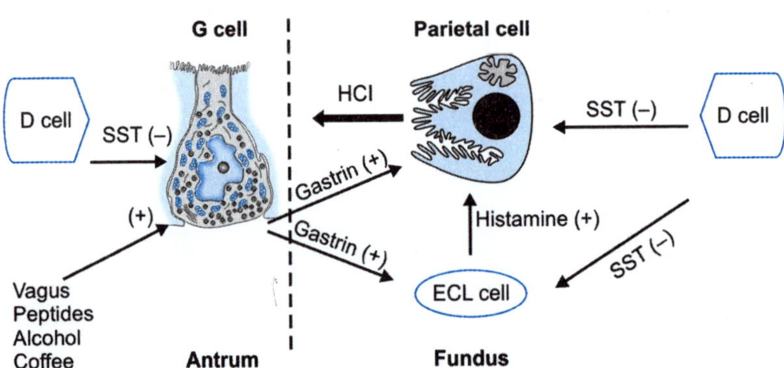

Fig. 9.7: Regulation of gastric acid secretion and role of D cells in stomach. ECL cells: Enterochromaffin-like cells; SST: Somatostatin

gastric and intestinal motility, constricts the pyloric sphincter and lower esophageal sphincter.

ROLE OF GASTRIN IN MUCOSAL GROWTH

One of the most important functions of gastrin is to regulate gastric mucosal proliferation. Gastrin has been recognized as the single most important trophic hormone of the stomach.[11] The regulatory effect of gastrin on gut mucosal growth was initially identified more than three decades ago. Studies by Johnson[12] revealed that administration of a synthetic gastrin analog, pentagastrin, increases parietal cell mass in rats. Removal of endogenous gastrin by antral resection causes mucosal atrophy that can be overcome by exogenous administration of pentagastrin. In the stomach, the oxyntic, acid-secreting mucosa and enterochromaffin-like cells are particularly sensitive to the trophic actions of gastrin. Subsequently, increased fundic mucosal proliferation was also demonstrated in rodent models following the administration of H_2-receptor antagonists, which results in hypergastrinemia.[11] Consistently, patients with Zollinger-Ellison syndrome are associated with an increase in gastric mucosal growth due to high levels of gastrin in the circulation. In the small intestine, only duodenal mucosa seems to be sensitive to trophic action of gastrin.[11-13]

II. Intestinal Endocrines

1. Cholecystokinin (CCK)

The original discovery of CCK in 1928 was based upon the observation that a substance within intestinal extracts stimulated gallbladder contraction in dogs (hence the name cholecystokinin). In 1943, a similar extract (which was called pancreozymin) was noted to stimulate pancreatic enzyme secretion. However, purification of the hormone and determination of its amino acid sequence showed that the actions on the gallbladder and pancreas were due to the same hormone. CCK possesses a five amino acid sequence at the carboxyl terminus that is identical to that of gastrin (Fig. 9.8). The carboxyl terminus confers the biologic activity of CCK; as a result, gastrin has weak CCK-like activity and CCK has weak gastrin-like activity.

CCK is synthesized principally in the open-type **I cells** of the duodenum and (to a lesser extent) in proximal 2/3 of the jejunum. CCK exists in multiple molecular forms, including molecules containing 8, 22, 33, 39 and 58 amino acids. CCK-8 and CCK-33 are thought to play a key role in gastrointestinal physiology.

The principal stimuli for CCK release from the duodenal and jejunal I cells are digestion products of fat and protein in lumen of the intestine. Fatty acids with carbon chain lengths of eight or longer or their monoglycerides are the most potent stimulus for CCK release. Peptides and aromatic (particularly tryptophan and phenylalanine) or aliphatic (particularly valine and methionine) amino acids, are also strong stimulants. CCK secretion is also induced by other intraluminal factors such as CCK-releasing peptide (CCK-RP) that is released from enterocytes or monitoring peptide that is released by pancreatic acinar cells. Both peptides stimulate CCK release by I cells of the duodenal and jejunal mucosa.[14] In addition, release of CCK is stimulated by acetylcholine (released by the parasympathetic nerve fibers of the vagus nerve).[14]

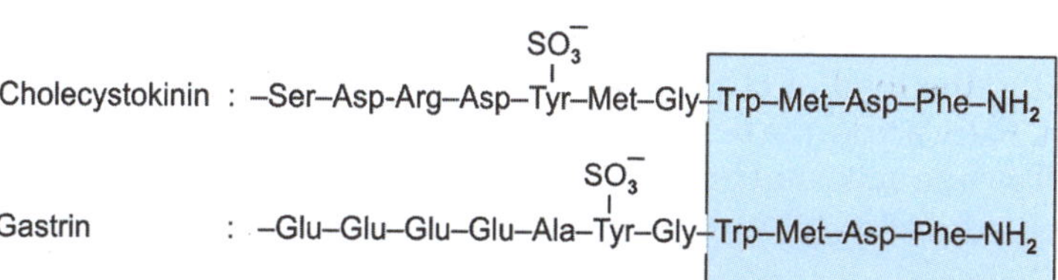

Fig. 9.8: Similarity in terminal amino acid sequence between CCK and gastrin

CCK has numerous physiological actions. It is a strong stimulus for pancreatic enzyme secretion. It also potentiates secretin-induced bicarbonate production in the pancreas. CCK stimulates pancreatic enzyme secretion by acinar cells mainly *via* activation of CCK1 receptors on vagal afferent terminals. Activation of CCK receptors triggers vagal efferents from motor nuclei in the brainstem leading to an increase in acetylcholine release and stimulation of enzyme secretion from the pancreatic acinar cells[15] (*see* Chapter 11 for details).

An increasing body of evidence demonstrates the trophic effects of CCK on the pancreas. Chronic administration of CCK leads to increase in pancreatic DNA synthesis, pancreatic content of DNA, RNA and protein and pancreatic weight. Secretin potentiates this trophic effect.[16] Inactivation of CCK receptors by CR 1409 completely abolishes the trophic effects of exogenous CCK-8. In addition, treatment with CR 1409 alone also decreases pancreatic weight, DNA, and protein content.[17]

CCK is a potent stimulus for bile expulsion from the gallbladder by causing the smooth muscle of the gallbladder to contract and the sphincter of Oddi to relax. CCK also has been shown to delay gastric emptying. CCK stimulates insulin secretion from α cells in the pancreas.

2. Secretin

Secretin is produced by **S cells** that are most abundant in the duodenum. Some S cells are present in the jejunal and ileal mucosa also. S cells are open-type cells. Secretin is a 27-amino acid peptide. The most potent stimulus for secretin release is acidification of the duodenum, a consequence of emptying of the stomach. Secretin release begins when duodenal pH falls below 4.0. Secretin release is directly related to the load of H⁺ delivered to the duodenum. Fatty acids, particularly the oleic acid, as well as bile salts and ethanol are weak stimulants for secretin release.

The primary action of secretin is the stimulation of the secretion of bicarbonate from the pancreatic and biliary ducts, and duodenal Brunner's glands. Secretin potentiates CCK-induced pancreatic enzyme secretion, inhibits gastric acid secretion, and inhibits gastric motility and gastric emptying.

3. Glucose-dependent Insulinotropic Polypeptide (GIP)

Initially, GIP was shown to suppress gastrin release and acid secretion and to inhibit gastric and intestinal motility, hence named gastric inhibitory peptide (GIP). However, it was subsequently observed that the effects on gastric acid secretion are seen only at very high concentrations of GIP, outside the physiological range. It is now generally accepted that the major physiological action of GIP is the incretin activity leading to stimulation of insulin release in response to food ingestion. Therefore, this peptide is currently called glucose-dependent insulinotropic polypeptide; acronym remains the same, GIP.

GIP is secreted by the **K cells** in the duodenal and jejunal mucosa. GIP is a linear 42-amino acid peptide. The complete molecule is required for full biological activity. The primary stimulus for GIP release is food, especially glucose (Fig. 9.9), as well as long-chain fatty acids and their monoglycerides in the lumen of the small intestine. Certain amino acids (arginine, histidine, isoleucine,

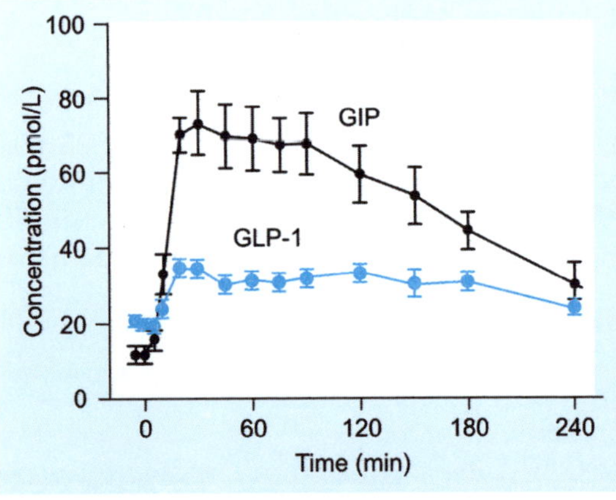

Fig. 9.9: Effect of oral ingestion of 75 g of glucose on plasma GIP and GLP-1 levels

leucine, lysine and threonine) are effective in releasing GIP. Acidification of the upper small intestine has no role in GIP release. Its release may be inhibited by high levels of insulin or glucagon.

Many decades ago, oral glucose load was shown to produce a much greater insulin response than intravenous injection of glucose.[18] Now it is confirmed that the effect is attributable two polypeptides called incretins (GIP and GLP-1) which are released from the upper gut after ingestion of glucose or nutrients and stimulate insulin secretion from pancreatic β cells. GIP secretion from K cells is enhanced in response to ingestion of meals. As compared to ingestion of 75 g of glucose, a mixed meal produces greater increase in plasma GIP levels. Moreover, ingestion of proteins produces more rapid and robust GIP secretion than that of fats.[19] GIP also promotes β cell proliferation and reduces β cell apoptosis.

In adipose tissue, GIP stimulates the insulin-dependent glucose uptake, fatty acids synthesis and lipogenesis, as well as reduces glucagon-stimulated lipolysis. However, unlike GLP-1, GIP does not inhibit, but stimulates glucagon secretion in humans. GIP administration in healthy subjects increases food intake and decreases energy expenditure. For all these reasons, GIP seems to be *useless as a potential remedy in the treatment of diabetes.*[20]

4. Glucagon-like Peptide 1 (GLP-1)

GLP-1, formerly called enteroglucagon, is a 30-amino acid peptide hormone, produced by intestinal endocrine **L cells**. The open-type L cells are most abundant in the distal ileum and colon, but are also present in the duodenum.[21] GLP-1 is released in response to food, especially glucose (Fig. 9.9), fatty acids, proteins and bile salts in the bowel.[21] Like GIP, GLP-1 is an incretin hormone. GLP-1 stimulates secretion of endogenous insulin and inhibits glucagon release. Administration of GLP-1 leads to decrease in blood glucose levels both in healthy individuals and in patients with type 2 diabetes.[22]

As mentioned above, GLP-1 inhibits glucagon secretion. Since in patients with type 2 diabetes there is fasting hyperglucagonemia as well as exaggerated glucagon responses to meal ingestion, and since it is likely that the hyperglucagonemia contributes to the hyperglycemia of the patients, this glucagon-inhibitory effect of GLP-1 may be as important clinically as the insulinotropic effects. Indeed, in patients with type 1 diabetes and complete lack of β cell activity (C-peptide negative), GLP-1 is still capable of lowering fasting plasma glucose concentrations, presumably as a consequence of a powerful lowering of the plasma glucagon concentrations. Other important effects of GLP-1 in the gut include the inhibition of pancreatic and gastric exocrine secretion, the reduction of gastric emptying, and suppression of hepatic glucose production.

Animal and human experimental studies mentioned above have indicated that GLP-1 may be useful in the therapy of type 2 diabetes and obesity. However, GLP-1 has a very short circulating half-life of 5–6 min. This problem has been solved by use of chemically modified degradation-resistant agonists of GLP-1 receptors. Administration of GLP-1 receptor agonists reduces fasting and postprandial blood glucose level without developing hypoglycemia, decreases concentration of HbA1c and lowers body weight in diabetic patients.[23]

Unabsorbed nutrients in the ileum inhibit the motility of upper gut ("ileal brake") (see below). PYY, and possibly GLP-1, participate in the ileal brake.

5. Somatostatin

Somatostatin exists in two main molecular forms: One composed of 14-amino acids (SS14) and the second one composed of 28-amino acids (SS28). Both SS14 and SS28 derived from their common precursor, a 116-amino acid presomatostatin peptide, and both have identical fourteen C-terminal amino acid residues. Somatostatin is produced and released by neurons in the central and peripheral nervous systems, as well as by

D cells in the gastrointestinal mucosa and pancreatic islets. D cells may be both open-type and closed-type cells, depending on their location.

Somatostatin release is stimulated by dietary components of a meal, especially fat and proteins. Low pH of the gastric contents increases somatostatin release from pyloric D cells. Somatostatin release is also induced by numerous hormones and neuropeptides that inhibit hydrochloric acid secretion in the stomach, including glucagon, secretin, GLP-1, CCK, GRP, and VIP.[3] Gastrin upregulates somatostatin synthesis and release. This suggests that somatostatin plays a role in the prevention of hypergastrinemia through a negative feedback mechanisms.[3] Moreover, somatostatin inhibits gastrointestinal motility, decreases visceral blood flow, as well as cell growth and cell renewal in the gastrointestinal tract.[24]

6. Peptide YY (PYY)

Peptide YY consists of 36-amino acids. PYY is secreted by **L cells** in the ileum and colon in response to food intake. Fats and other products of food digestion, such as carbohydrates are primary stimulants for PYY release. It is also supposed that neural mechanisms, including GRP, are involved in PYY secretion. PYY inhibits both gastric and intestinal motility and suppresses gastric and pancreatic exocrine secretion. Reduced intestinal motility may increase efficiency of food digestion and nutrient absorption. This phenomenon called "ileal brake" seems to be an effect of GLP-1 and PYY.[25]

7. Ghrelin

Ghrelin is a 28-amino acid peptide, predominantly secreted in the stomach. However, it was also found in other body organs such as intestine, pancreas, kidneys, pituitary gland and hypothalamus. After the discovery of ghrelin, it was realized that the stomach is an important organ not only for digestive function but also for the regulation of energy metabolism and the secretion of GH. Two important functions of ghrelin include:

i. **Regulation of appetite and food intake:** Ghrelin is believed to influence not only a short-term but also a long-term regulation of food intake. Ghrelin controls energy balance of the body. This peptide increases food intake and fat accumulation in adult humans and animals. Fasting plasma ghrelin level is inversely correlated with body mass index (BMI). Therefore, plasma ghrelin level is low in patients with obesity and high in those with anorexia. Food has significant effect on circulating ghrelin levels which are high during periods of fasting and low after eating.[26]

Ghrelin acts as a hunger signal from peripheral tissues. Intravenous injections of ghrelin increase food intake; likewise, peripherally injected ghrelin stimulates hypothalamic neurons and food intake. The factors involved in the regulation of ghrelin secretion have not yet been identified. Blood glucose level may be a most probable candidate: Oral or intravenous administration of glucose decreases plasma ghrelin concentration.[27] Because gastric distension by water intake does not change ghrelin concentration, mechanical distension of the stomach alone does not seem to decrease ghrelin release after food intake.

Human studies have shown that ghrelin could be useful in the treatment of diseases associated with negative energy balance. Administration of exogenous ghrelin and its analogs improves appetite and food intake in patients with cancer. Antagonists of ghrelin, as anorexigenic agents, could be valuable in the treatment of obesity and type 2 diabetes.

ii. **Regulation of growth hormone secretion:** Another important action of ghrelin involves its strong GH-releasing activity. The maximal stimulation effected by ghrelin is two to three times greater than that of GHRH, in both rats and human. Ghrelin stimulates GH release from anterior pituitary gland by a direct action on somatotrophs. A synergistic effect of ghrelin and GHRH is also important. Co-administration of ghrelin and GHRH results in more GH release than does either GHRH or ghrelin alone.[28]

8. 5-Hydroxytryptamine

Enterochromaffin (EC) cells are the most abundant enteroendocrine cells (EECs) of the GI tract and are distributed widely, populating the gastric antrum, duodenum, jejunum, ileum and appendix *as well as the colon and rectum. EC cells have* been shown to make up over 70% of the EEC population in the proximal large bowel, which falls to around 40% in the rectum. Originally named because of their affinity to bind chromium salts, EC cells' main secretory product is serotonin (5-HT). Ninety-five percent of the body's 5-HT content is within the GI tract, the majority of which is contained within the secretory granules of EC cells. Furthermore, 5-HT content is said to be highest in the rectum.[29]

The primary function of EC cells is to synthesize and secrete serotonin (5-HT) for modulation of gastrointestinal neurons. In the ENS, serotonin is an essential modulator of sensory transduction and mucus secretion. Release of serotonin from EC cells can be triggered by a multitude of stimuli, particularly luminal distension, parasympathetic innervation or changes in osmotic concentrations in intestinal contents.[30] 5-HT causes increased gut motility and accelerated intestinal transit. It has been suggested that bidirectional interaction between EC cells and the enteric nervous system is partly responsible for this effect whereby 5-HT released from EC cells binds to 5-HT 4 receptors on the nerve terminals of intrinsic primary afferent neurons (IPANs), which in turn results in the activation of excitatory cholinergic neurons that innervate smooth muscle to initiate peristaltic activity.[31]

9. Neurotensin

Neurotensin, a 13-amino acid peptide, is produced and released by open-type endocrine **N cells** in the gastrointestinal tract. N cells are predominantly expressed in the ileum, though such cells are also present in small number in the jejunal, duodenal and colonic mucosa. In the gut, it is released after a meal, particularly those containing high lipid levels. It has a range of paracrine and endocrine functions modulating vascular smooth muscle activity, gastrointestinal motility, and pancreaticobiliary secretions.[32]

10. Motilin

Motilin is a 22-amino acid peptide produced by **M cells** that are found primarily in the duodenum and in smaller number in the more distal portions of the small bowel. Plasma motilin level is low during the digestive phase of gastrointestinal activity and its concentration increases, and varies cyclically during the interdigestive phase, with peak in the phase III of this period.

Motilin is considered to be a unique hormone playing a role in the interdigestive period, rather than, as in the case of most gastrointestinal hormones, in the postprandial period. In the interdigestive state, GI motility is characterized by cycling interdigestive migrating contractions (IMCs) originating in the stomach and propagating along the small intestine. IMCs are assumed to have an important housekeeping role by forcefully pushing the content of the gut forward while cleaning the bowel of debris and bacteria that would otherwise accumulate and lead to bacterial overgrowth and compromise nutrient absorption from the small intestine.[33]

References

1. Bayliss WM, Starling EH. The mechanism of pancreatic secretion. J Physiol. 1902; 28: 325–353.
2. Starling EH. The Croonian Lectures on the chemical correlation of the function of the body. Lecture 1. Lancet. 1905; 2: 339–41.
3. Ceranowicz P, Warzecha Z, Dembinski A. Peptidyl hormones of endocrine cells origin in the gut—their discovery and physiological relevance. J Physiol Pharmacol. 2015; 66: 11–27.
4. Moran GW, Leslie FC, Levison SE, et al. Enteroendocrine cells: Neglected players in

gastrointestinal disorders? Therapeutic Advances in Gastroenterology. 2008; 1: 51–60.

5. Bertrand PP. The cornucopia of intestinal chemosensory transduction. Front. Neurosci., 04 December 2009 | https://doi.org/10.3389/neuro.21.003.2009

6. Furness JB, Rivera LR, Cho HJ, et al. The gut as a sensory organ. Nature Reviews Gastroenterology & Hepatology. 2013; 10, 729–40.

7. Barker N. Adult intestinal cells: critical drivers of epithelial homeostasis and regeneration. Nature Reviews Molecular Cell Biology. 2014;15: 19–33.

8. Levant JA, Walsh JH, Isenberg JI. Stimulation of gastric secretion and gastrin release by single oral doses of calcium carbonate in man. N Engl J Med. 1973; 289: 55–558.

9. Hersey SJ, Sachs G. Gastric acid secretion. Physiol Rev. 1995; 75: 155–89.

10. Dockray DJ.Gastrin and gastric epithelial physiology. J Physiol. 1999; 518: 315–24.

11. Thomas RP, Hellmich MR, Townsend CM, et al. Role of gastrointestinal hormones in the proliferation of normal and neoplastic tissues. Endocrine Rev. 2003; 24: 571–99.

12. Johnson LR. The trophic action of gastrointestinal hormones. Gastroenterology 1976; 70: pp. 278–88.

13. Liddle RA. Regulation of cholecystokinin secretion by intraluminal releasing factors. Am J Physiol. 1995; 269: G319–27.

14. Chey WY, Chang T. Neural hormonal regulation of exocrine pancreatic secretion. Pancreatology. 2001; 320–35.

15. Fried GM, Ogden WD, Sakamoto T, et al. Experimental evidence for a vagally mediated and cholecystokinin-independent enteropancreatic reflex. Ann Surg. 1985; 202: 69–74.

16. Petersen H, Solomon T, Grossman MI. Effect of chronic pentagastrin, cholecystokinin, and secretin on pancreas of rats. Am J Physiol. 1978; 234: E286–93.

17. Watanapa P, Egan M, Deprez PH, et al. Role of cholecystokinin in pancreatic adaptation to massive enterectomy. Gut. 1992; 33: 959–64.

18. Erlick H, Stimmler L, Hlad CJ, et al. Plasma insulin response to oral and intravenous glucose administration. J Clin Endocrinol Metab. 1964; 24: 1076–82.

19. Carr RD, Larsen MO, Winzell MS, et al. Incretin and islet hormonal responses to fat and protein ingestion in healthy men. Am J Physiol Endocrinol Metab. 2008; 295: E779–84.

20. Mentis N, Vardarli I, Kothe LD, et al. GIP does not potentiate the antidiabetic effects of GLP-1 in hyperglycemic patients with type 2 diabetes. Diabetes. 2011; 60: 1270–76.

21. Tolhurst G, Reimann F, Gribble FM. Nutritional regulation of glucagon-like peptide-1 secretion. J Physiol. 2009; 587: 27–32.

22. Nathan DM, Schreiber E, Fogel H, et al. Insulinotropic action of glucagon like peptide-1 (7–37) in diabetic and nondiabetic subjects. Diabetes Care. 1992; 15: 270–76.

23. Tasyurek HM, Altunbas HA, Balci MK, et al. Incretins: their physiology and application in the treatment of diabetes mellitus. Diabetes Metab Res Rev. 2014; 30: 354–71.

24. Low MJ. The somatostatin neuroendocrine system: physiology and clinical relevance in gastrointestinal and pancreatic disorders. Clin Endocrinol Metab. 2004; 18: 607–22.

25. Lin HC, Zhao XT, Wang L, et al. Fat-induced ileal brake in the dog depends on peptide YY. Gastroenterology. 1996; 110: 1491–95.

26. Kojima M, Hosoda H, Date Y, Nakazato M, et al. Ghrelin is a growth-hormone releasing acylated peptide from stomach. Nature. 1999; 402: 656–60.

27. Shiiya T, Nakazato M, Mizuta M, et al. Plasma ghrelin levels in lean and obese humans and the effect of glucose on ghrelin secretion. J Clin EndocrinolMetab. 2002; 87: 240–244.

28. Kojima M, Hosoda H, Date Y,et al. Ghrelin is a growth-hormone releasing acylated peptide from stomach. Nature. 1999; 402: 656–60.

29. Spiller R. Serotonin and GI clinical disorders. Neuropharmacology. 2008; 55, 1072–80.

30. Gunawardene AR, Corfe BM, Staton CA. Classification and functions of enteroendocrine cells of the lower gastrointestinal tract Int. J. Exp. Path. 2011; 92: 219–31.

31. Costedio MM, Hyman N, Mawe GM. Serotonin and its role in colonic function and in gastrointestinal disorders. Dis. Colon Rectum. 2007; 50, 376–88.

32. Qiu S, Pellino G. Fiorentino F, et al. A review of the role of neurotensin and its receptors in colorectal cancer. Gastroenterology Research and Practice Volume 2017, Article ID 6456257, 8 pages https://doi.org/10.1155/2017/6456257.

33. Ohno T, Mochiki E, Kuwano H. The Roles of Motilin and Ghrelin in Gastrointestinal Motility. International Journal of Peptides Volume 2010, Article ID 820794, 6 pages http://dx.doi.org/10.1155/2010/820794.

Chapter 10

Lower Esophageal Sphincter

INTRODUCTION

In the undergraduate textbooks, physiology of lower esophageal sphincter is usually described in one small paragraph. Gastric physiology is discussed in greater detail in the context of peptic ulcer disease. However, nowadays gastroesophageal reflux disease (GERD) is reported to be one of the most prevalent diseases of the upper gastrointestinal tract. In a study in USA, the prevalence of heartburn and acid regurgitation was found to be 42% and 45%, respectively.[1] Another study showed two distinct and opposing patterns among upper gastrointestinal diseases. On the one hand, the prevalence of peptic ulcer and gastric cancer decreased; on the other hand, that of gastroesophageal reflux disease (GERD) and esophageal adenocarcinoma increased continuously during the period 1970–1995[2] (Fig. 10.1). In patients with gastroesophageal reflux, a greater duration of exposure of the esophagus to a pH <4 leads to increases the severity of esophagitis and its complications. **Barrett's esophagus** is a well-recognized premalignant condition for the development of esophageal adenocarcinoma, and it results from chronic gastroesophageal reflux disease. That is why the sphincter mechanism at the esophagogastric junction has been an area of intense investigation for last few decades.

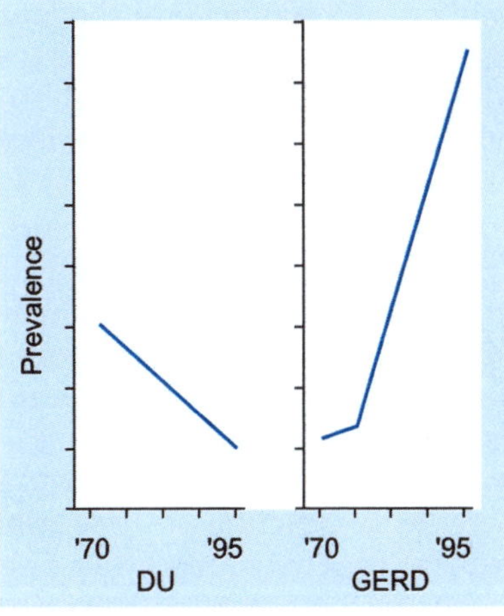

Fig. 10.1: Trends of prevalence of duodenal ulcer (DU) and GERD during 1970 to 1995

After a full meal, if a subject is turned upside down on a tilt table, food does not regurgitate from the stomach into the esophagus and mouth. It is obvious that there is a valve or sphincter-like mechanism between the lower end of esophagus and the stomach. A high pressure zone was described by Fyke et al in 1956[3] at the lower end of esophagus which was believed to prevent regurgitation of gastric contents into the esophagus. The average lower esophageal sphincter (LES) pressure later was found to be an average of 29 mmHg in reference to the intragastric pressure

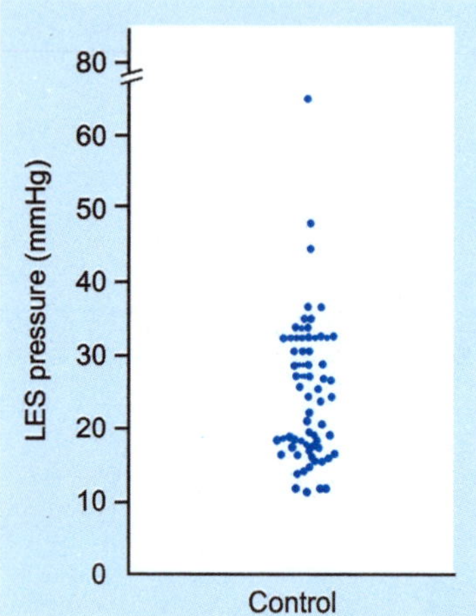

Fig. 10.2: LES pressure in normal subjects[4]

(0 mmHg)[4] (Fig. 10.2). The anatomy and functioning of the sphincter at the lower end of the esophagus is fairly complex. It is now clear that the LES is more than tonic contraction of the smooth muscle of lower end of esophagus.

Smooth muscles of the lower end of the esophagus and diaphragmatic sphincter constitute two active sphincters mechanism at the lower end of the esophagus and both contribute to the high pressure zone identified in humans at the lower end of esophagus.[5] The neuromuscular mechanisms that maintain tonic or reflex contraction of these sphincters are essential for their antireflux behavior. Impairment of these mechanisms promotes GERD. The phrenoesophageal ligament anchors the lower end of the esophagus to the diaphragmatic sphincter and plays a key role in the maintenance of an intact antireflux barrier.[6]

1. Contribution of Lower Esophageal Tone to LES Action

a. Anatomic Considerations

The smooth muscle of the esophagus is organized into two distinct layers, the circular and the longitudinal, and these layers continue into the LES. If one examines *autopsy specimens, both of these muscle layers are not thicker in the LES as compared to rest of the esophagus.* In living state, the thickness of the LES muscles changes in a dynamic fashion with the change in LES tone, that is, it increases and decreases with the increase and decrease in the LES pressure, respectively. It is for this reason that autopsy studies fail to show LES as a distinct anatomic structure, because with the loss of muscle tone in the autopsy specimen the LES looks no different than the esophagus.[6]

Anatomic descriptions of the LES are limited by the fact that the LES is best identified functionally on manometry, and the anatomic dissections can only guess the identity of the LES (Fig. 10.3). Nonetheless, anatomic studies suggest that LES muscle cells are not completely circular and in the lower part of the LES they are arranged in the

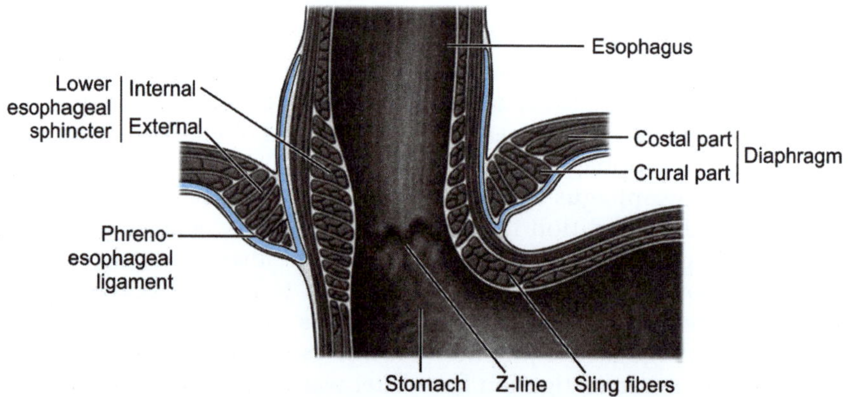

Fig. 10.3: Anatomy of gastroesophageal junction (diagrammatic), after Mittal et al[6]

form of incomplete "C-shaped" fibers arising from the left and the right sides that clasp each other. On the left side, the "C-shaped" fibers intermingle with the gastric sling fibers. The gastric sling fibers loop around the gastro-esophageal junction on the left side, and its two limbs continue on the anterior and posterior walls of the stomach, respectively, parallel to the lesser curvature of stomach. The gastric sling fibers form the oblique muscle layer of the stomach and may play an important role in the formation and modulation of angle of His.[6]

b. Lower Esophageal Tone

The difference between LES and the proximal esophageal muscle is that the former maintains higher basal tone than the latter, and in the *in vivo* situation the LES in humans is recognized as a zone of high pressure, 2 to 4 cm in length. What is responsible for this tone? A multitude of myogenic, neural, and neurohumoral factors can either increase or decrease LES tone (Table 10.1). Ingestion of food is often associated with a drop in LES pressure, resulting in large part from the secretion of hormones such as secretin and cholecystokinin (CCK), particularly with fat intake. LES pressure is decreased by chocolate, alcohol, and caffeine. During pregnancy, LES pressure is low because of high concentrations of circulating estrogen and progesterone, particularly the latter. Pharmacological agents such as dopamine, nitroglycerin, anti-cholinergics and β-adrenergic agonists also decrease LES tone. All these medications can potentially predispose to gastroesophageal reflux.

From an *electrophysiological point of view, the LES muscle is in a state of greater depolarization than the esophageal muscle*, as evidenced by a higher resting membrane potential than the esophageal muscle.[7] The depolarized state of the sphincter smooth muscle is suggested to be due to the resting chloride conductance. Periodic spike bursts or increase in the depolarization result in an increase in the LES tonic activity. Tonic LES contraction is both spike dependent and spike independent.[6]

The relative contribution of myogenic tone to the LES pressure differs in different species. In humans, neural cholinergic drive contributes significantly to the basal LES tone. Atropine, reduces LES pressure by 50 to 70% in humans.[8] Immunohistochemical staining of the myenteric plexus of the LES shows the presence of acetylcholine and substance P, the two likely candidates for excitatory neuro-transmissions. These nerve cells are likely to be located in the parasympathetic pathway to the LES.

The LES pressure increases in response to increases in intra-abdominal pressure. However, considerable controversy exists as to whether the rise in LES pressure during abdominal compression is due to reflex LES contraction or merely a passive transmission of the increased intra-abdominal pressure. It is unlikely to be a passive transmission because a contracted diaphragm muscle is relatively noncompliant. LES abdominal compression reflex is believed to be a true reflex that is mediated via the vagus nerve and

Table 10.1: Neurohumoral factors influencing LES pressure

	Increase	Decrease
Hormones	Gastrin	Secretin
	Motilin	Cholecystokinin
	Substance P	GIP
	Vasopressin	VIP
	Glucagon	Progesterone
Drugs	Adrenergic agonist:	Adrenergic antagonist:
	Norepinephrine	Isoproterenol
	Phenylephrine	
	Cholinergic:	Anticholinergic:
	Bethanechol	Atropine
	Methacholine	
	Betazole	
	Metaclopramide	Theophylline
Miscellaneous	Prostaglandin F_2	Prostaglandin E_1, E_2, A_2
	Protein meal	Nicotine
	Gastric alkalinization	Ethanol
		Fat meal
		Chocolate
		Gastric acidification

is dependent on cholinergic neurons.[9] In the human, increases in abdominal pressure during abdominal compression or other physical maneuvers result in reflex contraction of not only the LES but also the diaphragmatic sphincter.

The brainstem contains the central control mechanism for the LES, which is closely integrated with the swallow pattern generator. There is a topographic representation of neurons representing LES in the dorsal motor nucleus of the vagus nerve (DMV). The rostral cells are involved in the excitatory and caudal cells in the inhibitory innervation to the LES. The afferent information from the sensory nucleus of the tractus solitarius (NTS) is relayed to the DMV motor neurons via the interneurons. Glutamate is the neurotransmitter of the sensory afferents. The postganglionic neurons of vagus nerve release excitatory acetylcholine/substance P or inhibitory NO/VIP (Fig. 10.4).[10–12]

2. Contribution of Diaphragmatic Sphincter to LES Action (Fig. 10.5)

a. Anatomic Considerations

The mammalian diaphragm should be considered as two separate muscles consisting of

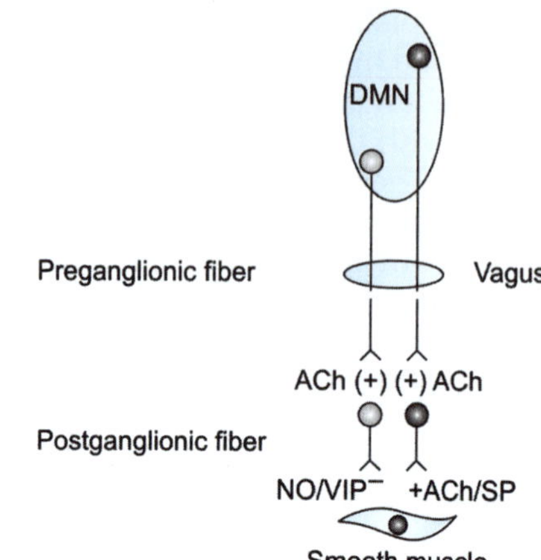

Fig. 10.4: Motor innervation of lower esophageal sphincter. DMN: Dorsal motor nucleus of vagus

the crural and the costal diaphragms. The costal diaphragm is a respiratory muscle, but the crural diaphragm has two functions, respiratory and gastrointestinal. Respiratory function relates to ventilation, and gastrointestinal to the sphincter like action at the lower end of the esophagus. The costal diaphragm originates primarily from the ribs, and the crural diaphragm from the lumbar vertebra, as two pillars (right and left crura). The esophageal hiatus is formed primarily by the right crus. The fibers of the crus are

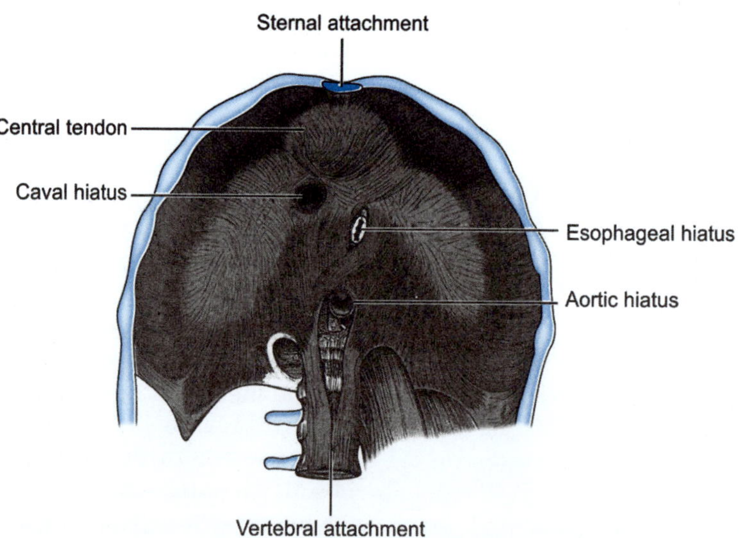

Fig. 10.5: Anatomy of diaphragmatic sphincter

oriented in the craniocaudal direction. Phrenic nerves formed by the branches of C5, C6, and C7 nerve roots provide both motor and sensory innervation to the crural and costal diaphragms.[6]

b. Diaphragmatic Contraction

Measuring the contribution of diaphragmatic sphincter pressure to the LES pressure is problematic in humans. The LES and diaphragmatic sphincter are anatomically superimposed on each other and, therefore, it is difficult to discern whether the intraluminal pressure is related to LES or diaphragmatic sphincter contraction.

Contraction of diaphragmatic sphincter provides a powerful sphincter mechanism at the lower end of the esophagus. It contributes to both tonic (sustained) and phasic pressure increases at the level of the LES. It is widely believed that end-expiratory pressure at the lower end of the esophagus is due to tonic contraction of the LES, and the increase in pressure with inspiration is due to the contribution from the diaphragmatic sphincter. The best evidence for the tonic contraction of the diaphragmatic sphincter is provided by studies in patients with a completely absent LES (the latter resected because of cancer at the distal esophagus or proximal stomach). Klein et al[13] found a high pressure zone that had both end-expiratory and inspiratory pressure oscillations in these patients. In humans, tidal inspirations cause a 15- to 20-mmHg increase in the LES pressure, and with forceful inspiration the increase in LES pressure can be 100 to 150 mmHg. The diaphragmatic sphincter also contracts reflexively during all those physiologic maneuvers that are associated with an increase in intra abdominal pressure, such as coughing and defecation.

LOWER ESOPHAGEAL SPHINCTER RELAXATION

For the transport of ingested contents into the stomach, relaxation of both the LES and the diaphragmatic sphincter is essential. Deglutition and distension of the esophagus are the two major stimuli that induce esophagogastric junction relaxation. Deglutition-induced LES relaxation starts within 2 seconds of the onset of swallowing and lasts 6 to 10 seconds (Fig. 10.6). Lower esophageal sphincter relaxation is terminated by the arrival of esophageal peristaltic contraction at the LES and is followed by an after-contraction that may last up to 10 seconds. In the upright position, the swallowed bolus may reach the LES very quickly owing to gravity, and in that situation the bolus may be transiently held at the LES before passage into the stomach. The LES normally relaxes to a pressure very close to the intragastric pressure. The LES relaxation associated with primary peristalsis is mediated by vagal efferent nerves and abolished by bilateral cervical vagal section or vagal cooling.

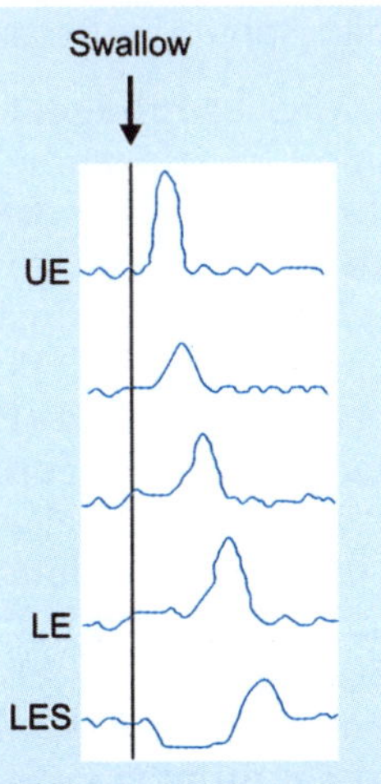

Fig. 10.6: Manometric recording of swallow-induced fall in LES pressure (diagrammatic representation). UE: Upper esophagus; LE: Lower esophagus; LES: Lower esophageal sphincter

During a swallow, stimuli from the oropharyngeal region traverse via the vagal afferent pathways into the nucleus tractus solitarius, which then relays information into groups of cells, rather loosely organized (reticular formation). These cells are felt to be the pattern generator for the deglutition reflex and communicate with the motor neurons of the DMV. The rostral and caudal cells in the DMV are responsible for the excitation and inhibition of LES respectively. Following its activation, myenteric plexus neurons release nonadrenergic noncholinergic (NANC) neurotransmitter or neurotransmitters that cause relaxation of the smooth muscles of LES. There are several possible candidate NANC transmitters: Adenosine triphosphate (ATP), vasoactive intestinal peptide (VIP), pituitary adenylate cyclase-activating peptide (PCAP), nitric oxide (NO), and carbon monoxide (CO) in the smooth muscle of the GI tract. However, the predominant one in the LES is NO. Deglutition-induced, vagal efferent stimulation-induced, esophageal distention-induced, and direct myenteric plexus neuron stimulation-induced relaxation are all blocked by NO antagonists.[14]

Diaphragmatic Sphincter Relaxation

Deglutition and esophageal distension, besides causing LES relaxation, also induces selective inhibition of the diaphragmatic sphincter muscle. Deglutition-induced inhibition of diaphragmatic sphincter is usually not complete, and as a result flow across the esophageal gastric junction (EGJ) is interrupted if the subject inspires during the period of deglutition-induced LES relaxation. In fact, alteration in the breathing pattern can cause complete arrest of the bolus in the esophagus above the EGJ. Vagotomy abolishes esophageal distension-mediated inhibition of the diaphragmatic sphincter, suggesting that vagal afferents inhibit the brainstem medullary neurons responsible for diaphragmatic sphincter contraction. Nitric oxide suppresses the excitatory motor nerve junction potentials at the crural diaphragm motor end plates.[6]

Transient Lower Esophageal Sphincter Relaxation (TLESR)

Weakness of LES and diaphragmatic sphincters can be demonstrated only in some of the patients suffering from gastroesophageal reflux disease. In the majority, however, especially in mild to moderate reflux disease,

the LES and diaphragmatic sphincter pressure is normal. A large body of data indicates that transient relaxation of the LES and diaphragmatic sphincter (TLESR) is the major mechanism of reflux in patients with reflux disease.[15] The term *TLESR* may not even be appropriate because deglutition-induced LES relaxation is also transient, and furthermore there is also concurrent and simultaneous relaxation of the diaphragmatic sphincter during TLESR.[6] In fact, similar to deglutition-induced LES relaxation being a physiologic mechanism for the antegrade flow, gastric distension-induced TLESR is a physiologic mechanism for the retrograde flow of stomach contents into the esophagus. Belching, vomiting, and physiologic reflux in normal subjects occur through the mechanism of TLESR.[16] Transient LES relaxation is seen as an abrupt fall in LES pressure to the level of intragastric pressure that is not triggered by deglutition. Transient LES relaxation is typically of longer duration than the swallow-induced LES relaxation, lasting 10 to 45 seconds (Fig. 10.7). The optimal criteria for the definition of TLESR are:

- Independent of swallowing
- Not accompanied by esophageal peristalsis
- Accompanied by diaphragmatic inhibition

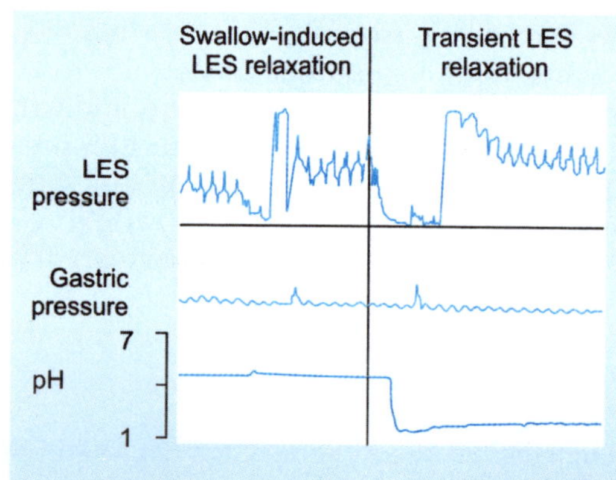

Fig. 10.7: Swallow-induced fall in LES pressure (left), as compared to transient LES relaxation (right). Note the prolonged decrease on LES pressure and severe fall in esophageal pH during TLESR

- Persists longer than swallow-induced LES relaxation (>10 seconds).

Gastric distension is a potent stimulus for TLESR, which is not surprising given that TLESR is the mechanism by which gas is vented from the stomach during belching. Approximately 15 ml of air is delivered to the stomach with each swallow,[16] and without an in-built venting mechanism, uncontrolled gastrointestinal bloating would occur. Studies in which the stomach has been partitioned surgically have shown that the subcardiac region of the stomach is primarily responsible for triggering TLESRs. In humans, a volume of 750 to 1000 ml causes a fourfold increase in the rate of TLESRs within the first 10 minutes. In some studies, a similar effect has been reported after meals.[17]

The production of TLESRs by gaseous gastric distension is almost totally suppressed in the supine posture. In patients with reflux disease, TLESRs occur less frequently in the supine and lateral recumbent positions compared to the sitting position. Transient LES relaxation does not occur during stable sleep; reflux episodes that do occur during the night-time sleep periods are totally confined to periods of arousal during sleep that may last for only 10 seconds.[6]

References

1. Richter JE, Friedenberg FK. Gastroesophageal reflux disease. In: Feldman M, Friedman LS, Brandt LJ, eds. *Sleisenger and Fordtran's Gastrointestinal and Liver Disease.* 10th ed. Philadelphia, PA: Elsevier Saunders; 2016.
2. El-Serag HB, Sonnenberg A. Opposing time trends of peptic ulcer and reflux disease. *Gut.*1998; 43: 327–33.
3. Fyke FE, Code CF, Schlegel JF. The gastro-esophageal sphincter in healthy human beings. Gastroenurologia. 1956; 86: 135–50.
4. https://www.hon.ch/OESO/books/Vol_4_Prim_Motility/Articles/ART032.HTML
5. Mittal RK, Balaban DH. The esophagogastric junction. N Engl J Med.1997; 336: 924–932.
6. Mittal RK, Goyal RK. Sphincter mechanisms at the lower end of the esophagus. GI Motility online (2006) doi: 10.1038/gimo14
7. Papasova M, Milousheva E, Bonev A, et al. On the changes in the membrane potential and the contractile activity of the smooth muscle of the lower esophageal and ileo-caecal sphincters upon increased K in the nutrient solution. Act a Physiol Pharmacol Bulg. 1980; 6: 41–49.
8. Dodds WJ, Dent J, Hogan WJ, et al. Effect of atropine on esophageal motor function in humans. Am J Physiol. 1981; 240: G290–96.
9. Lind JF, Warrian WG, Wankling WJ. Responses of the gastroesophageal junctional zone to increases in abdominal pressure. Can J Surg. 1966; 9: 32–38.
10. Rossiter CD, Norman WP, Jain M, et al. Control of lower esophageal sphincter pressure by two sites in dorsal motor nucleus of the vagus. Am J Physiol. 1990; 259: G899–906.
11. Hyland NP, Abrahams TP, Fuchs K, et al. Organization and neurochemistry of vagal preganglionic neurons innervating the lower esophageal sphincter in ferrets. J Comp Neurol. 2001; 430: 222–34.
12. Wiedner EB, Bao X, Altschuler SM. Localization of nitric oxide synthase in the brainstem neural circuit controlling esophageal peristalsis in rats. Gastroenterology.1995; 108: 367–75.
13. Klein WA, Parkman HP, Dempsey DT, et al. Sphincter like thoracoabdominal high pressure zone after esophagogastrectomy. Gastroenterology. 1993; 105: 1362–69.
14. Ergun G, Kahrilas P, Lin S, et al. Shape, volume, and content of the deglutitive pharyngeal chamber imaged by ultrafast computerized tomography. Gastroenterology. 1993; 105: 1396–403.
15. Ryan JP, Snape WJ Jr, Cohen S. Influence of vagal cooling on esophageal function. Am J Physiol. 1977; 232: E159–E164.
16. Mittal RK, Holloway RH, Penagini R, et al. Transient lower esophageal sphincter relaxation. Gastroenterology. 1995; 109: 601–610.
17. Penagini R, Bartesaghi B, Conte D, et al. Rate of transient lower oesophageal sphincter relaxations of healthy humans after eating a mixed nutrient meal: time course and comparison with fasting. Eur J Gastroenterol Hepatol. 1992; 4: 35–38.

Chapter 11

Exocrine Pancreas

INTRODUCTION

The role of the pancreas in digestion was discovered for the first time by the French physiologist, Claude Bernard. The discovery was based on an observation on the kitchen table. While trying to clean a rabbit before cooking, Bernard noticed that the lacteals were filled with a white fluid (absorbed fat) only up to some distance from the pylorus, in contrast to dogs in which the lacteals were full of absorbed fat right up to the pylorus. He correlated the difference with the entry point of the pancreatic duct in the intestine in the rabbit and dog. He felt that pancreatic juice was somehow important in the digestion of fats in the intestine. This fact was confirmed when he observed splitting of neutral fats when incubated with crushed pancreatic tissue. Moreover, he demonstrated that obstruction of pancreatic duct in a dog resulted in loss of large amount of fat in the stools of the animal. This discovery was made in 1846, and that earned him a prize (the Grand Prix, 1849) in experimental physiology from the French Academy of Sciences. In 1902, Bayliss and Starling discovered secretin, an intestinal hormone which increases pancreatic exocrine secretion.[1] Subsequently, a large number of gastrointestinal hormones were discovered (*see* Chapter 9).

STRUCTURE OF EXOCRINE PANCREAS

The exocrine pancreas is a highly lobulated gland, consisting of clusters of secretory cells forming acini and a highly branched ductal system. The pancreatic acini consist of pyramidal shaped cells with their apexes projecting towards the lumen of a minute duct. The acinar cells are typical protein secreting cells. Highly basophilic basal region contains the nucleus surrounded by extensive endoplasmic reticulum (Fig. 11.1). The Golgi apparatus is located in supranuclear position. Apexes of the cells are filled with acidophilic zymogen granules. The smallest excretory ducts starting from the acini are lined with centroacinar cells. The secretion is drained into ducts of progressively increasing size and finally into the main pancreatic duct. The main pancreatic duct joins the common bile duct to form ampulla of Vater, which opens into the duodenum. The ductal epithelium consists of cuboidal cells that are devoid of zymogen granules and have very little endoplasmic reticulum. The centroacinar cells and cell lining the smaller ducts contain carbonic anhydrase, the enzyme required for synthesis of bicarbonate. The pancreas is heavily innervated by sympathetic and parasympathetic peripheral nerves and contains a dense network of blood vessels.

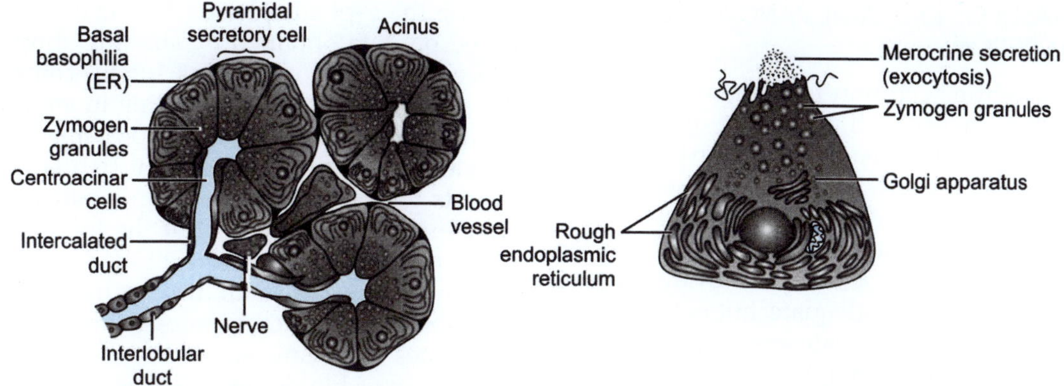

Fig. 11.1: Histological structure of exocrine pancreas. ER: Endoplasmic reticulum

The autonomic innervation regulates pancreatic blood flow and modulates pancreatic secretion.

Mechanism of Exocrine Pancreatic Secretion

The pancreatic enzymes (organic component of pancreatic juice) are synthesized in the acinar cells. The enzymes are synthesized in the rough endoplasmic reticulum and packaged into zymogen granules to be released into ductal lumen by exocytosis. Of the various pancreatic enzymes, proteases (e.g. trypsin, carboxypeptidases, chymotrypsin, and elastase) are the most abundant (according to weight) comprising around 90% of the enzymes found in human pancreatic fluid; amylase, lipase, and nucleases are relatively less abundant. In *in vitro* studies, receptors for secretagogue hormones (e.g. CCK) or neurotransmitters (e.g. ACh) can be found on the basolateral membrane of the isolated acinar cells. Following production and storage of inactive proenzymes (termed zymogens) and active enzymes in the acinar cells, they are secreted into a network of pancreatic ductules and ducts in response to a meal.

The inorganic components of the pancreatic juice, especially bicarbonate and water are chiefly contributed by the ductal cells. Although the exact mechanism of secretion of inorganic components is not clear; a tentative hypothesis is given below.

The mechanism of bicarbonate secretion is basically similar to that proposed for hydrochloric acid secretion in the oxyntic cells of the stomach. However, the crucial difference is that in the pancreas, H^+ is added to the blood and HCO_3^- to the lumen of the duct. The bicarbonate is derived from hydration of carbon dioxide in the presence of carbonic anhydrase. Bicarbonate is transferred to the ductal lumen by a HCO_3^-/Cl^- antiport operating at luminal border of the cells (Fig. 11.2).[2]

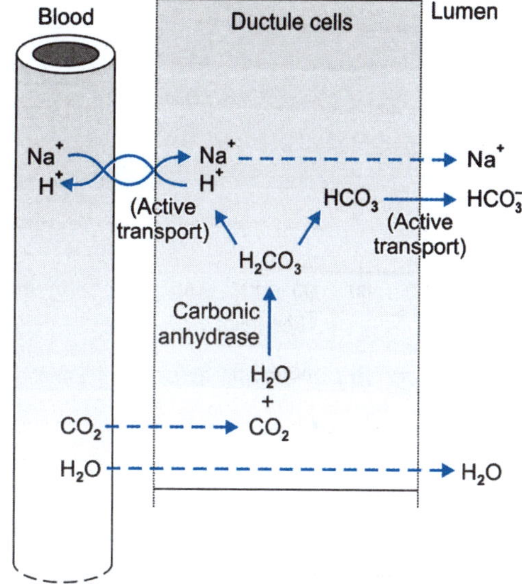

Fig. 11.2: Mechanism of bicarbonate generation in pancreatic acinus

PHASES OF EXOCRINE PANCREATIC SECRETION

Pancreatic secretion in response to a meal occurs in three-distinct but overlapping phases which are named based on the location of ingested food. The three phases of pancreatic secretion are cephalic, gastric and intestinal. Considerable crosstalk and inter-regulation is associated within the phases, thereby ensuring adequate, but not excessive, enzyme and bicarbonate secretion. Each phase is regulated by a complex network of neural, humoral, and paracrine feedback mechanisms which help to maintain an optimal environment for food digestion and absorption. Delivery of pancreatic juice into the duodenum increases rapidly after a meal, and reaches maximal values within the first postprandial hour or even within 20–30 minutes postprandially. Following peak output, enzyme secretion decreases to almost stable secretory rates at lower levels until about 3–4 hours postprandially depending on the size of the meal. The interdigestive range is reached again at the end of the digestive period (Fig. 11.3).

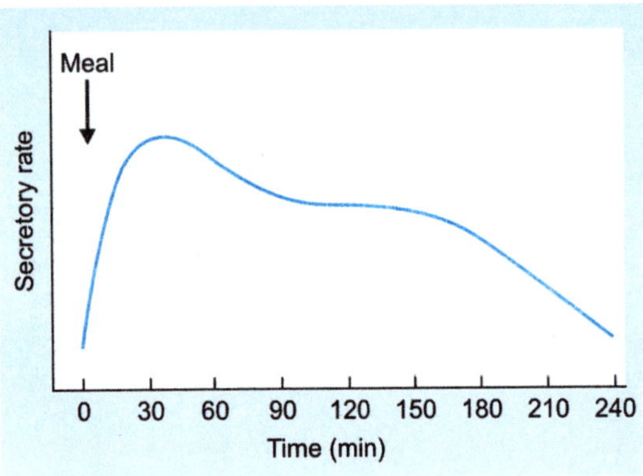

Fig. 11.3: Exocrine pancreatic response to a mixed meal[19]

a. Cephalic Phase

Sensory inputs such as mastication, taste, smell, or even sight of food initiate the first phase of pancreatic secretion known as the cephalic phase. Approximately 20–25% of the total pancreatic exocrine secretion occurs during the cephalic phase.[3] This estimate is based on data obtained by sham feeding, a process by which, in humans, food is chewed but not swallowed. Sham feeding in animals such as dogs, has been made possible by a surgically prepared esophageal fistula that diverts food from the esophagus, allowing chewing and swallowing but the food does not reach the stomach (Fig.11.4). The pancreatic response to sham feeding in humans lasts approximately 60 minutes. Sham feeding stimulates pancreatic secretion which is low in bicarbonate but rich in enzymes, suggesting that pancreatic acinar, rather than ductal cells are stimulated in this phase.[4]

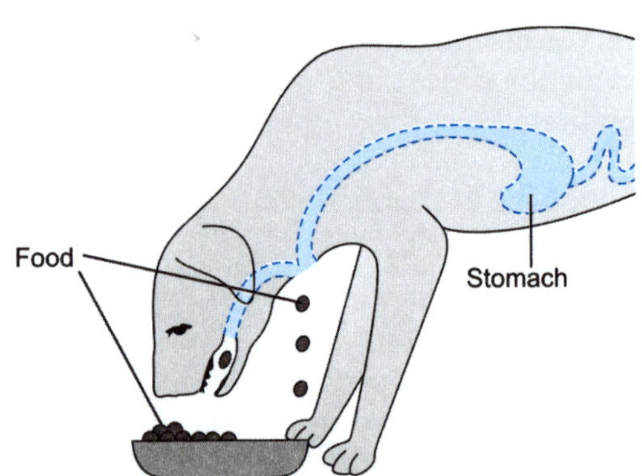

Fig. 11.4: Sham feeding experiment in dog

The cephalic phase of exocrine secretion is under the control of the vagus nerve. Sensory inputs arising from mastication of food in the mouth are integrated in the dorsal vagal complex (located in the brainstem) and transmitted to the exocrine pancreas via the vagus nerve. Cholinergic agonists produce secretory responses similar to cephalic stimulation while vagotomy blocks the cephalic responses, suggesting that acetylcholine released by vagal efferents is the primary mechanism by which sensory inputs lead to exocrine secretion. During vagal stimulation, acetylcholine is believed to be the main neurotransmitter which regulates the cephalic phase.[5,6]

b. Gastric Phase

Experiments in which food was put directly into the stomach but prevented from entering into the duodenum demonstrated that the gastric phase accounted for only approximately 10% of pancreatic secretion. Secretions induced during this phase consist mainly of enzymes with minimal release of bicarbonate suggesting that acinar cells are primarily involved in the induction of this phase.[7]

The vagus nerve plays an important role in the gastric phase of pancreatic secretion. Early experiments in anesthetized cats demonstrated that stimulation of the antrum resulted in vagal stimulation of pancreatic amylase release. Antral distension in dogs also increased pancreatic secretion by long route vagal pathways.[7] An antro-pancreatic short reflex pathway which is blocked by hexamethonium as well as atropine mediates this phase.[8] In addition, vagotomy blocks the gastric phase providing further evidence that gastric contributions to pancreatic secretion are mediated by vagovagal cholinergic reflexes that originate in the stomach and terminate in the pancreas.[9]

The role of gastrin in this phase of pancreatic secretion remains unclear. When exogenous gastrin was administered to dogs, the amount required to stimulate exocrine pancreatic secretion was much larger than normal postprandial gastrin levels, suggesting that gastrin did not have a physiological role.[10] These findings are supported by other studies demonstrating that gastrin release is not required for pancreatic enzyme secretion during this phase.[6]

Though gastric phase does not substantially contribute to exocrine pancreatic secretion, the stomach plays an important role in intestinal phase of pancreatic secretion. In the stomach, pepsin hydrolyses proteins into peptides, while salivary amylase contributes to the continued digestion of carbohydrates. Peptic digests of proteins are effective in stimulating the intestinal phase. Thus, when gastric chyme enters the duodenum, it stimulates the intestinal phase of pancreatic secretion. Therefore, the rate of gastric emptying regulates the discharge of nutrients into the intestine and consequently the activation of the intestinal phase through neural and hormonal pathways.

c. Intestinal Phase

Digestion of food in the stomach is followed by release of acidic chyme into the duodenum, which initiates the intestinal phase of pancreatic secretion. A majority of the pancreatic secretory response (50–80%) occurs during the intestinal phase and is regulated by hormonal and neural mechanisms.

The intestinal phase is more easily studied than the gastric phase as food can be instilled directly into the intestinal lumen without concern for gastric emptying. Stimulation of both pancreatic acinar and ductal cells results in the production of enzyme and bicarbonate-rich secretion. In the duodenum, the high volume of bicarbonate released neutralizes the acidity of gastric chyme, while pancreatic enzymes hydrolyse partially digested food into molecules that are easily absorbed by intestinal enterocytes.

In the intestinal phase, pancreatic response is regulated primarily by the hormones *secretin* and *cholecystokinin (CCK)*, and by neural influences including the enteropancreatic reflex. The enteropancreatic reflex is mediated by the enteric nervous system and amplifies the pancreatic secretory response. Entry of low pH gastric chyme into the intestine stimulates release of secretin from mucosal S cells into the blood. The main action of secretin is to stimulate bicarbonate release from pancreatic duct cells, but it also has a direct effect on acinar cells and potentiates enzyme secretion. CCK is released by mucosal I cells in response to luminal proteins and fats and their partial digestion products: Peptides and fatty acids. Experiments in dogs with chronic pancreatic fistulae have shown that CCK antagonism diminishes pancreatic protein response to a meal suggesting that CCK plays an important role in this phase. Similar results were also obtained in humans, where CCK receptor antagonism reduced pancreatic enzyme secretion during the intestinal phase.[11]

Cholinergic regulation plays a critical role during this phase of pancreatic secretion. Atropine inhibited pancreatic enzyme secretion, implicating cholinergic mechanisms.[12]

Vagovagal enteropancreatic reflexes mediated by M1 and M3 muscarinic receptors play an important role in the intestinal phase of secretion.[13] The vagovagal enteropancreatic reflexes are modulated by input from the dorsal motor nucleus of the vagus projecting into the pancreas.

Atropine also partially inhibits pancreatic bicarbonate secretion stimulated by low pH due to acidic chyme in the duodenum. In addition, the amount of bicarbonate produced by infusion of secretin is lower than that released by entry of food into the duodenum suggesting that other factors contribute to meal-stimulated pancreatic bicarbonate secretion.

AGENTS MODULATING EXOCRINE PANCREATIC SECRETION

1. Neural Mechanisms

The pancreas is innervated by parasympathetic nerve fibers, postganglionic sympathetic neurons, as well as a network of intrapancreatic nerves. Together these nerves regulate pancreatic exocrine function by releasing neurotransmitters such as acetylcholine and neuropeptides such as VIP, GRP, serotonin, and neuropeptide Y (NPY).

Intrapancreatic postganglionic neurons are activated by central input during the cephalic phase and by vagovagal responses initiated during the gastric and intestinal phases of stimulation. They stimulate enzyme and bicarbonate secretion primarily by releasing acetylcholine, which activates muscarinic receptors located on acinar and duct cells.

Information relayed by sensory vagal afferent nerves innervating the pancreas is first processed in the nucleus of the solitary tract, which then projects onto the preganglionic motor neurons of the DMV. Parasympathetic preganglionic efferent vagal nerves innervating the pancreas originate primarily from the DMV and terminate in the pancreatic ganglia. Electrical and chemical stimulation of the DMV induces rapid pancreatic secretion, and this response is inhibited by vagotomy or blockade of muscarinic receptors by atropine. *It has been suggested that vagal cholinergic neurons mediate pancreatic secretion during low loads of intestinal stimulants, whereas hormones mediate the response during high loads of intestinal stimuli.*[14]

CCK increases exocrine pancreatic secretion through both a direct effect on pancreatic acinar cells and an indirect effect on the vagus nerve. CCK is released from I cells of the small intestine and diffuses into the blood stream where it is carried to the pancreas. CCK binds to receptors on acinar cells to stimulate pancreatic enzyme secretion. In addition, secreted CCK also diffuses through the paracellular space and *binds to CCK receptor-bearing vagal afferents* in the submucosa. Vagal afferent signals are integrated in the dorsal vagal complex which also receives signals from other regions of the brain (e.g. hypothalamus). Vagal efferent fibers transmit cholinergic signals to the pancreas to stimulate pancreatic secretion (Fig. 11.5).

2. Stimulatory Hormonal Mechanisms

i. **Cholecystokinin:** CCK is released from specialized enteroendocrine cells (I cells) located mainly in the upper small intestine. The major stimulants of CCK release are dietary fats and proteins. Various molecular forms of CCK, ranging in size from CCK-8 to CCK-58, have been described in dogs, rats, and humans. The actions of CCK-8, CCK-33 and CCK-58 appear to be functionally identical suggesting that CCK-8 retains the biological activity ascribed to this hormone.[6]

ii. **Secretin:** Secretin is a 27-amino acid hormone released by S cells of the small intestine. Secretin release is stimulated during the intestinal phase upon entry of gastric acid and ingested fatty acids into the duodenum. It augments fluid and bicarbonate secretion and is one of the most potent stimulators of pancreatic secretion.

Large increases in pancreatic fluid and bicarbonate secretion have been demonstrated with secretin infusions in extremely

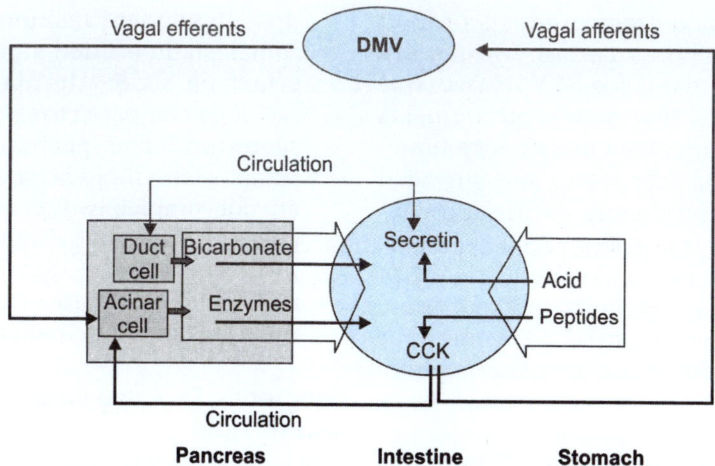

Fig. 11.5: Neural and hormonal regulation of exocrine pancreatic secretion during intestinal phase. DMV: Dorsal motor nucleus of vagus

low concentrations (Fig.11.6). Although secretin is believed to be the single most powerful stimulator of pancreatic bicarbonate secretion, infusion of exogenous secretin equivalent to postprandial blood levels only produced 10% of the maximal pancreatic bicarbonate secretory response suggesting that other hormones and neurotransmitters play important roles in postprandial pancreatic bicarbonate secretion in humans.[15]

Secretin receptors have been localized in acinar and duct cells in the rat pancreas.

Secretin stimulates the release of fluid and bicarbonate by ductal cells, and to a lesser extent, enzymes from acinar cells by a cholinergic mechanism. Perfusion of acetic and lactic acids in the duodenum of anesthetized rats increased fluid and protein output from the pancreas concomitant with elevation of plasma secretin levels. In addition, treatment of rats with atropine decreased plasma secretin levels and inhibited fluid secretion, indicating that the fluid secretion is partly dependent on cholinergic input.

Several hormones and peptides modulate the effects of secretin on pancreatic secretion. CCK augments secretin-induced pancreatic fluid secretion and protein output by stimulating acetylcholine release and this effect can be blocked by atropine. Venous drainage from pancreatic islets bathes the exocrine pancreas with high concentrations of islet hormones. Several of these hormones have potent effects on pancreatic secretion. Insulin enhances secretin-mediated fluid and enzymes secretion while glucagon has an inhibitory effect.[6]

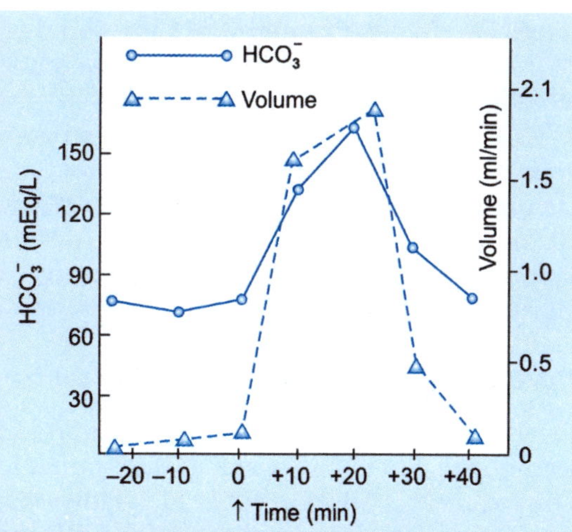

Fig. 11.6: Pancreatic response to secretin

3. Inhibitory Hormonal Mechanisms

i. **Peptide YY:** Peptide YY consists of 36-amino acids. PYY is secreted by L cells in the ileum and colon in response to food intake. Fats and other products of food digestion, such as carbohydrates are primary stimulants for PYY release. It is also supposed that neural mechanisms and GRP are involved in PYY secretion.

PYY inhibits both gastric and intestinal motility and suppresses gastric and pancreatic exocrine secretion. Reduced intestinal motility may increase efficiency of food digestion and nutrient absorption. Thus, along with GLP-1, PYY seems to be involved in "ileal brake" phenomenon. PYY levels in blood are elevated postprandially and following instillation of fatty acids into the distal small intestine. Physiological concentrations of PYY in the circulation inhibit both meal- and hormone-stimulated pancreatic secretion.[16]

ii. **Glucagon-like peptide-1 (GLP-1):** GLP-1, is a 30-amino acid peptide hormone, produced by intestinal endocrine L cells. The open-type L cells are most abundant in the distal ileum and colon, but are also present in the duodenum. GLP-1 is released in response to food, especially glucose, fatty acids, proteins and bile salts in the bowel .Unabsorbed nutrients in the ileum inhibit the motility of upper gut ("ileal brake"). PYYand GLP-1 participate in the ileal brake. Inhibition of exocrine pancreas is another action of GLP-1.

iii. **Pancreatic polypeptide:** Secretion by PP cells of islets is also stimulated by ingesting a meal. The effect can be reproduced by intraduodenal infusion of acid, aromatic amino acids, or fatty acids. Like PYY, PP attenuates secretin- and CCK-mediated exocrine pancreatic secretion.[17]

iv. **Somatostatin:** Somatostatin is composed of 14- or 28-amino acids and is produced by δ cells of pancreatic islets. It is also secreted by certain intestinal D cells. It is released into the blood after a meal but functions primarily through a paracrine mechanism. It has broad inhibitory actions on the release of several hormones and their target organs. Somatostatin and its analogs inhibited secretin- and CCK-induced pancreatic enzyme secretion in a dose-dependent fashion. Low doses of somatostatin exerted a greater inhibitory effect on CCK-stimulated pancreatic secretion compared to secretin-stimulated secretion.[6] The mechanisms by which somatostatin inhibits pancreatic secretion are not completely understood. However, it is believed that somatostatin has an inhibitory effect on the release of hormones and neurotransmitters that normally stimulate pancreatic secretion.

Pancreatic Exocrine Response to Gastric Acid

Intraduodenal infusion of hydrochloric acid elicited a concentration-dependent increase in both the amount of bicarbonate and volume of pancreatic secretion. Secretion was similar to that attained with intravenous infusion of exogenous secretin suggesting that pH changes resulting from entry of acidic contents into the duodenum are important in inducing pancreatic secretion. Acid infusion in both the duodenum and upper jejunum elicited pancreatic secretion suggesting that the proximal small intestine responds to this stimulus.

The pH of a liquid gastric meal also plays a significant role in pancreatic bicarbonate secretion; in cats and dogs, pH >4.5 resulted in little pancreatic bicarbonate secretion, while at pH <4.0 secretion increased substantially suggesting that a pH threshold of <4.5 is critical for stimulation of pancreatic secretion.[18]

This evidence implies that gastric acid is an important regulator of pancreatic bicarbonate secretion which neutralizes the acid to create an alkaline environment optimal for the action of pancreatic enzymes and continued digestion of food.

PANCREATIC RESPONSE TO VARIOUS DIETS

The basal (interdigestive) and postprandial secretion of pancreatic enzymes in response to a mixed meal is given in Table 11.1. However, exocrine pancreas response varies with the type of diet.

Table 11.1: The basal (interdigestive) and postprandial secretion of pancreatic enzymes in response to a mixed meal[19]

	Interdigestive	Mean postprandial
Amylase (U/min)	50–250	500
Lipase (U/min)	1000	2000–4000
Trypsin (U/min)	50–100	150–500

i. **Pancreatic response to protein diet:** The protein component of a meal contributes *markedly* to induction of the digestive pancreatic enzyme response. However, the stimulatory effect of dietary protein on pancreatic exocrine secretion appears to depend on adequate protein digestion and stimulatory potency is limited to certain essential amino acids: Phenylalanine, valine, methionine, and tryptophan increased trypsin secretion, whereas isoleucine, leucine, lysine, threonine, and other amino acids did not.[19]

ii. **Pancreatic response to lipid diet:** Lipids are the *strongest stimulants* of pancreatic enzyme secretion. Duodenal free fatty acids rather than triglycerides appear to be responsible for the release of cholecystokinin (CCK) and subsequent stimulation of pancreatic enzyme output. Duodenal perfusion of fatty acids increases pancreatic enzyme output in a dose-dependent manner. Apart from the dose of fatty acids delivered to the duodenum, fatty acid chain length has a major influence on pancreatic enzyme response: Long chained fatty acids augment pancreatic responses, medium chained fatty acids do not.[20]

iii. **Pancreatic response to carbohydrate diet:** Carbohydrates are *weaker stimulants* of pancreatic enzyme secretion compared with proteins and lipids (Fig. 11.7). Most investigators report a significant increase in enzyme output induced by carbohydrate meals *followed by a fast decline.* 100 g glucose dissolved in 500 ml water

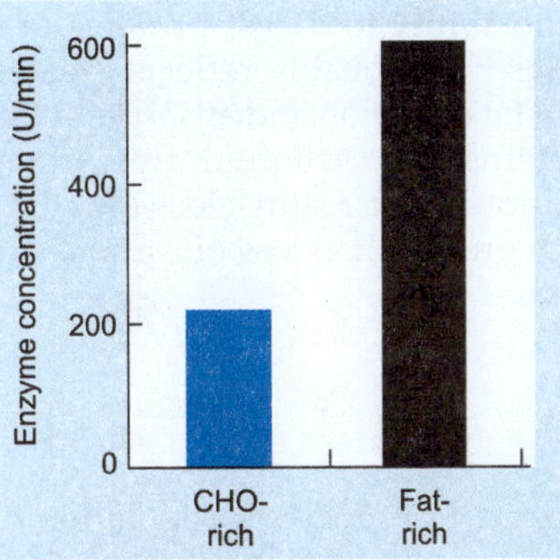

Fig. 11. 7: Pancreatic enzyme secretion in response to carbohydrate-rich and fat-rich diets

increased basal lipase output (800 U/min) about threefold (2600 U/ min). However, one hour postprandially lipase output had already returned to pre-prandial values.[21]

Enzyme response to a solid meal is more sustained compared with an identical meal which has been homogenised. This is partly due to slower gastric emptying of the solid compared with the homogenised meal which leads to prolonged stimulation of enzyme output. In addition, increased gastric acid secretion and, consequently, increased duodenal acid delivery in response to a solid meal may also cause higher pancreatic secretory response.

Ratios of Pancreatic Enzymes

In general, output of various pancreatic enzymes into the duodenum occurs in parallel. This becomes evident from most studies which analysed two or more enzymes. Overall, postprandial lipase: Amylase ratio and lipase: Trypsin ratio are approximately 3–6/1 and 5–10/1, respectively. This means that in terms of catalytic units as defined under optimised conditions *in vitro*, a three- to six-fold excess of lipolytic activity in comparison with amylase and a five to tenfold excess of lipolytic activity in comparison with trypsin is released into the duodenum. It does not automatically imply that the ability of human intestinal contents to digest lipids is higher than the ability to digest proteins or carbohydrates.[22] It probably reflects greater difficulty in fat digestion and absorption than that of proteins and carbohydrates. In patients with chronic pancreatitis, loss of fat in faeces is much greater than loss of carbohydrates or proteins.

Feedback Regulation of Pancreatic Secretion

The concept of feedback regulation of pancreatic secretion originated from a series of studies demonstrating that (1) instillation of trypsin inhibitor into the upper small intestine or (2) surgical diversion of the bile–pancreatic duct removing bile and pancreatic juice from the duodenum of rats stimulated pancreatic enzyme secretion. Conversely, infusion of trypsin into the duodenum during bile-pancreatic juice diversion suppressed pancreatic enzyme release. Thus, the protease concentration in the upper small intestine appears to be intimately linked to pancreatic secretion through a negative feedback system in which active proteases within the duodenum limit pancreatic secretion but reduced protease activity stimulates pancreatic secretion.[23]

POSTPRANDIAL TERMINATION OF EXOCRINE PANCREATIC SECRETION

Late postprandial pancreatic enzyme output declines to interdigestive values, although the proximal small intestine is still exposed to substantial nutrient concentrations. This is probably not due to inhibitory mechanisms induced by duodenal nutrient exposure—for example, release of somatostatin or pancreatic polypeptide. The liberation of these inhibitory mediators coincides with the release of stimulatory hormones. Therefore, duodenal hormonal inhibitors are generally assumed to be modulators of the magnitude of the digestive response rather than its duration.

Inhibitory mechanisms activated predominantly in the late postprandial period are more likely to mediate the termination of digestive pancreatic enzyme secretion even in the presence of duodenal nutrient exposure. Candidate hormones are peptide YY (PYY) and glucagon-like peptide-1 (GLP-1). These peptides are released from the distal intestinal mucosa in response to a meal or intra-ileal nutrients. Ileal perfusion of lipids inhibits secretory and motor functions of upper gastrointestinal organs both in the inter-digestive state and during weak or moderate endogenous stimulation.

Under physiological circumstances, considerable amounts of nutrients pass the small intestine unabsorbed. This phenomenon is referred to as "physiological malabsorption". Therefore, the ileum appears to contribute to the regulation of the transition from the fed to the subsequent fasting state. Stimulatory and inhibitory mechanisms induced by nutrients both in the proximal and distal small bowel may interact to co-regulate pancreatic exocrine secretion during the functional interface between the late postprandial and the subsequent interdigestive period.[19]

References

1. Marya RK. History of Medicine. Chapter 48. JP publishers, New Delhi, 2009.
2. Marya RK. Medical Physiology, 4th Edn, CBS Publishers, New Delhi, 2015.
3. Defilippi C, Solomon TE, Valenzuela JE. Pancreatic secretory response to sham feeding in humans. Digestion. 1982; 23: 217–23
4. Anagnostides A, Chadwick VS, Selden AC, et al. Sham feeding and pancreatic secretion. Evidence for direct vagal stimulation of enzyme output. Gastroenterology.1984; 87: 109–14.
5. Pandol SJ. The Exocrine Pancreas. San Rafael (CA): Morgan & Claypool Life Sciences, 2010.
6. Chandra R, Liddle RA. Regulation of Pancreatic Secretion. Pancreapedia: Exocrine Pancreas Knowledge Base, 2015.
7. Cargill JM, Wormsley KG. Effect of gastric distension on human pancreatic secretion. Acta Hepatogastroenterol (Stuttg). 1979; 26: 235–38.
8. Furukawa N, Okada H. Effects of antral distension on pancreatic exocrine secretion in dogs: evidence for a short reflex. Jpn J Physiol. 1987; 37: 671–85.
9. White TT, Lundh G, Magee DF. Evidence for the existence of a gastropancreatic reflex. Am J Physiol Gastrointest Liver Physiol. 1960; 198: 725–28.

10. Kohler E, Beglinger C, Eysselein V, et al. Gastrin is not a physiological regulator of pancreatic exocrine secretion in the dog. Am J Physiol Gastrointest Liver Physiol. 1987; 252: G40–44.

11. Hildebrand P, Beglinger C, Gyr K, et al. Effects of a cholecystokinin receptor antagonist on intestinal phase of pancreatic and biliary responses in man. J Clin Invest. 1990; 85: 640–46.

12. Bozkurt T, Adler G, Koop I, et al. Effect of atropine on intestinal phase of pancreatic secretion in man. Digestion. 1988; 41: 108–15.

13. Singer MV, Niebergall-Roth E. Secretion from acinar cells of the exocrine pancreas: role of enteropancreatic reflexes and cholecystokinin. Cell Biol Int. 2009; 33: 1–9.

14. Niebergall-Roth E, Singer MV. Central and peripheral neural control of pancreatic exocrine secretion. J Physiol Pharmacol. 2001; 52: 523–38.

15. Schaffalitzky DOB, Fahrenkrug J, Matzen P, et al. Physiological significance of secretin in the pancreatic bicarbonate secretion. II. Pancreatic bicarbonate response to a physiological increase in plasma secretin concentration. Scand J Gastroenterol. 1979; 14: 85–90.

16. Pappas TN, Debas HT, Goto Y, et al. Peptide YY inhibits meal-stimulated pancreatic and gastric secretion. Am J Physiol Gastrointest Liver Physiol. 1985; 248: G118–123.

17. Beglinger C, Taylor IL, Grossman MI, et al. Pancreatic polypeptide release: role of stimulants of exocrine pancreatic secretion in dogs. Gastroenterology. 1984; 87: 530–36.

18. Dembinski A, Konturek SJ, Thor P. Gastric and pancreatic responses to meals varying in pH. J Physiol. 1974; 243: 115–28.

19. Keller J, Layer P. Human pancreatic exocrine response to nutrients in health and disease. Gut 2005;54(Suppl VI): vi1–28.

20. Guimbaud R, Moreau JA, Bouisson M, et al. Intraduodenal free fatty acids rather than triglycerides are responsible for the release of CCK in humans. Pancreas. 1997; 14: 76–82.

21. Ekelund K, Johansson C. Output of bilirubin and pancreatic enzymes in response to different liquid test meals in man. Scand J Gastroenterol. 1975; 10: 507–11.

22. Czako L, Hajnal F, Nemeth J, et al. Effect of a liquid meal given as a bolus into the jejunum on human pancreatic secretion. Pancreas. 1999; 18: 197–202.

23. Green GM, and Lyman RL. Feedback regulation of pancreatic enzyme secretion as a mechanism for trypsin inhibitor-induced hypersecretion in rats. Proc Soc Exp Biol Med. 1972; 140: 6–12.

Chapter 12

Molecular Basis of Smooth Muscle Contraction and Relaxation

INTRODUCTION

Although, muscle protein myosin was isolated in 1864, the mechanism of muscle contraction remained elusive for more than a century. Muscle protein actin was isolated only in 1942. In the same year, Szent-Györgyi recorded the contraction of " actomyosin thread" on addition of ATP resulting in "superprecipitation" of the muscle proteins (Fig. 12.1). At that time, skeletal muscle contraction was believed to result from some chemical reaction between the two muscle proteins, actin and myosin, in the presence of ATP.[1] By 1950s, electron microscopy was available, which revealed actin and myosin filaments in sarcomeres. The sliding filament theory was based on the observations of constancy of the length of the A-band and the shortening of the I band during a contraction observed by interference microscopy in frog's muscle.

Fig. 12.1A and B: Contraction of "actomyosin thread" on addition of ATP. Shown are the same thread (A) before and (B) after addition of boiled muscle juice (a source of ATP) (Szent-Györgyi)

Thus, the ultrastructure and molecular basis of skeletal muscle contraction was worked by early 1950s. The ultrastructure of smooth muscle is also known since many decades, but the *molecular basis of smooth muscle contraction is yet to be completely elucidated.* Ultrastructure of smooth muscle and molecular basis of its contraction can be best explained in the background knowledge of skeletal muscle ultrastructure and microfilament mechanism of skeletal muscle contraction.

SKELETAL MUSCLE ULTRASTRUCTURE

A skeletal muscle is made up of numerous skeletal muscle fibers. Muscle fibers are long cylindrical multinucleated single cells. Each muscle fiber is surrounded by a membranous sarcolemma which insulates it from the adjacent fibers. Each muscle fiber contains up to several thousand myofibrils, stacked longitudinally throughout the length of the muscle fiber. The myofibrils are aligned along the longitudinal axis of the muscle fiber.

A characteristic feature of skeletal muscle fiber is the well-developed cross-striations (Fig. 12.2). Cross-striations are produced by a difference in the refractive index of different parts of myofibrils. Some components are *isotropic* while others are *anisotropic* to polarized light. These components, visible as

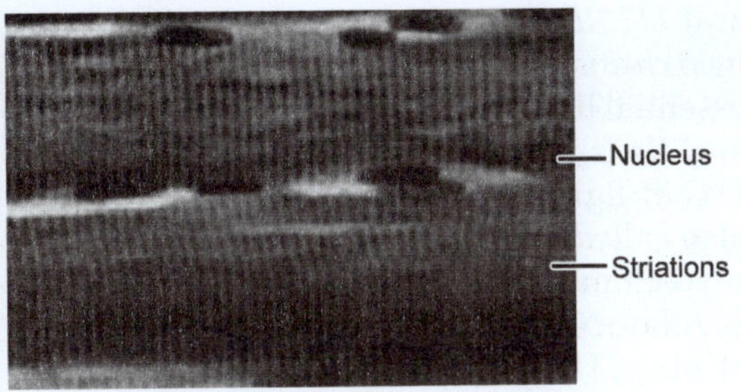

Fig. 12.2: Skeletal muscle fibers

light and dark bands of the myofibrils, are known as *I bands* and *A bands* respectively. Since I bands and A bands of adjacent myofibrils are in register with each other, the cross-striations seem to run across the muscle fiber. At a higher magnification, the I bands are seen to be divided by dark Z lines and the dark A bands have lighter H zones in their center. The area between two adjacent Z lines is called a *sarcomere*, which constitutes a unit of skeletal muscle (Fig. 12.3). Each muscle fiber is composed of thousands of sarcomeres.[2]

Under an electron microscope, each myofibril is seen to contain about *1,500 thick myosin filaments* and *3,000 thin actin filaments*. Actin and myosin filaments partly interdigitate. The light I bands contain only thin filaments. The dark A band is seen where myosin filaments are overlapped by actin filaments. When the muscle is relaxed, the actin filaments attached to the opposite Z lines do not meet each other and a small gap is left where the actin filaments do not overlap the myosin filaments. The gap is seen as H zone (Fig. 12.3).

The myofibrils are suspended in a matrix called *sarcoplasm*. The sarcoplasm of skeletal muscle has all cytoplasmic organelles, but it is myofibrils that occupy most of the available space, pushing the numerous nuclei at the periphery of the muscle fiber. The sarcoplasm is extraordinarily rich in the mitochondria; indicating high energy requirement of the tissue. The sarcoplasm also contains a large amount of glycogen, and enzymes of Embden Meyerhoff pathway. The enzymes of citric acid cycle pathway are present in the mitochondria.[2]

The molecular basis of muscle contraction is related to the enzymatic and physical properties of thick and thin filaments. The proteins of the thick and thin filaments can be separated into myosin, actin and accessory proteins or protein complexes. The accessory proteins/complexes include the tropomyosin and the troponin complex in skeletal muscle. Smooth muscle fibers do not have troponin, but have tropomyosin and recently discovered proteins such as caldesmon and calponin.

Actin Molecule

The actins were initially characterized as being either α (alpha), β (beta), or γ (gamma) actins. The α-actins are the predominant actins of the contractile apparatus of muscle cells. The β- and γ-actins are involved in the regulation of cell motility functions.

The main body of the thin filament is made up of a double strand of F-actin protein molecule. One end of the actin filament is inserted into Z-disc while the other free end lies in the space between the myosin filaments. At regular intervals, actin filament possesses binding sites for the myosin heads. *In vitro* actin molecules exist in two forms: G actin and F-actin. *G actin* is a monomeric globular protein with a molecular weight of 46,000. Each monomer contains binding sites for other actin monomers and for myosin, tropomyosin, troponin I and ATP. *F-actin* is a fibrous polymer, consisting of approximately 300 monomers of G actin. Thin filament of a sarcomere consists of two strands of F-actin polymers, intertwined in the conformation of a double-stranded helix.

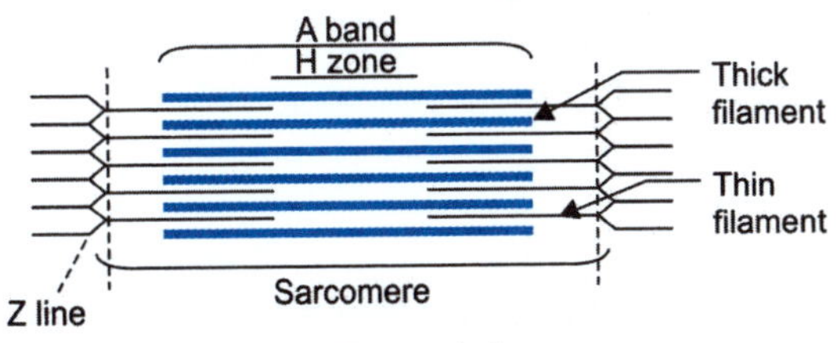

Fig. 12.3: Microfilaments in a sarcomere

Myosin Molecule

Each myosin molecule is made up of six subunits, two very large, heavy chains (HC), and two pairs of non-identical light chains (LC) (Fig. 12.4). Within the myosin molecule, it is the HC proteins that possess the ATPase-mediated motor function. In a given muscle fiber the two large subunits are identical. Heavy chains contain a long linear C-terminal α-helical domain (1,300-amino acids) (tail) and a prominent globular N-terminal domain of about 800-amino acids (head). The two HC α-helical domains are inter-wound, giving the molecules a long, rigid superhelical structure with two globular headpieces. A complete myosin molecule also contains the two pairs of LC in the region known as neck, which are associated with the globular headpieces.[2]

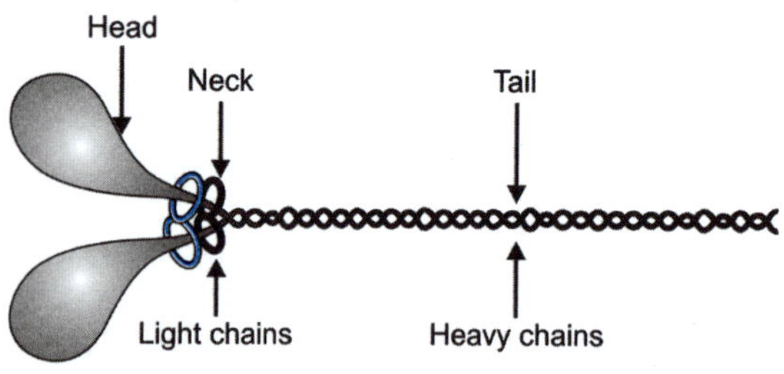

Fig. 12.4: A myosin molecule

The function of LC is well known in smooth muscle only. The LC proteins were originally characterized by their chemical sensitivities and so are also referred to as *alkali light chains* and *DTNB (5,5'-dithiobis/2-nitrobenzoate) light chains.* Two alkali light chains (also called essential light chains) are not phosphorylated and their function is not yet clear. The two DTNB light chains are phosphorylated and also called the regulatory light chains of the myosin molecule (also known as LC20). The regulatory LC proteins bind Ca^{2+} with high affinity. The phosphorylation of the LC proteins is catalyzed by one of four myosin light chain kinases (MLCK). The myosin light chain kinases are also regulated by Ca^{2+} binding to their calmodulin subunits. Phosphorylation of myosins by MLCK serves to regulate the ATPase activity of the myosin molecule in smooth muscle.[3] *The function of light chains shall be discussed later in this chapter when mechanism of smooth muscle contraction is discussed.*

A functionally important landmark exists on the myosin molecule. Near the midpoint of the long linear superhelical region is a site defined by its ready susceptibility to proteolytic trypsin digestion. Trypsin cleaves myosin complexes into two portions: One containing both globular headpieces and some superhelical region, and the other consisting of the remaining superhelical portion of the carboxy terminus. The portion containing the headpiece is referred to as heavy meromyosin (HMM; molecular weight 350,000). The C-terminal fragment is referred to as light meromyosin (LMM; molecular weight 125,000) (Fig. 12.5).

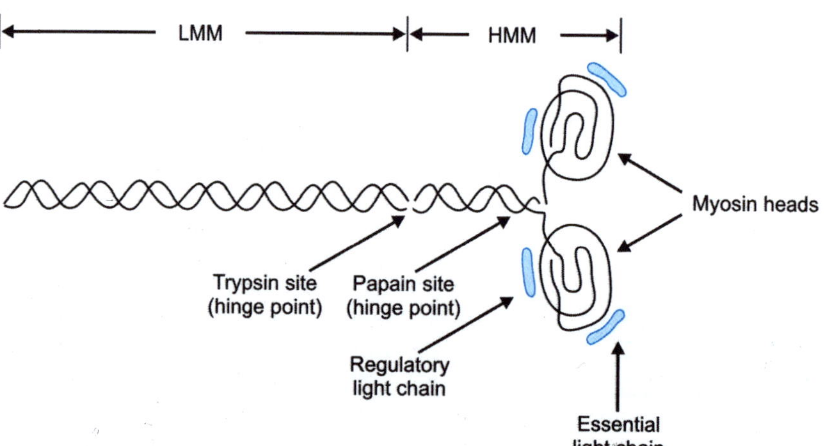

Fig. 12.5: Details of myosin molecule showing light chains and trypsin site

The significance of the trypsin site is that it reflects an interruption in the otherwise rigid superhelix, allowing this site to act as one of the hinge points involved in mechanical events of contraction and relaxation.[3] At the trypsin hinge point, the heavy meromyosin angles sharply outward from the main axis of the thick filament (Fig. 12.6). This extension of the heavy meromyosin away from the main axis of the thick filament helps bring the headpiece into close proximity to the actin thin filaments lying between the thick filaments. Papain site is another hinge point on the myosin molecule (Figs 12.5 and 12.6). The protease papain cleaves myosin at the base of the heads, releasing the long rod-shaped tail and two myosin subfragments. This site also acts as a hinge point. The molecular event underlying muscle contraction is the regulated binding of the myosin headpieces to actin thin filaments, followed by rapid myosin conformational changes about its hinge points with the bound actin being translocated towards M line.[4]

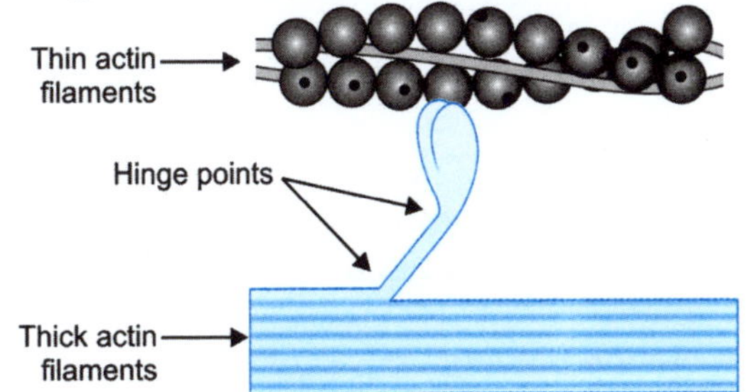

Fig.12.6: Role of hinge points on the myosin molecule

Thick Filaments

A thick filament is composed of approximately 400 myosin molecules, 200 arrayed on either side of the M line (M band). The myosin molecules are most tightly packed in the regions represented by the LMM portion of the molecules. At the trypsin hinge point the heavy meromyosin angles sharply outward from the main axis of the thick filament. This extension of the heavy meromyosin away from the main axis of the thick filament helps bring the headpiece into close proximity to the actin thin filaments lying between the thick filaments.

Thin Filaments

Thin filaments are composed of many subunits of the globular protein G actin (G for globular: 42 kD) and several accessory proteins. In thin filaments, G actin is polymerized into long fibrous (filamentous) arrays known as F-actin (F for filamentous). Each G actin molecule contains a high-affinity myosin head-binding site. In skeletal muscle, accessory proteins of the thin filament physically regulate the availability of this site for binding myosin. Thus, the accessory proteins control contractile events. The main thin filament accessory proteins are the tropomyosin and the troponin complex.

Tropomyosin is a long, rod-like molecule. In relaxed muscle, tropomyosin molecules cover the myosin binding sites of G actin residues, preventing interaction between actin and myosin and thus maintaining the relaxed state. In the absence of calcium, tropomyosin, along with the troponin complex, prevent myosin access to its binding site on actin. In the presence of calcium, the tropomyosin-troponin complexes allow the myosin-actin interactions to occur. This onset of contractile activity involves the calcium-mediated activation of the troponin complex.

Troponin is a heterotrimeric complex of troponin C (TnC), troponin I (TnI), and troponin T (TnT). TnI component binds troponin tightly to actin filaments. The TnC protein of the complex is the Ca^{2+}-binding subunit, with similarity to calmodulin, whose role is to effect the Ca^{2+}-dependent regulation of muscle contraction. When TnC binds calcium, the whole troponin complex undergoes the conformational changes that move the attached tropomyosin away from the myosin binding sites on actin. The conformational change permits nearby myosin heads to interact with myosin binding sites on actin, and contractile activity ensues.

The TnT protein binds troponin complex with tropomyosin, thereby regulating troponin complex interaction with thin filaments. The entire troponin complex is attached to one end of each tropomyosin molecule and to actin, physically linking these later two proteins.[2]

Thin Filament Activation

Events on the thin filament can be summarized as follows: Prior to the appearance of free calcium in the sarcoplasm, tropomyosin covers the myosin binding sites on actin. The appearance of calcium in the sarcoplasm (cytoplasmic compartment of muscle cells) leads to calcium binding to the TnC subunit of the troponin complex. The resulting conformational changes in the troponin complex move the attached tropomyosin molecule more deeply into the helix groove of F-actin, uncovering the myosin binding sites on G actin subunits. The exposed sites are then available to interact with myosin headpieces. Removing calcium from the sarcoplasm restores the original conformational states of the troponin complex and tropomyosin, preventing interaction between actin and myosin and leading to the relaxed state.

SLIDING FILAMENT MECHANISM OF SKELETAL OF MUSCLE CONTRACTION

The contraction of the skeletal muscle is initiated by the spread of an action potential from the nerve to the sarcolemma, and then through T-tubules to the terminal cisterns. The Ca^{2+} released from the terminal cisterns bind with troponin C. The troponin C-Ca^{2+} complex undergo a conformational change leading to lateral displacement of tropomyosin strand.

As a result, the binding sites on the actin filaments are exposed. The myosin heads attach to these exposed binding sites on the actin filaments, forming crossbridges which bend immediately. The ATP attached to the crossbridge splits into ADP and energy released is used for mechanical bending of the crossbridges. *The inward bending of the crossbridges produces sliding of actin filaments inwards.* Almost immediately the myosin heads bind another ATP molecule, release actin filament and return to the perpendicular position with respect to the tail of the myosin filament. The cycle of binding, bending, and release is repeated once again. As long as the sarcoplasm contains Ca^{2+}, the process is repeated until the sliding movements of actin filaments bring the Z lines closer and the muscle shortens[2] (Fig. 12.7). The sequence of events leading to contraction of skeletal muscle is summarized below.

Summary of steps of skeletal muscle contraction:

1. Stimulation of motor nerve.
2. Neuromuscular transmission
3. Release of acetylcholine at motor end plate
4. Action potential in sarcolemma
5. Action potential in T tube
6. Release of Ca^{2+} from terminal cisterns

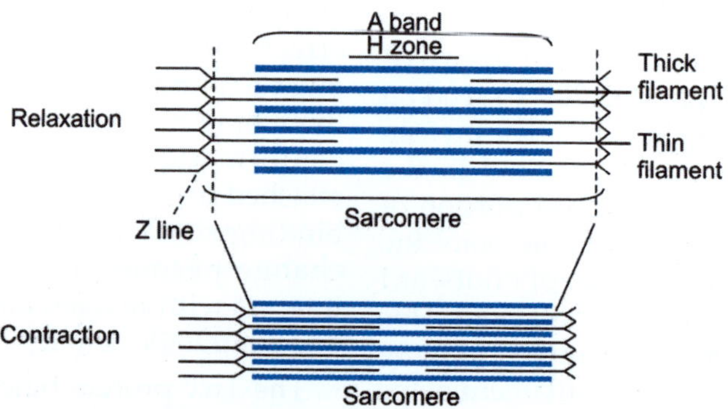

Fig.12.7: Skeletal muscle shortening

7. Increased cytosolic Ca^{2+} concentration
8. Ca^{2+}-troponin C binding
9. Actin binding sites uncovered
10. Myosin heads activated
11. Crossbridges formed
12. Sliding of thin filaments over thick filaments
13. Muscle shortens

Skeletal Muscle Relaxation

Within a few milliseconds of Ca^{2+} release, the Calcium-pump operating at the surface of the longitudinal sarcoplasmic reticulum lowers the sarcoplasmic of Ca^{2+} concentration to such an extent that the interaction between actin and myosin filaments stops and the muscle *relaxes* once again. *ATP is consumed not only to produce contraction but also to produce relaxation of the skeletal muscle.* Energy is required for mechanical bending of the crossbridges to produce contraction and for active transport of Ca^{2+} by the longitudinal sarcoplasmic reticulum to produce relaxation.[2]

SMOOTH MUSCLE

Smooth muscle cell is fusiform in shape: Wide in the middle with tapered ends and only a single nucleus in central position. These cells do not contain the sarcomeres found in skeletal and cardiac muscle. Smooth muscle cells are seen unstriated because there is no regular arrangement of actin and myosin filaments (Fig. 12.8A). Cytoplasmic reticulum is poorly developed. Instead, smooth muscle cells use extracellular calcium as the source of calcium ions which initiates contraction. Smooth muscle contains both actin and myosin filaments having chemical characteristics similar to but not exactly same as in skeletal muscle. The actin filaments are attached to *dense bodies* (and not to Z discs).

The myosin myofilaments are dispersed throughout the muscle cell cytoplasm. Actin filaments are held together in contractile bundles of 12–15. The actin filaments of contractile units are attached to *dense bodies* (Fig. 12.8A). Dense bodies are also attached to intermediate filaments and thereby appear to serve as anchors from which the thin filaments can exert force. Dense bodies may coordinate tensions from both the contractile machinery and the cytoskeleton.

Each actin bundle overlaps the myosin filament on one side, and converges to attach to a dense body on the other side. A single myosin filament is overlapped by two seperate actin contractile bodies, each of which are attached to a different dense body. The actin–myosin filament ratio is 10:1 in smooth muscle as compared to 2:1 in skeletal muscle (Fig. 12.8B).

The contractile filament bundles of actin and myosin are loosely arranged in a diagonal fashion, in different directions around the perimeter of the smooth muscle cell. This arrangement of fibers causes the muscle cell to become globular upon contraction.

In addition to the thick myosin and thin actin filaments, smooth muscles possess noncontracting intermediate filaments. The intermediate fibers attach the dense bodies, that are scattered throughout the sarcoplasm to the sarcolemma. During contraction, the movements of myosin and actin are transferred to intermediate fibers, which pull the dense bodies; these in turn pull the muscle cells together. In this way, the dense bodies function similarly to the Z discs in striated muscles. The intermediate filaments are connected to other intermediate filaments via dense bodies, which eventually are attached to adherens junctions in the cell membrane of the smooth muscle cell, called the sarcolemma. The adherens junctions consist of large number of proteins including α-actinin, vinculin and cytoskeletal actin.[5] The adherens junctions are scattered around *dense bands* that encircle the smooth muscle cell in a rib-like pattern.[5]

Excitation–Contraction Coupling

While the sliding filament model adequately describes the basic mechanism of contraction in all muscle types, there are significant differences between striated and smooth

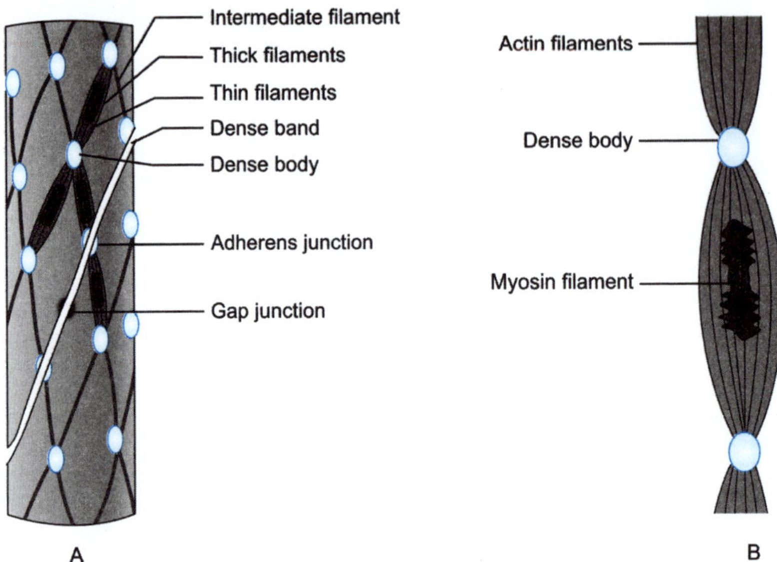

Fig. 12.8: (A) Smooth muscle ultrastructure; (B) Detailed arrangement of actin filaments, myosin filament and dense body

muscles. An appreciation of these differences stems from the observation that although smooth muscle lacks the troponin complex, its contractile activity is still regulated by cytoplasmic calcium levels. This was explained when it was discovered that a Ca^{2+}/calmodulin was involved in regulating the movement of smooth muscle tropomyosin on and off the myosin binding sites of thin filaments. Alterations in smooth muscle cytosolic calcium levels occur via both voltage-dependent activation processes and by receptor-mediated processes.

To initiate smooth muscle contraction, activation of plasma membrane may occur by two types of Ca^{2+} channels; voltage-gated and ligand-gated calcium channels (Fig. 12.9). The voltage-gated calcium channels are of the L-type. They are opened by an action potential in sarcolemma. On the other hand, ligand-gated calcium channels are activated by a neurotransmitter, e.g. norepinephrine, ligands such as vasopressin, angiotensin II or simply by a stretch of smooth muscle fiber.

The voltage-gated ion channels transfer Ca^{2+} from ECF into sarcoplasm. The ligand-gated receptor is coupled to a G protein which leads to production of inositol triphosphate (IP_3) and diacylglycerol (DAG). The IP_3 binds to receptors on the endoplasmic reticulum leading to release of stored Ca^{2+} into the sarcoplasm. DAG opens calcium channels that increase Ca^{2+} from extracellular fluid into the smooth muscle cell (Fig. 12.9). However, unlike skeletal muscle, endoplasmic reticulum is poorly developed in smooth muscle. Therefore, in smooth muscle contraction, ECF is the chief source of Ca^{2+}.

The Ca^{2+} ions bind to calmodulin. (Smooth muscle cells do not contain troponin, like in skeletal muscle cells, so Ca^{2+} binds to calmodulin.) This calcium-calmodulin activates an enzyme called myosin light chain kinase (MLCK). Activated MLCK is responsible for phosphorylating the regulatory light chains at the neck of myosin filaments. Phosphorylated myosin can form crossbridges with actin filaments. The sliding of actin filaments over myosin filaments results in smooth muscle contraction[6] (Fig. 12.9).

In striated muscle cells, the overall sliding of thick and thin filaments pulls the two ends towards the center of the cell decreasing the total length of the fiber. Smooth muscle cells have been observed contracting in a spiral corkscrew fashion, and contractile proteins

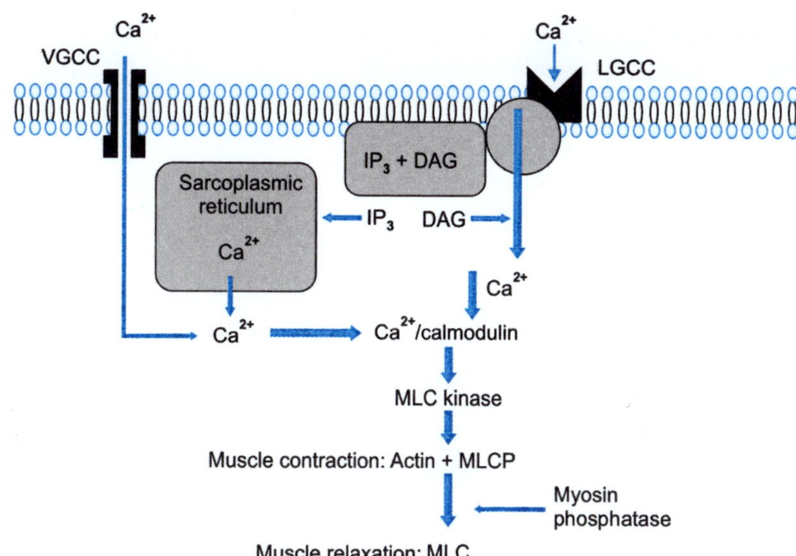

Fig. 12.9: Mechanisms of smooth muscle contraction. VGCC: Voltage-gated calcium channel; LGCC: Ligand-gated calcium channel; IP₃: Inositol triphosphate; DAG: Diacylglycerol; MLC kinase: Myosin light chain kinase; MLCP: Myosin light chain phosphate; MLC: Myosin light chain (dephosphorylated)

have been observed organizing into zones of actin and myosin along the axis of the cell (Fig. 12.10).

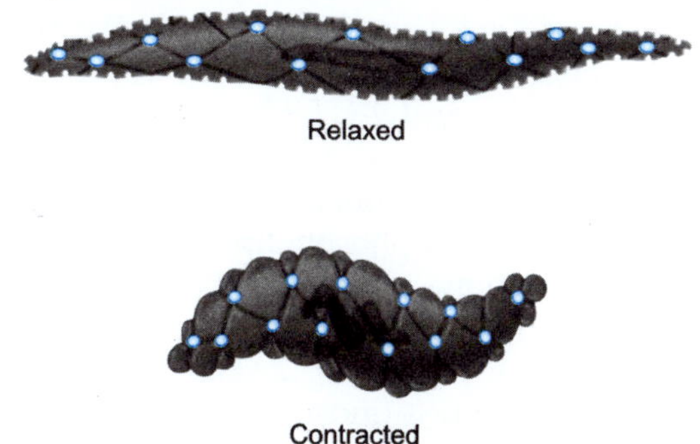

Fig. 12.10: Smooth muscle fiber in relaxed and contracted state

Thin Filament-associated Regulatory Proteins

These days, the role of two thin-filament associated regulatory proteins in smooth muscle contraction is being investigated, i.e. caldesmon and calponin.

Caldesmon

Smooth muscle thin filaments contain tropomyosin, but, unlike skeletal and cardiac muscle, they lack troponin. Because tropomyosin plays an auxiliary role by conferring cooperativity between the troponin complexes spaced apart along the thin filament, additional thin filament regulatory proteins analogous to troponin might be expected to be present in smooth muscle. Caldesmon is one such protein under intense investigations. In the smooth muscle cells, Ca^{2+} binds to calmodulin (Fig. 12.11). It is now believed that calcium-calmodulin complex removes the protein caldesmon from the actin sites where myosin will attach.[3] Calcium-calmodulin complex also activates an enzyme called a myosin light chain kinase (MLCK). Activated

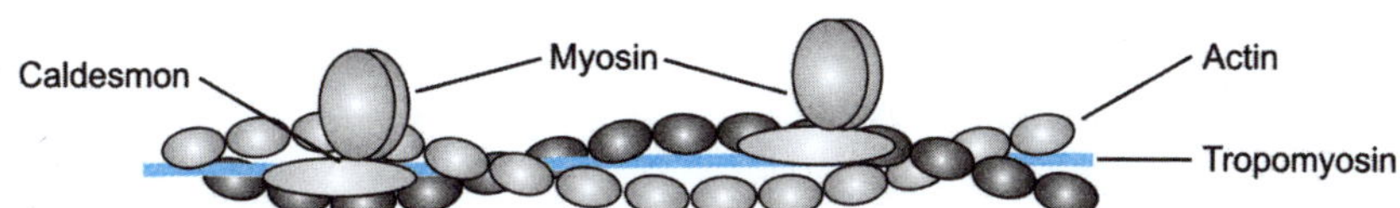

Fig. 12.11: Position of actin filament-associated proteins tropomyosin and caldesmon in smooth muscle

MLCK is responsible for phosphorylating light chains at the neck of myosin filaments. Phosphorylated myosin can form cross-bridges with actin filaments (Fig. 12.10). The sliding of myosin filaments over actin filaments results in smooth muscle contraction. In essence, caldesmon in smooth muscle cells replaces the troponin complex as a calcium dependent regulator of the location of the tropomyosin complex within the context of the thin filaments.[7] In short, it seems, in smooth muscle, increased concentration of Ca^{2+} in the cytosol leads to activation of both myosin and actin molecules (Fig. 12.12), in contrast to skeletal muscle where only actin reactive sites are activated by Ca^{2+}.

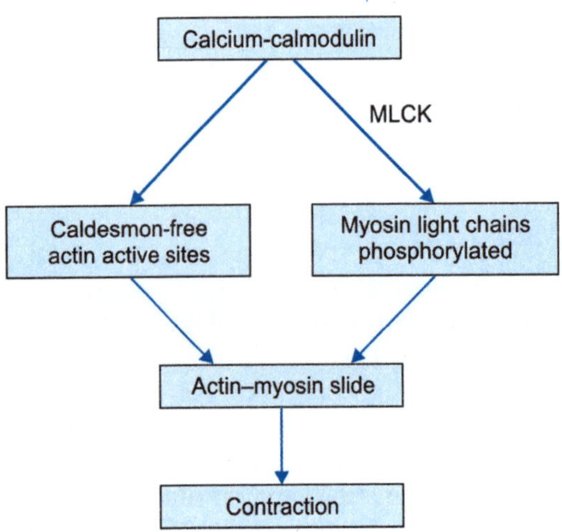

Fig. 12.12: Co-activation of actin and myosin filaments by calcium-calmodulin complex in smooth muscle contraction

Calponin

The demonstrated capacity of protein calponin to bind actin and inhibit actomyosin ATPase *in vitro* indicates a possible involvement of calponin in thin filament regulation. Some studies support the view that calponin is a thin filament regulatory protein and that its inhibitory activity is regulated by phosphorylation. However, role of calponin in smooth muscle contraction is yet to be confirmed.[3]

Steps of Smooth Muscle Contraction[8]

1. An action potential in the sympathetic motor neuron travels through the axon and reaches the synaptic terminal.
2. The action potential causes activation of voltage-gated Ca^{2+} channels on the presynaptic nerve terminal inducing influx of Ca^{2+} ions into the cytoplasm.
3. The increase concentration of Ca^{2+} will promote exocytosis of synaptic vesicles promoting the expulsion of the neurotransmitter norepinephrine into the interstitial space.
4. Norepinephrine reaches the smooth muscle cell membrane where it binds to a G protein coupled ligand-gated Ca^{2+} channel receptor.
5. Once the transmitter-receptor complex is formed, an underlying protein undergoes a conformational change that activates the G protein.
6. The inactive G protein consists of 3 heterogeneous α, β- and γ-subunits with GDP bound on the α-subunit. Activation occurs when GDP is substituted with GTP which promotes dissociation of the G protein into α-, β- and γ-GTP individual components.
7. The γ-GTP component binds to a specific enzyme called phospholipase C inducing its activation.
8. The activated phospholipase C cleaves phospholipids into DAGs (diacylglycerols) and IP_3s (inositol triphosphates).
9. The DAG and IP_3 act as second messengers that will activate Ca^{2+} channels. DAG binds on plasma membrane Ca^{2+} receptors opening a channel allowing Ca^{2+} influx. IP_3 binds on receptors on the sarcoplasmic reticulum and opens channels promoting Ca^{2+} efflux from the reticulum into the cytosol.
10. The Ca^{2+} accumulated inside the smooth muscle cell binds with calmodulin giving rise to the Ca^{2+}-calmodulin complex.
11. The Ca^{2+}-calmodulin complex binds and activates myosin light chain kinase (MLCK).

12. MLCK phosphorylates the myosin light chains enabling the myosin crossbridge to bind to the actin filament and allows contraction to begin.

Smooth Muscle Relaxation

The mechanism of skeletal muscle relaxation described earlier is rather simple. The skeletal muscle is normally relaxed. In response to alpha motor neuron stimulus, it contracts to be immediately followed by relaxation. In contrast, smooth muscle contraction may occur in a variety of forms. Smooth muscle relaxation occurs either as a result of removal of the contractile stimulus, or smooth muscle may remain in a continuous tonic contraction and relax only in the presence of a neural stimulus (e.g. lower esophageal sphincter) or a ligand (vascular smooth muscle relaxation in response to NO).

Smooth muscle relaxation occurs when the myosin is dephosphorylated by myosin phosphatase (MP) removing it from actin. This process is relatively slow because it relies on the diffusion of calcium ions over large distances. Hence, maximum contraction is often nearly a second long and uses very little ATP. This is important so that smooth muscle does not fatigue during sustained periods of activity. The rate of ATP splitting determines the rate of muscle contraction. In smooth muscle myosin, the rate of ATPase activity is 10 to 100 times slower than in skeletal muscle myosin.

In general, the relaxation of smooth muscle is induced by cell-signaling pathways that (Fig. 12.9):

i. increase the myosin phosphatase activity, or

ii. hyperpolarize the smooth muscle.

Increase in myosin phosphatase activity can occur in response to a decrease in cytosolic calcium concentration or by cGMP mechanism. Decrease in calcium ion concentration in the cytosol of smooth muscle may occur by a variety of mechanisms:

1. Closure of voltage-operated or ligand-operated calcium channels.

2. Transfer of calcium back into sarcoplasmic reticulum.

3. Extrusion of calcium by Na/Ca exchanger.

4. Extrusion of calcium by a Ca, Mg-ATPase pump.

The decrease in cytosolic calcium concentration increases myosin phosphatase activity.[8]

Unsolved Mysteries of Smooth Muscle Contraction

i. **Greater power generation than skeletal muscle:** One of the more intriguing puzzles in the field of muscle physiology is the fact that smooth muscle is capable of generating more force per cross-sectional area than striated muscle, despite the fact that smooth muscle cells contain only roughly one-fifth of the myosin content of striated muscle.[9] This difference in the two muscle types persists even at the single-cell level. It has been suggested that geometric factors might explain the difference. The longer filaments in smooth muscle result in greater numbers of crossbridges arranged in parallel, and insertion of multiple filaments along the cell membrane results in a greater number of total contractile units arranged in parallel.[10] However, Vanburen et al[11], obtained an estimate of the force generated per myosin head in smooth and skeletal muscle (0.6 and 0.2 units, respectively), thus demonstrating that the smooth muscle myosin molecule itself (separate from geometric factors and accessory proteins) can generate greater force than the striated muscle myosin molecule.

ii. **Latch effect:** Smooth muscle is known to enter a state where $[Ca^{2+}]_i$ levels decline to low or, in some cases, basal levels despite the fact that vascular tone is maintained. The phenomenon of Ca^{2+}-dependent force maintenance without detectable elevations in light chain phosphorylation was termed latch by Dillon et al[12] to indicate mechanical similarities with molluscan catch muscles. The fundamental reason in the latch phenomenon is the decreased rate of crossbridge cycling and ATP utilization. Hai and Murphy[13] have proposed that smooth muscle LC20 phosphorylation is obligatory for crossbridge attachment, and that the only regulatory mechanism that needs to be considered in smooth muscle is phosphorylation and dephosphorylation of LC20. The key assumption of the model is that the sole effect of dephosphorylating an attached (previously phosphorylated) crossbridge is the reduction in detachment rate to form a "latchbridge."

References

1. Szent-Györgyin AG.Milestone in Physiology: The early history of the biochemistry of muscle contraction. Published Online: 1 June, 2004 | Supp Info: http://doi.org/10.1085/jgp.200409091
2. Marya RK. Medical Physiology. 4th Edn. CBS publishers, New Delhi, 2015.
3. Horowitz A, Menice CB, Laporte R, et al. Mechanisms of smooth muscle contraction. Physiol Rev. 1996; 76: 967–1003.
4. https://themedicalbiochemistrypage.org/muscle.php 1996–2018
5. Aguilar HN, Mitchell BF. Physiological pathways and molecular mechanisms regulating uterine contractility. Hum Reprod Update. 2010; 16: 725–44.
6. Schwartz SM, Robert P.The vascular smooth muscle cell: molecular and biological responses to the extracellular matrix. Editors: Stephen M Schwartz, Robert P Mecham. Publisher: Academic Press, 1995.
7. Ansari S, Alahayan M, Marston SB, et al. Role of caldesmon in the Ca^{2+} regulation of smooth muscle thin filaments: Evidence for a cooperative switching mechanism. Jour Biological Chem. 2008; 283: 47–56.
8. Webb RC. Smooth muscle contraction and relaxation. APS refresher course report..01 DEC 2003https://doi.org/10.1152/advan.00025.2003
9. Murphy RA, Herlihy JT, Megerman J. Force-generating capacity and contractile protein content of arterial smooth muscle. J. Gen. Physiol 1974; 64: 691–705 .
10. Warshaw DM, McBride WJ, Work SS. Corkscrew like shortening in single smooth muscle cells. Science. Wash. DC 1987; 236: 1457– 59.
11. Vanburen PS, Work SS, Warshaw DM. Enhanced force generation by smooth muscle myosin in vitro. Proc Natl Acad. Sci. USA 1994; 91: 202–5.
12. Dillon PF, Aksoy MO, Driska SP, et al, Myosin phosphorylation and the crossbridge cycle in arterial smooth muscle. Science, Wash. DC 1981; 211: 495–97.
13. Hai CM, Murphy RA. Crossbridge phosphorylation and regulation of latch state in smooth muscle. Am J Physiol. 1988; 254 (Cell Physiol. 23): C99–106.

Chapter 13

Neuroglia

INTRODUCTION

The brain consists of two major types of cells, neurons and glial cells. Glial cells are as numerous as neurons. Neuroglia were initially described by Rudolf Virchow as a *connective tissue* that binds nervous elements together. At that time, glia were suggested to solely function as so-called "Nervenkitt" (the German word for nerve glue). This is also reflected in the name "glial cell" derived from the ancient Greek word *"glía"* meaning "glue" in English. With time, scientists started to speculate about additional possible roles of these cells. At present, glial cells are believed to play many roles crucial for normal neuronal function. In spite of huge number of researches, full functions of glial cells remain unresolved.[1]

While neurons have been studied for more than a century, extensive glial research has developed only over the last few decades. For the last two decades glial cell research has considerably increased, and there are a number of reasons for this development. It could be shown that there is not a single pathological process in the brain which occurs without participation of glial cells, specifically microglia and astrocytes. Another aspect is the evidence of glial excitability. This form of communication is much slower than the neuronal response. However, it has become evident that glial cells, and astrocytes in particular, modulate neuronal activity and thus brain function. We can assume that the combination of the activities of glial cells and neurons is crucial for all brain functions.[2]

We can distinguish three types of glial cells in the central nervous system: Astrocytes, oligodendrocytes and microglial cells (Fig. 13.1). Schwann cells can be found in the peripheral nervous system. Astrocytes are a rather heterogeneous cell population which interact with neurons and blood vessels. These cells detect neuronal activity and modulate neuronal networks. Oligodendrocytes in the central nervous system and Schwann cells in the peripheral nervous system produce myelin and are thus responsible for the high speed information processing in the axons of vertebrates. Microglial cells are the immune cells of the central nervous system and react with a so-called activation to various types of damage in the nervous system. They are therefore also called the pathological sensors of the brain. They migrate to the site of damage, proliferate and become phagocytes as well as interact with the peripheral immune system by antigen presentation. Today, we conceive the brain as an organ which can achieve its function due to the interaction of all these cell types only.[2] Microglia are of mesodermal origin, whereas rest of the neuroglia including

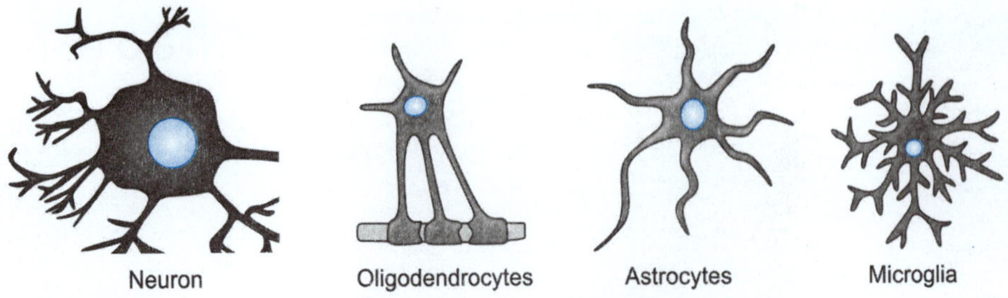

Fig. 13.1: A neuron and various types of neuroglia

Schwann cells are of ectodermal origin (Fig. 13.2).

1. Astrocytes

a. Morphology

Morphology of astrocytes ("star-shaped cells") is highly heterogeneous. Some astrocytes do have a star-like appearance, with several primary processes originating from the soma. Two types of astrocytes can be identified: Protoplasmic astrocytes and fibrous astrocytes (Fig. 13.3). *Protoplasmic astrocytes* are present in gray matter. They have many fine processes (approximately ~50 µm long), which are extremely elaborate and complex. The processes of protoplasmic astrocytes contact blood vessels, forming so-called 'perivascular' end-feet, and form multiple contacts with neurons. Some protoplasmic astrocytes also send processes to

Fibrous astrocyte

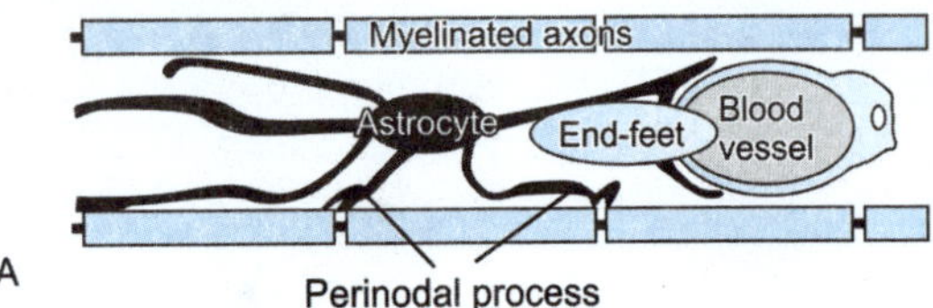

Protoplasmic astrocyte

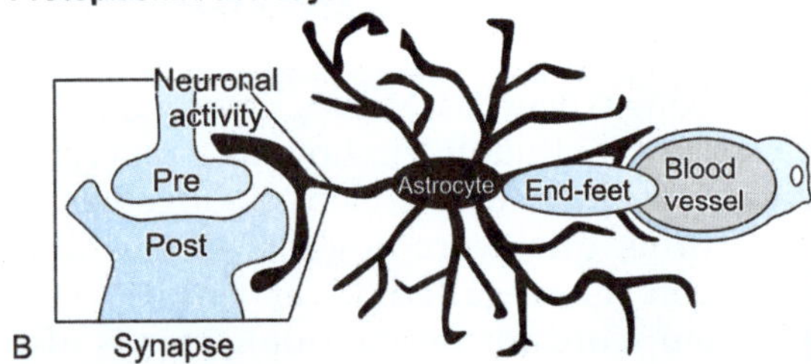

Fig. 13.3: A fibrous astrocyte (A) and a protoplasmic astrocyte (B)

the pial surface, where they form 'subpial' end-feet. Protoplasmic astrocyte processes

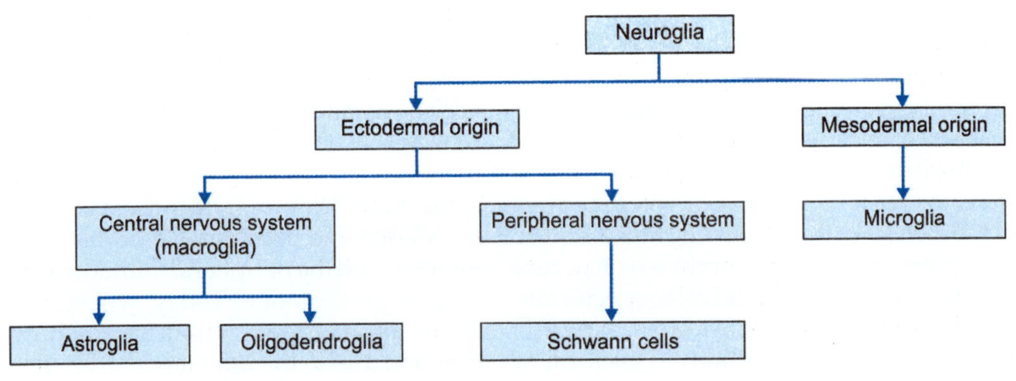

Fig. 13.2: Classification of glial cells and their embryonic origin

cover most of neuronal membranes within their reach.

Fibrous astrocytes are present in white matter. Their processes are long (up to 300 μm), though much less elaborate as compared to protoplasmic astrocytes. The processes of fibrous astrocytes establish several perivascular or subpial end-feet. Fibrous astrocyte processes also send numerous extensions ('perinodal' processes) that contact axons at nodes of Ranvier.

b. Functions of Astrocytes

Astrocytes act in a variety of ways to support neuronal function:

i. **Ion transport:** One of the well-established functions of astrocytes is the regulation of extracellular K^+ concentration of the brain. During neuronal activity, an increase in astrocyte K^+ ion concentration parallels loss of K^+ from active neurons. Astrocytes expedite the removal of impulse-evoked increase in extracellular K^+ concentration. Neuronal activity rapidly increases extracellular K^+ of the brain from 3 mM up to 10–12 mM. Therefore, homeostatic control of brain's extracellular K^+ concentration is a physiological necessity.[3] Increased extracellular K^+ concentration can influence transmitter release, cerebral blood flow and even neuronal activity.[4–6] Astrocytes remove extracellular K^+ by (i) an increase in Na^+ pump activity, (ii) an anion transporter that cotransports K^+ with Na^+ and Cl^- and (iii) Donnan forces that propel KCl into glial cells in the face of elevated K^+ concentration[3] (Fig. 13.4).

ii. **Glutamate-glutamine cycle:** Glutamate is the most abundant excitatory neurotransmitter in the brain. It is also a potential neurotoxin. It can be derived from neuronal glucose metabolism, but carbon labeling experiments have shown that astrocyte-derived glutamine is principal precursor of neuronal glutamate.[7,8] After release from the presynaptic neuron ter-

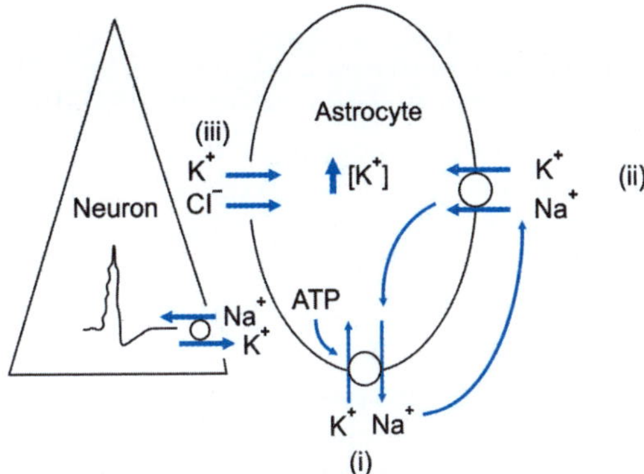

Fig.13.4: Mechanisms of K^+ uptake by astrocytes (*see* text) (modified from Magistretti and Ranson[3])

minal, a very small amount of glutamate combines with glutamate receptors on the postsynaptic membrane. However, most of it is taken up by the nearby astrocytes. In the glial cell, glutamate is converted to glutamine through an enzymes glutamine synthetase. Glutamine released from the astrocytes reaches the neuron through specific uptake carriers.[9] In the presynaptic neuron, glutamine is converted to glutamate through the enzyme glutaminase and packaged into microvesicles (Fig.13.5). The glutamine-glutamate cycle

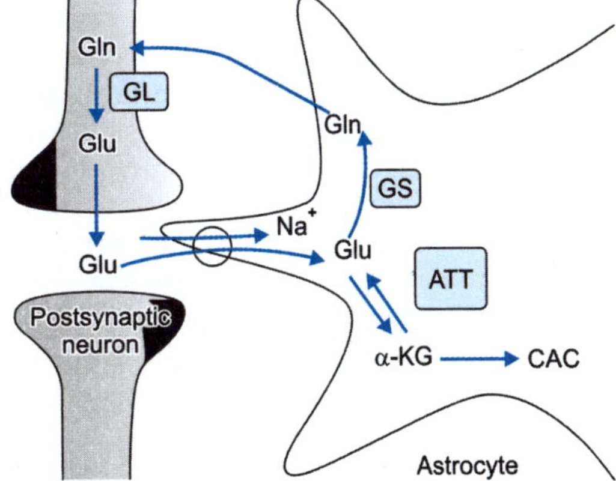

Fig. 13. 5: Glutamate-glutamine cycle. Glu: Glutamate; Gln: Glutamine; α-KG: α ketoglutarate; AAT: Aspartate aminotransferase; GS: Glutamine synthetase; GL: Glutaminase; CAC: Citric acid cycle

is one of the examples of cooperation between neurons and astrocytes. The cycle not only removes excess glutamate from the extracellular space but also supplies the neuron with an inert precursor for re-synthesis of glutamate. Thus, metabolic burden of neurotransmitter synthesis is off-loaded on the adjacent astrocyte.[3]

iii. **Tripartite synapse:** In the gray matter, astrocytes are closely associated with neuronal membranes and specifically in synaptic regions, so that astroglial membranes completely or partially enwrap presynaptic terminals and postsynaptic structures (Fig. 13.6B). The distance between terminal structures of astrocytes and neuronal pre- and postsynaptic membranes is as close as 1 μm. The very intimate morphological apposition of astrocytes and synaptic structures allow the former to be exposed to the neurotransmitters released from the synaptic terminals. Functionally, the processes of astroglial cells are endowed with neurotransmitter receptors. Astrocytes have a complement of receptors very similar to that of their neuronal neighbor. Close morphological relations between astrocytes and synapses as well as functional expression of relevant receptors in the astroglial cells prompted the concept of the *tripartite synapse*. According to this concept, synapses are built from three equally important parts, the presynaptic terminal, the postsynaptic neuronal membrane and the surrounding astrocyte (Fig.13.6B). A neurotransmitter released from the presynaptic terminal activates receptors in both the postsynaptic neuronal membrane and the perisynaptic astroglial membrane. This results in the generation of a postsynaptic potential in the neuron and a Ca^{2+} signal in the astrocyte. Astrocyte Ca^{2+} elevation stimulates release of different neuroactive substances ("gliotransmitters") which regulate neuronal excitability and synaptic transmission.[10]

The question of whether astrocytes actively participate in the ongoing synaptic transmission remains open. Astroglial signals are on a much slower time scale; in the second or even minute range as compared to the rapid signaling of neurons which occurs within milliseconds. Astrocytes therefore can be considered as integrators or modulators of synaptic transmission. At the same time the intimate coverage of synaptic structures by astroglial membranes can have another important role: The astroglial membranes may effectively isolate synapses and prevent neurotransmitter spillover (e.g. of glutamate and GABA); thereby increasing spatial precision of synaptic transmission.[3]

iv. **Energy support to neurons (astrocyte-neuron lactate shuttle):** Until recently it has been believed that due to their high energy demands neurons synthesize energy primarily by the oxidative metabolism of glucose (Krebs cycle and respiratory chain) using glucose as fuel. However, a lot of evidence suggests that neurons can efficiently utilize lactate and, in addition, have a preference for lactate if both glucose and lactate are present.[11] In addition, astrocyte processes possess not only receptor and reuptake sites for neurotransmitters, including glutamate, but *also astrocyte end-feet are rich in specific glucose transporter 1 (GLUT-1). Astrocytes convert glucose to lactate which is transported to the nearby neuron* (Fig. 13.7).

Experimental evidence demonstrates that the amount of glucose that astrocytes actually take up is disproportionally high in comparison to their energy requirements. In the resting rat brain, some studies have shown that astrocytes are responsible for approximately half of glucose uptake, and that this proportion increases even further upon functional activation. How can these data be reconciled with the fact that neurons—not astrocytes—have the highest energy needs? The transfer of energy substrates

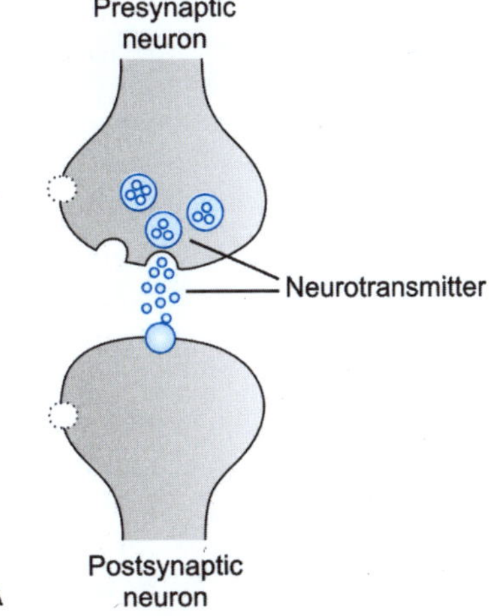

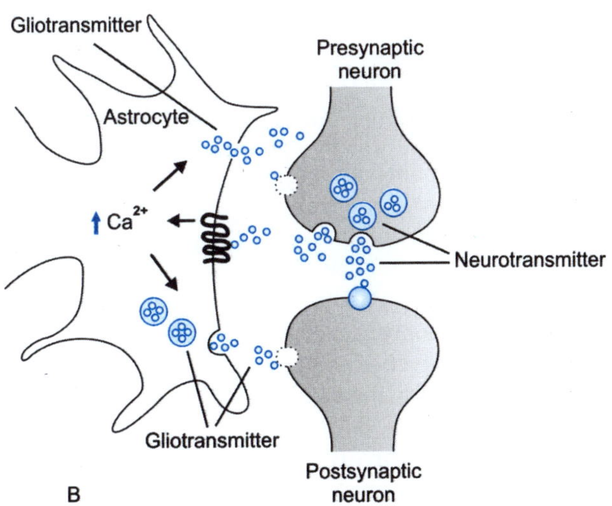

Fig. 13.6: Traditional view of a synapse (A) and a tripartite synapse (B)

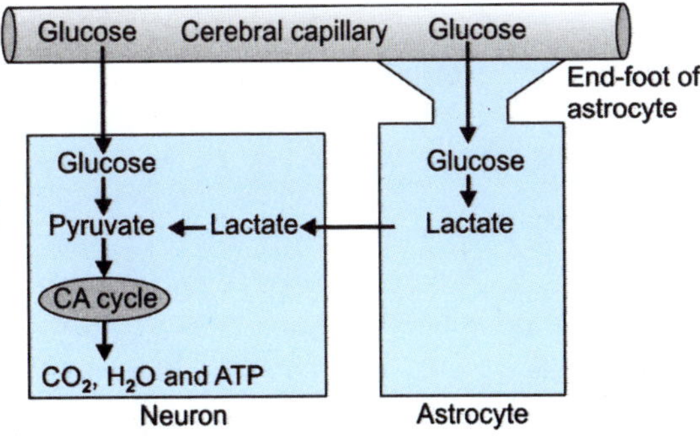

Fig.13.7: The astrocyte–neuron lactate shuttle (CA cycle: Citric acid cycle)

from astrocytes to neurons could provide a simple explanation for this apparent paradox. This is a central point of the astrocyte–neuron lactate shuttle (ANLS) model proposed over a decade ago by Pellerin and Magistretti.[12] The essence of this model is that (1) neuronal activity increases extracellular glutamate (via glutamatergic neurotransmission), which is avidly taken up via a Na^+-dependent mechanism by specific glial glutamate transporters; (2) the resulting increase in

$[Na^+]_i$ activates the Na^+/K^+ ATPase, glucose uptake, and glycolysis in astrocytes; (3) this in turn leads to a large increase in the production of lactate which is released in the extracellular space; and (4) lactate can be used as an energy substrate for neurons for oxidative-derived ATP production.[13]

Thus, to summarize, it seems, there is a mechanism by which astrocytes contribute to the uptake of glucose from the circulation into brain parenchyma in register with synaptic activity: The release of glutamate from synaptically active neurons stimulates glucose uptake in nearby astrocytes. The extent of glucose uptake would be proportional to the extent of activity, thereby "coupling" neuronal activity to glucose utilization. The astrocytes possess the necessary features to sense neuronal synaptic activity (Figs 13.5 and 13.6B) and couple it with greater glucose transport into the cell. In this manner, the increase neuronal synaptic activity is coupled with greater energy source in the form of astrocytes-derived lactate.[3]

v. **Glycogen store:** Glycogen, the storage form of glucose, is the largest energy reserve of the brain, and it is almost exclusively localized to astrocytes. The presence of glycogen in the brain, although lower than in the liver or muscles, indicates its essential role in neuronal activity. When glucose supply in the active neuron is insufficient, glycogen stored in the astrocytes can be converted to glucose and then lactate for transfer to the adjacent neuron.[3] Astrocytic glycogen is more than a simple emergency reserve, and plays an important and active role in complex brain physiological functions, in particular through an astrocyte-to-neuron transfer of energy metabolites in the form of lactate. It appears that both glucose- and glycogen-derived lactate are important to sustain neuronal function.[13]

vi. **Defense against oxidative stress:** CNS is particularly vulnerable to oxidative injury. This can be explained by the brain's high rate of oxidative energy metabolism (which inevitably generates ROS), combined with a relatively low intrinsic antioxidant capacity. Compared with neurons, astrocytes display a much more effective artillery against ROS. Accordingly, cooperative astrocyte–neuron defense mechanisms against oxidative stress seem to be essential for neuronal viability. This neuroprotective capacity of astrocytes may derive from the fact that they possess significantly higher levels of a variety of antioxidant molecules (including glutathione, ascorbate, and vitamin E) and display greater activities for ROS-detoxifying enzymes (including glutathione S-transferase, glutathione peroxidase, and catalase).[15]

Glutathione is the most important antioxidant molecule found in the brain. However, neurons are highly dependent on astrocytes for their glutathione synthesis. Astrocytes synthesize and release glutathione in the extracellular space, where it is cleaved by the astrocytic gamma-glutamyl transpeptidase (GGT) to produce cystein, which can then be taken up by neurons and reconverted to glutathione[15] (Fig. 13.8).

The recycling of ascorbate is another example of cooperation between astrocytes and neurons for antioxidant defense. Ascorbate

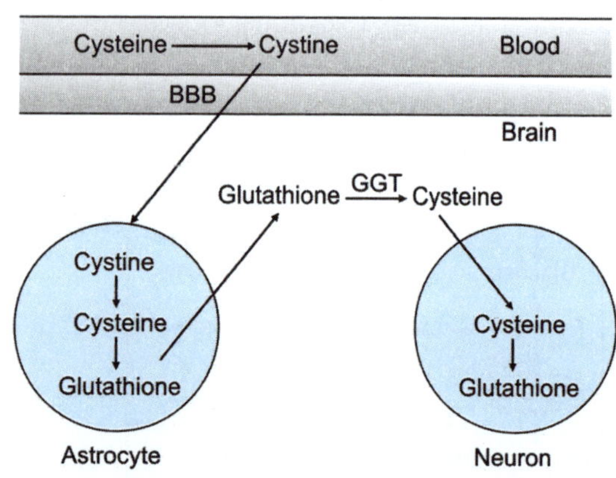

Fig. 13.8: Role of astrocyte in glutathione production in the neuron (GGT: Gamma-glutamyl transpeptidase)

can directly scavenge ROS, and is also an important cofactor for the recycling of oxidized vitamin E and GSH. Astrocytes are responsible for the uptake of the oxidation product of ascorbate, dehydroascorbic acid, from the extracellular space and its recycling back to ascorbic acid. The latter can then either be used intracellularly in astrocytes, or released into the extracellular space to be utilized by neurons for their own antioxidant defense.[15]

2. Microglia

Microglia constitute approximately 10–20% of the total population of glia cells in the adult. Microglia represent the endogenous brain defense and immune system, which is responsible for CNS protection against various types of pathogenic agents. Microglial cells are derived from progenitors that have migrated from the periphery during embryonic life. They are of mesodermal/ mesenchymal origin. After invading the CNS, microglial precursors disseminate relatively homogeneously throughout the neural tissue and acquire a specific phenotype, which clearly distinguish them from their precursors, the blood-derived monocytes.[16]

Until recently, microglia (Fig. 13.9) of the healthy adult CNS were considered quiescent, i.e. functionally dormant, due to their supposedly immobile morphology.[17] In reality, however, these "resting" cells actively scan their environment. The resting microglial cell is characterized by a small cell body and much elaborated thin processes, which send multiple branches and extend in all directions. The processes of resting microglial cells are constantly moving through its territory.

Considering the velocity of this movement, the brain parenchyma can be completely scanned by microglial processes every several hours. The "resting" microglia thus actively surveys the tissue, being ready to rapidly transform to "activated" states (Fig.13.9) upon appearance of signs indicating a threat to the CNS. When a brain insult is detected by microglial cells, they launch a specific program that results in the gradual transformation of resting, ramified microglia into an ameboid form; this process is generally referred to as 'microglial activation' and proceeds through several steps. Some microglial cells return into a proliferative mode, and microglial numbers around the lesion site start to multiply. Microglial cells become motile, and using amoeboid-like movements they gather around sites of insult. If the damage persists and CNS cells begin to die, microglial cells undergo further transformation and become phagocytes. The transition from the surveillance mode of a "resting" cell to the executive states thereby represents more of a shift in activities rather than "activation" *per se*. This view acknowledges that microglia have a constant functional importance. There are no periods of inactivity.[18]

3. Oligodendrocytes

The major function of oligodendrocytes and Schwann cells is the formation of myelin. Myelin acts as an insulator of axonal segments and is a prerequisite for the high velocity of nerve impulse conduction.

All white matter tracts contain oligodendrocytes to form myelin (Fig. 13.10). The myelin forming oligodendrocytes have several processes (up to 40) which connect to one myelin segment. Each of these segments is several hundred micrometers long and is also termed the internode. Segments are interrupted by structures known as node of Ranvier which spans for less than 1 micron. At the node, as compared to the internodal region, the axon is not enwrapped by myelin. In addition, astrocyte processes contact the axonal membrane at the nodal region.

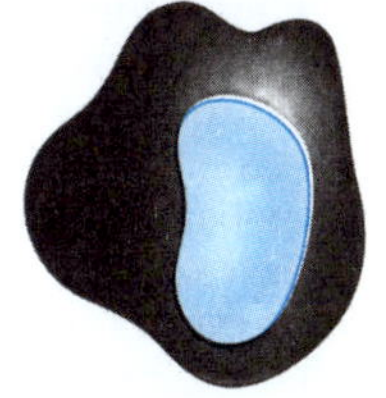

Fig. 13.9: Microglia, resting and activated

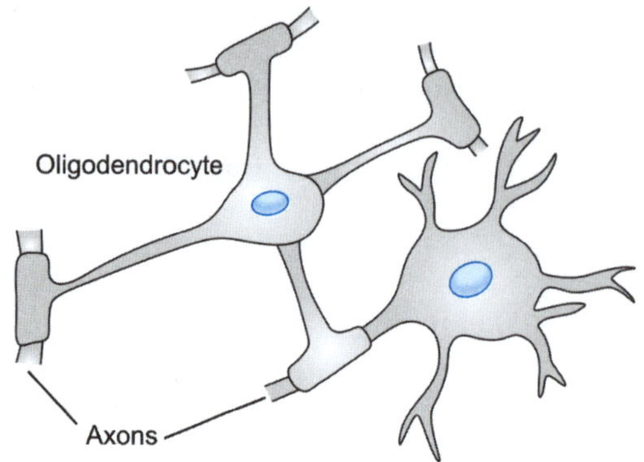

Fig.13.10: An oligodendrocyte myelinating multiple axons

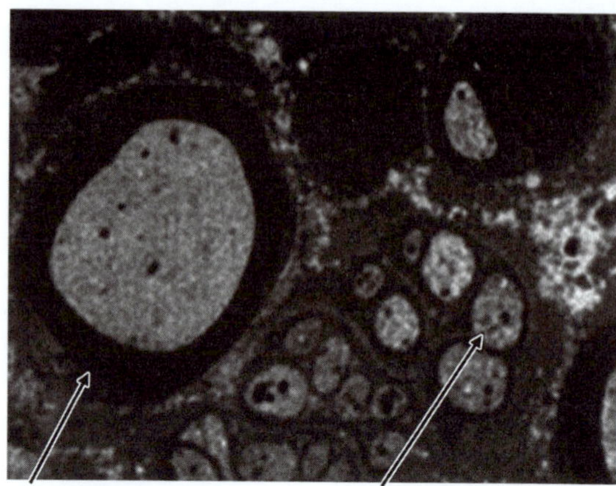

(a) A Schwann cell surrounding a myelinated nerve fiber

(b) A Schwann cell surrounding many unmyelinated nerve fibers

Fig.13.11: Myelinating Schwan cell (a) and non-myelinating Schwann cell (b)

Besides white matter, oligodendrocytes are also found in gray matter. Oligodendrocytes found in the gray matter are known as satellite oligodendrocytes. Satellite oligodendrocytes are functionally distinct from other oligodendrocytes found in the white matter. They are not attached to neurons and, therefore, do not serve an insulating role. They remain juxtaposed to neurons and, like astrocytes, seem to regulate the ionic composition of extracellular fluid.

4. Schwann Cells

Schwann cells are the cellular counterparts to oligodendrocytes in the peripheral nervous system. Similar to oligodendrocytes, they form the myelin sheath. In contrast to the oligodendrocyte, each Schwann cell is associated with only one axonal segment. While the myelin structure formed by oligodendrocytes and Schwann cells has a similar ultrastructure, it is not composed of an identical set of proteins.

During development, Schwann cells are derived from undifferentiated migrating neural crest cells. The immature Schwann cells produce either myelinating or non-myelinating Schwann cells. Myelinating Schwann cell surrounds only one nerve fiber. Non-myelinating Schwann cell loosely enwraps several axons without forming myelin (Fig. 13.11).

References

1. Jakel S, Dikou L. Glial cells and their function in the adult brain: A journey through the history of their ablation.Front Cell Neurosci. 2017; 11: 24. Published online 2017 Feb 13. doi: 10.3389/fncel.2017.00024
2. *http://www.networkglia.eu/en*
3. Magistretti PJ and Ranson BR. Astrocytes Ch 10 in: Neuropsychopharmacology: The fifth generation of progress. Eds Davis KL & Charney D, et al. American college of Neupsychopharmacology, 2002.
4. Heinmann U, Lux HD. Ceiling of stimulus-induced rise in extracellular potassium concentration in the cerebral cortex of cat. Brain Res. 1977; 120: 231–49.
5. Balestrino M, Aitken PG, Somjen GG. The effects of moderate changes in extracellular K$^+$ and Ca^{2+} on synaptic and neuronal function in CA1 region of the hippocampal slice. Brain. 1986; 377: 229–39.
6. Ballanyi K, Grafe P, Bruggencate G. Ion activities and potassium uptake mechanisms of glial cells in guinea-pig olfactory cortex. J Physio (London). 1987; b382: 159–74.
7. Hamberger AC, Chiang, GH, Nylén ES, et al. Glutamate as a CNS transmitter.1. Evaluation of glucose and glutamine as a precursor for synthesis of preferentially released glutamate. Brain Res. 1979; 168: 513–30.
8. Stobart JK. Multifunctional role of astrocytes as gatekeepers of neuronal energy supply. Frontiers in CellularNeuroscience. .https://www.frontiersin.org/article/10.3389/fncel.2013.00038

9. Anderson CM, Swanson RA. Astrocyte glutamate transport: review of properties, regulation and physiological functions. Glia. 2000; 32: 1–14.

10. Perea G, Navarrete M, Araque A. Tripartite synapses: astrocyte processes and control of synaptic information. Trends Neurosci. 2009; 32: 421–31.

11. Hertz L, Peng L, Dienel GA. Energy metabolism in astrocytes: High rate of oxidative metabolism and spatiotemporal dependence on glycolysis/glycogenolysis. J. Cereb. Blood Flow Metab. 2007; 27: 219–49.

12. Pellerin L, Magistretti PJ. Glutamate uptake into astrocytes stimulates aerobic glycolysis: a mechanism coupling neuronal activity to glucose utilization. Proc Natl Acad Sci USA. 1994; 91: 10625–29.

13. Magistretti PJ. Role of glutamate in neuron-glia metabolic coupling. Am J ClinNutr. 2009; 90: S875–80.

14. Wilson JX. Antioxidant defense of the brain: a role for astrocytes. Can J Physiol Pharmacol. 1997; 75: 1149–63.

15. Belanger M, Magistretti PJ. The Role of Astrocyte-Neuron Interactions for the Defense against Oxidative Stress. Dialogues Clin Neurosci. 2009; 11: 281–95.

16. Kettenmann H, Hanisch UK,Noda M, et al. Physiology of microglia. Physiol Rev.2011; 91: 461–553.

17. Kreutzberg GW. Microglia: a sensor for pathological events in the CNS. Trends Neurosci. 1996; 19: 312–318.

18. Hanisch UK, Kettenmann H. Microglia: active sensor and versatile effector cells in the normal and pathologic brain. Nat Neurosci. 2007; 10: 1387–94.

Chapter 14

Motor and Nonmotor Functions of the Basal Ganglia

INTRODUCTION

The term "basal ganglia" refers to a group of subcortical nuclei involved in motor control. In this term,"basal" comes from the fact that most of its elements are located in the basal part of the forebrain. The term "ganglia" is a misnomer: In modern usage, neural clusters are called "ganglia" only in the peripheral nervous system; in the central nervous system they are called "nuclei". For this reason, the basal ganglia are also occasionally known as the "basal nuclei". Terminologia anatomica (1998), the international authority for anatomical naming, named them "nuclei basales", but this is not commonly used; the structures continue to be known as the basal ganglia.

The motor functions of basal ganglia are well known. Recent researches have highlighted many nonmotor functions of the basal ganglia such as motor learning, executive functions, behavior and emotions. In the strictest sense, the term basal ganglia refer to nuclei embedded deep in cerebral cortex, i.e. caudate nucleus, putamen, nucleus accumbens and globus pallidus. These structures are intimately connected to some other "related nuclei": Subthalamic nucleus (located in diencephalon) and substantia nigra (situated in mesencephalon). All these structures are discussed under the heading of"basal ganglia" (Fig. 14.1). Moreover, basal ganglia have been traditionally considered a part of extrapyramidal system. But nowadays, the terms

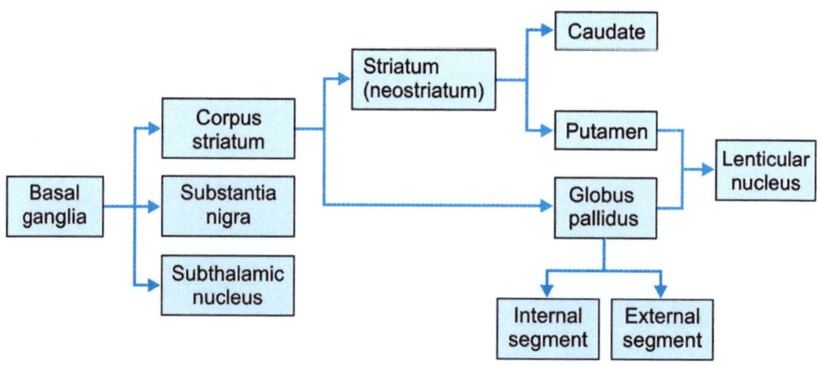

Fig. 14.1: Basal ganglia nomenclature

"pyramidal system" and "extrapyramidal system" are considered misleading since they are not exclusive of each other. The motor cortex and basal ganglia and related nuclei are intimately involved in all types of motor activity.[1]

STRUCTURE OF BASAL GANGLIA

The **caudate nucleus** is a C-shaped structure that is closely associated with the lateral wall of the lateral ventricle. It is largest at its anterior pole (the head), and its size diminishes posteriorly as it follows the course of the lateral ventricle (the body) all the way to the temporal lobe (the tail), where it terminates at the amygdaloid nuclei. The **putamen** is also a large structure that is separated from the caudate nucleus by the anterior limb of the internal capsule. The putamen is connected to the caudate head by bridges of cells that cut across the internal capsule. Because of the striated appearance of these cell bridges, the caudate and putamen are collectively referred to as the **striatum or neostriatum**, and the **nucleus accumbens** is often called the **ventral striatum**. Functionally, the caudate nucleus and the putamen are considered equivalent to each other; indeed, most mammals have only a single nucleus called the striatum. It is unclear whether there is any functional significance of the separation of the striatum into the caudate and putamen in primates. The putamen and the globus pallidus are collectively called the **lenticular nucleus, or lentiform nucleus**. The **globus pallidus** is divided into two segments: The internal (or medial) segment and the external (or lateral) segment (Fig. 14.2).

The **subthalamic nucleus** is part of the diencephalon; as its name implies, it is located just below the thalamus. The **substantia nigra** is a midbrain structure, composed of two distinct parts: The pars compacta and the pars reticulata. The substantia nigra is located between the red nucleus and the crus cerebri (cerebral peduncle) in the ventral part of the midbrain. The *pars compacta* is the source of a clinically important dopaminergic pathway to the striatum; loss of neurons in this area is the cause of Parkinson's disease. An area that is functionally analogous to the substantia nigra, pars compacta is the ventral tegmental area, which is located nearby and makes a dopaminergic projection to the nucleus accumbens. The neurons of *the pars reticulata* produce the neurotransmitter, gamma-

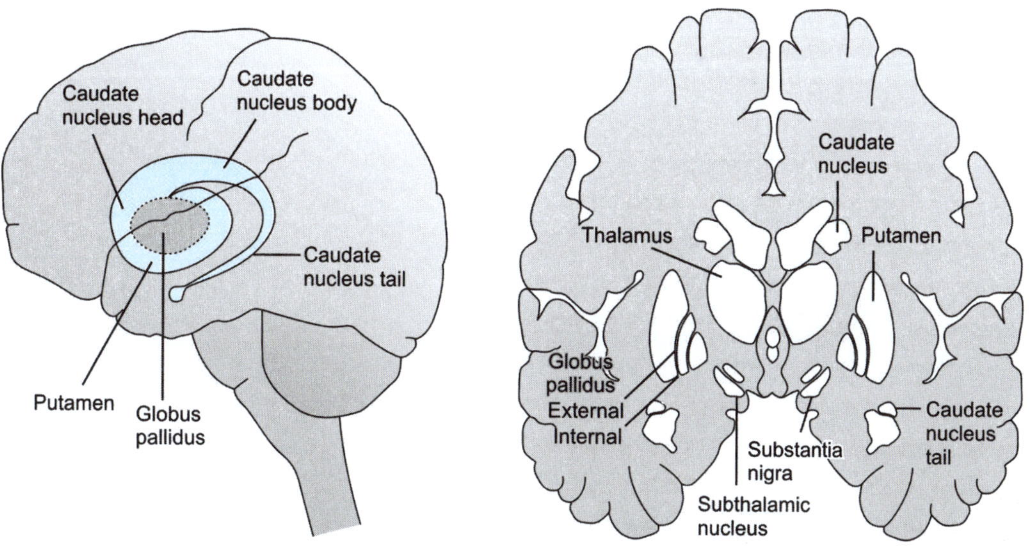

Fig. 14.2: Location of the basal ganglia

aminobutyric acid (GABA). The neurons of the pars reticulata, through the nigrothalamic bundle, send axons to nucleus ventralis anterior (VA) a part of the motor thalamus.

Basal Ganglia Afferents[2]

The **striatum** is the main recipient of afferents to the basal ganglia (Fig.14.3). These excitatory afferents arise from:

1. The entire cerebral cortex
2. The intralaminar nuclei of the thalamus (primarily the centromedian nucleus and parafascicularis nucleus).
3. The limbic cortex.
4. Substantia nigra (nigrostriate afferents)

The largest afferent source of the basal ganglia arises from the cerebral cortex. In fact, most regions of the cortex contribute projections to the basal ganglia. These include inputs from motor, sensory, association, and even limbic areas of the cerebral cortex. While the caudate nucleus and putamen serve as the primary target regions of afferent projections from the cortex, the source of cortical inputs to these regions of the basal ganglia differ. The principal inputs from the primary motor, secondary motor, and primary somatosensory regions of cortex are directed to the *putamen* (Fig. 14.3). These inputs to the putamen are somatotopically organized, which means that different regions of the putamen receive sensory and motor inputs that are associated with different parts of the body. The *caudate nucleus*, on the other hand, receives inputs from cortical association regions, frontal eye fields, and limbic regions of cortex. Thus, the putamen appears to be concerned primarily with motor functions, while the caudate nucleus, which appears to receive more varied and integrated cortical inputs, is likely involved with cognitive aspects of movement, eye movements, and emotional correlates of movement (relegated to the ventral aspect of the neostriatum). The *ventral striatum (nucleus accumbens)* receives direct input from multiple regions in the cerebral cortex and limbic structures such as the amygdala, thalamus, and hippocampus, as well as the entorhinal cortex and the inferior temporal gyrus. **The neostriatum also receives an indirect** source of cortical input. The source of this input is the centromedian nucleus of the thalamus. This nucleus receives afferent fibers primarily from the motor cortex and projects its axons topographically to the putamen as thalamostriate fibers (Fig. 14.4). *Substantia nigra (pars compacta)* is the source of a clinically important dopaminergic pathway to the striatum; loss of neurons in this area is the cause of Parkinson's disease.

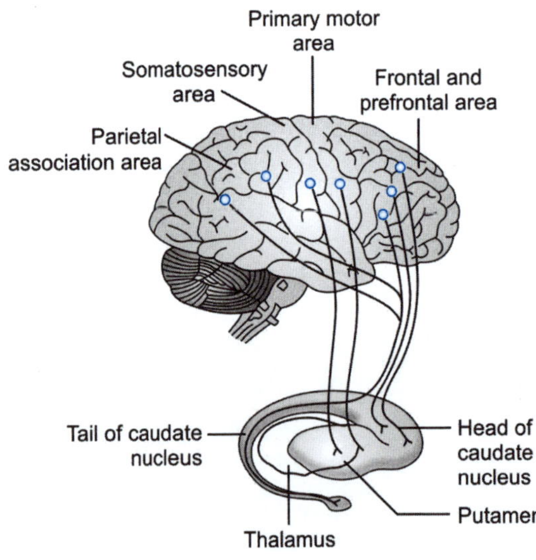

Fig. 14.3: Cortical afferents to striatum

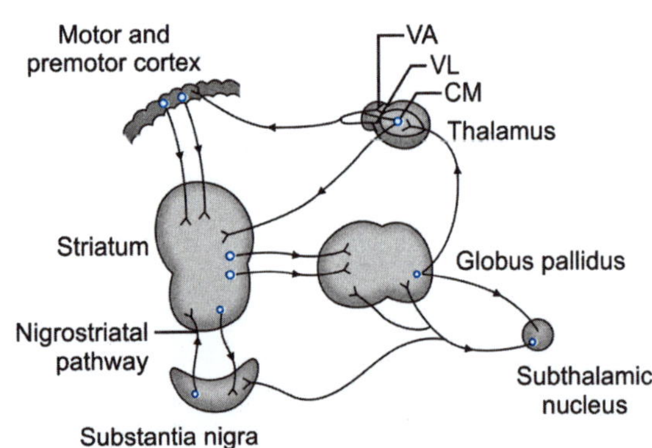

Fig. 14.4: Afferent and efferent connections of basal ganglia (VA: Ventral anterior nucleus; VL: Ventral lateral nucleus; CM: Centromedian nucleus)

Basal Ganglia Efferents[2]

The major output structures of the basal ganglia are the **globus pallidus** internal (GPi) segment and the **substantia nigra** (SNr) (pars reticulata) (Fig. 14.5). Both of these structures make GABAergic, inhibitory connections on their targets. The GPi projects to a number of thalamic structures by way of two fiber tracts: The **ansa lenticularis** and the **lenticular fasciculus**. The loop that processes sensorimotor information from the motor cortex and the somatosensory cortex projects to the ventral anterior (VA) and ventral lateral (VL) nuclei. The loop that processes other neocortical information projects to the dorsomedial nucleus (DM), intralaminar nuclei, and parts of the VA nucleus. The SNr projects to the **superior colliculus**, which is involved in eye movements, as well as to the VA/VL thalamic nuclei.

Substantia Nigra (Fig. 14.5)

Substantia nigra, consists of two components, pars compacta and pars reticulata. The dopaminergic neurons of the pars compacta project many of their axons along the nigrostriatal pathway to the dorsal striatum, where they release the neurotransmitter dopamine. Neurons in the pars reticulata are much less densely packed than those in the pars compacta. The neurons of the pars reticulata produce the neurotransmitter gamma-aminobutyric acid (GABA). The neurons of the pars reticulata through the nigrothalamic bundle send axons to nucleus ventral anterior (VA) a part of the motor thalamus. VA is the origin of one output of the basal ganglia system. It sends axons to the frontal and oculomotor cortex.

Basal Ganglia: Intrinsic Connections (Fig. 14.5)

1. The *striatopallidal pathway* is a GABAergic, inhibitory connection between the striatum and both segments of the globus pallidus.
2. The *striatonigral pathway* is a GABAergic, inhibitory connection between the striatum and the substantia nigra.
3. The *globus pallidus* external segment makes a GABAergic, inhibitory connection to *the subthalamic nucleus.*
4. The *subthalamic nucleus* makes glutamatergic, excitatory connections onto both segments of the *globus pallidus and the substantia nigra.* This pathway is the only purely excitatory pathway among the intrinsic pathways of the basal ganglia.
5. The *nigrostriatal* pathway makes a dopaminergic synapse onto striatal neurons. As we will see below, this is a mixed pathway, with excitatory effects on some striatal neurons and inhibitory effects on others.
6. The circuits within the basal ganglia by which signals are transmitted back to the cerebral cortex may be direct or indirect. The differences between the direct and indirect routes are discussed below.

TWO PATHWAYS PROCESS SIGNALS IN THE CEREBRAL CORTEX

Basal Ganglia and Cerebral Cortex Loop[2]

There are two distinct pathways that process signals in the cerebral cortex: Basal ganglia and cerebral cortex loop: The *direct pathway* and the *indirect pathway*. These two pathways

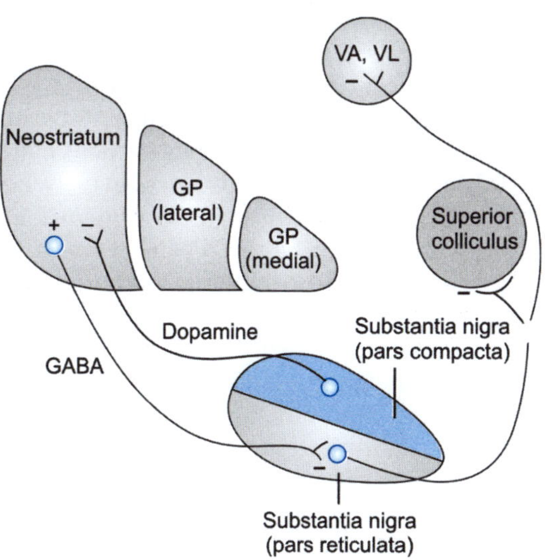

Fig. 14.5: Details of connections of substantia nigra (GP: Globus pallidus; VA: Ventral anterior nucleus; VL: Ventral lateral nucleus)

have opposite net effects on thalamic target structures. Excitation of the direct pathway has the net effect of exciting thalamic neurons (which in turn make excitatory connections onto cortical neurons). Excitation of the indirect pathway has the net effect of inhibiting thalamic neurons (rendering them unable to excite motor cortex neurons). The normal functioning of the basal ganglia apparently involves a proper balance between the activities of these two pathways. One hypothesis is that the **direct pathway** selectively facilitates certain motor programs in the cerebral cortex that are adaptive for the present task, whereas the **indirect pathway** simultaneously inhibits the execution of competing motor programs (Fig. 14.6). An upset of the balance between the direct and indirect pathways results in the motor dysfunctions that characterize the so-called *extrapyramidal syndromes.*

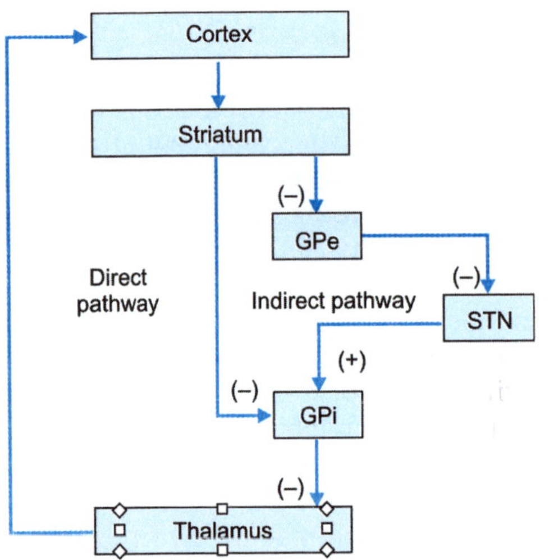

Fig. 14.6: Direct and indirect pathways in cortex: Basal ganglia and cortex loop (GPi: Globus pallidus internal; GPe: Globus pallidus external; STN: Subthalamic nucleus)

Direct pathway: The direct pathway starts with cells in the striatum that make inhibitory connections with cells in the globus pallidus internal (GPi). The GPi cells in turn make inhibitory connections on cells in the thalamus. Thus, the firing of GPi neurons *per se*

inhibits the thalamus, making the thalamus less likely to excite the neocortex. When the direct pathway striatal neurons fire, however, they inhibit the activity of the GPi neurons. This inhibition releases the thalamic neurons from inhibition (i.e. it disinhibits the thalamic neurons), allowing them to fire to excite the cortex. Thus, because of the "double negative" in the pathway between the striatum and GPi as well as the GPi and thalamus, **the *net result of exciting the direct pathway striatal neurons is to excite motor cortex.***

Indirect pathway: The indirect pathway starts with a different set of cells in the striatum. These neurons make inhibitory connections to the external segment of the globus pallidus external (GPe). The GPe neurons make inhibitory connections to cells in the subthalamic nucleus, which in turn make excitatory connections to cells in the GPi. (The subthalamic-GPi pathway is the only purely excitatory pathway within the intrinsic basal ganglia circuitry.) The GPi neurons make inhibitory connections on the thalamic neurons. To see the net effects of activation of the indirect pathway, let us work backwards from the GPi. When the GPi cells are active, they inhibit thalamic neurons, thus making cortex less active. When the subthalamic neurons are firing, they increase the firing rate of GPi neurons, thus increasing the net inhibition on cortex. Firing of the GPe neurons inhibits the subthalamic neurons, thus making the GPi neurons less active and disinhibiting the thalamus. However, when the indirect pathway striatal neurons are active, they inhibit the GPe neurons, thus disinhibiting the subthalamic neurons. With the *subthalamic neurons free to fire, the GPi neurons inhibit the thalamus, thereby producing a net inhibition on the motor cortex.*

Thus, as a result of the complex sequences of excitation, inhibition, and disinhibition, the net effect of the cortex exciting the direct pathway is to further excite the cortex (positive feedback loop), whereas the net effect of cortex exciting the indirect pathway is to inhibit the cortex (negative feedback loop) (Fig. 14.6).

The Nigrostriatal Projection

An important pathway in the modulation of the direct and indirect pathways is the dopaminergic **(nigrostriatal projection)** from the substantia nigra (pars compacta) to the striatum. Direct pathway striatal neurons have D1 dopamine receptors, which depolarize the cell in response to dopamine. In contrast, indirect pathway striatal neurons have D2 dopamine receptors, which hyperpolarize the cell in response to dopamine. The nigrostriatal pathway thus has the dual effect of exciting the direct pathway while simultaneously inhibiting the indirect pathway. Because of this dual effect, excitation of the nigrostriatal pathway has the net effect of exciting cortex by two routes, by exciting the direct pathway (which itself has a net excitatory effect on cortex) and inhibiting the indirect pathway (thereby disinhibiting the net inhibitory effect of the indirect pathway on cortex). The loss of these dopamine neurons in Parkinson's disease causes the poverty of movement that characterizes this disease, as the balance between direct pathway excitation of cortex and indirect pathway inhibition of cortex is tipped in favor of the indirect pathway, with a subsequent pathological global inhibition of motor cortex areas.

Neurotransmitters profile of basal ganglia is shown in Fig. 14.7.

FUNCTIONS OF THE BASAL GANGLIA

The basal ganglia were traditionally thought to function only during voluntary movement. Indeed, for some time it was believed that the basal ganglia sent their entire output to the motor cortex via the thalamus and thus act as a funnel through which movement is initiated by different cortical areas. It is now widely accepted that, through their interaction with the other areas cerebral cortex, the basal ganglia also contribute to a variety of behaviors other than voluntary movement, including oculomotor, cognitive, and even emotional functions.

The basal ganglia may be viewed as the principal subcortical components of a family of circuits linking the thalamus and cerebral cortex. These circuits are largely segregated, both structurally and functionally. Each circuit originates in a specific area of the cerebral cortex and engages different portions of the basal ganglia and thalamus. The thalamic output of each circuit is directed back to the portions of the frontal lobe from which the circuit originates.[2] Thus, the **motor circuit** begins and ends in the precentral

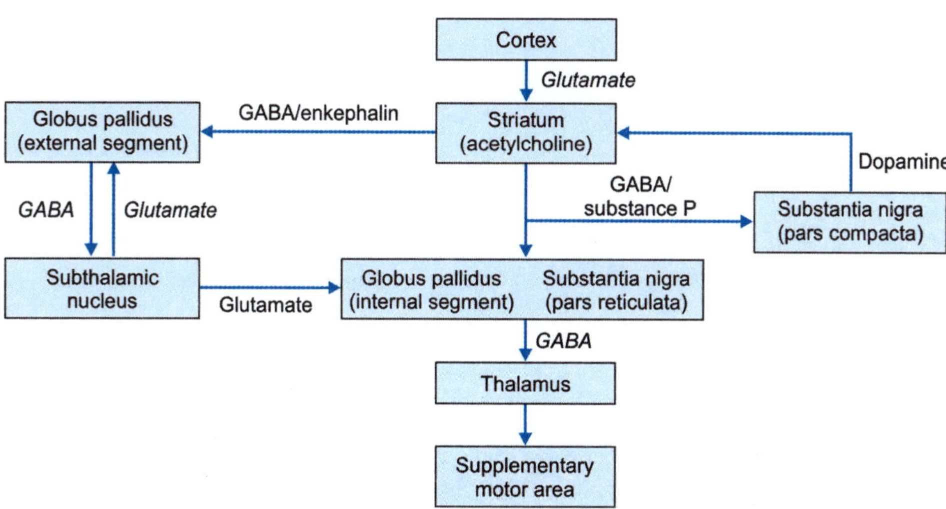

Fig. 14.7: Neurotransmitter profile of basal ganglia

motor fields (the premotor cortex, the supplementary motor area, and the motor cortex); the **oculomotor circuit,** in the frontal and supplementary eye fields; **the prefrontal circuits,** in the dorsolateral prefrontal and lateral orbitofrontal cortices; and the **limbic circuit,** in the anterior cingulate area and medial orbitofrontal cortex. Each area of the neocortex projects to a discrete region of the striatum and does so in a highly topographic manner.[3-6]

1. The Motor Circuit[5]

It appears that the basal ganglia are involved in the enabling of practiced motor acts and in gating the initiation of voluntary movements by modulating motor programs stored in the motor cortex and elsewhere in the motor hierarchy. Thus, voluntary movements are not initiated in the basal ganglia (they are initiated in the cortex); however, proper functioning of the basal ganglia appears to be necessary in order for the motor cortex to relay the appropriate motor commands to the lower levels of the hierarchy.

What is the function of the tonic inhibitory output of the basal ganglia? Stimulating the motor cortex of monkeys at various locations results in stereotyped sequences of move-ments, such as bringing the hand to the mouth or adopting a defensive posture. It appears that a number of "primitive" motor programs are stored in the cortex, and motor control may require the activation of these elemental motor programs in the precise temporal order to accomplish a sophisticated motor plan. It is important that only one motor program be active at a given time, however, such that one motor act (e.g. use hand to bring food to the mouth) should not compete with a conflicting motor act (e.g. use hand to shield face from dangerous object). It is thought that the basal ganglia are normally active in suppressing inappropriate motor programs, and that activation of the direct pathway temporarily releases one motor program from inhibition, enabling it to be executed by the organism (Fig. 14.8). Thus, the basal ganglia act as a gate that enables the execution of automatic programs in the motor hierarchy.[7]

2. The Oculomotor Circuit

The oculomotor circuit is involved in the control of saccadic eye movements. It originates in the frontal and supplementary motor eye fields and projects to the body of the caudate nucleus. The caudate nucleus in turn projects via the direct and indirect pathways

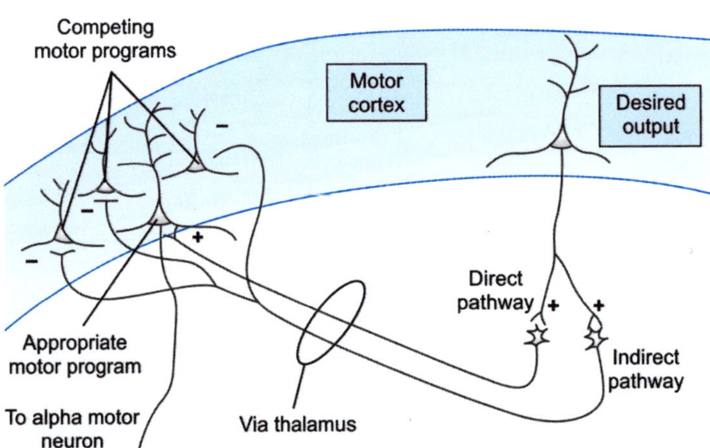

Fig. 14.8: Role of direct and indirect pathways in motor control. The basal ganglia enable the proper motor program stored in motor cortex circuits via the direct pathway and inhibit competing motor programs via indirect pathway[7]

to the lateral portions of the substantia nigra (pars reticulata), which projects back to the frontal eye fields as well as to the superior colliculus. Inhibition of tonic activity in the substantia nigra (pars reticulata) disinhibits output neurons in the deep layers of the superior colliculus whose activity is associated with saccades. Inactivation of neurons in the pars reticulata results in involuntary saccades to the contralateral side.[8]

A saccade is a rapid eye movement generated to shift visual attention onto a specific target area. Saccades are categorized as ballistic movements, meaning that after it is initiated, corrections are unable to be made. Saccades are initiated either through reflexive mechanisms or under voluntary control. Sudden attention to a visual, sensory, or auditory stimulus would be an example of an involuntary saccade. Conversely a voluntary saccade is initiated when desiring to look at a particular object of interest (e.g. reading).

The superior colliculus plays a prominent role in the initiation of saccades. It has two distinct functional areas: The superficial (sensory) layers and the deep (motor) layers. The superior colliculus receives direct retinal projections and information from the frontal eye field regarding visual input. This information comes into the superficial layers, and corresponds to a specific portion of the visual field, maintaining organization. In addition to the visual input, the superior colliculus receives auditory and somatosensory information. This varied input allows the superior colliculus to generate saccades in response to these differing stimuli. For example, a tap on the shoulder or loud noise that catches your attention will excite a certain part of the superior colliculus, generating a saccade oriented toward that stimulus.

The activity level of the superior colliculus is monitored by regions of the basal ganglia. The substantia nigra has a tonic inhibition on the superior colliculus, until just prior to a saccade. As visual input is processed, information will eventually reach the frontal eye field. The frontal eye field then sends exci-

tatory signals to the superior colliculus and the caudate nucleus of the basal ganglia. The caudate nucleus then sends inhibitory projections to the substantial nigra (pars reticulata), which releases its inhibition on the superior colliculus and allows for the generation of a saccade (Figs 14.9 and 14.10).[8]

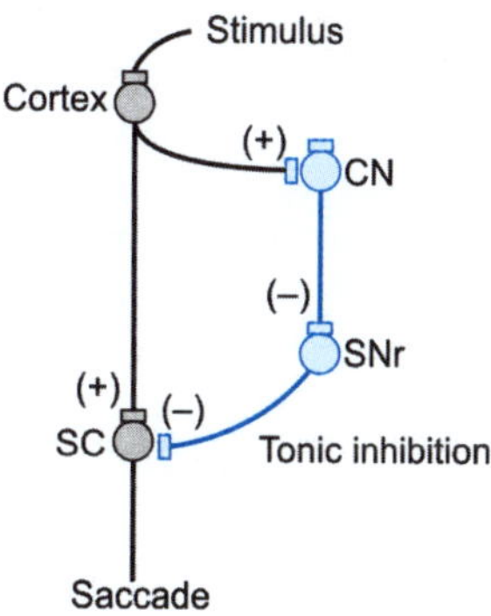

Fig. 14.9: A simplified view of saccade control system (CN: Caudate nucleus; SNr: Substantia nigra; SC: Superior colliculus)

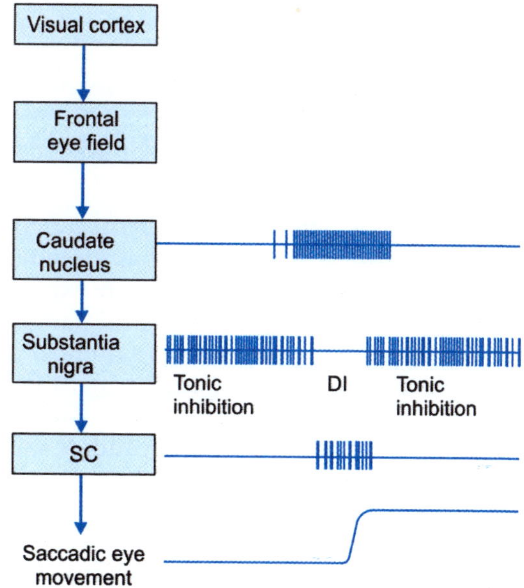

Fig. 14.10: Mechanism of saccade movement (SC: Superior colliculus; DI: Disinhibition)[8]

3. Cognitive Function (Executive Function)

The basal ganglia, besides motor function, have a role in cognition. Executive function is the ability to selectively attend to, work with, and plan for specific information. This means that executive function is deciding what information, cognitions, or stimuli are relevant, worth holding and working with that information, and then planning what to do with it. As such, executive function is largely the role of planning and organization. Executive function also involves being able to select the appropriate response or behavior while at the same time inhibiting inappropriate responses or behavior.[9] Many animal experimental research publications support the view that basal ganglia play an important role in cognition.[10] There exist a number of cortical loops through the basal ganglia that involve prefrontal association cortex and limbic cortex. Through these loops, the basal ganglia are thought to play a role in cognitive function that is similar to their role in motor control. That is, the basal ganglia are involved in selecting and enabling various cognitive, executive, or emotional programs that are stored in these other cortical areas. Moreover, the basal ganglia appear to be involved in certain types of learning. For example, in rodents the striatum is necessary for the animal to learn certain stimulus-response tasks (e.g. make a right turn if stimulus A is present and make a left turn if stimulus B is present).[11]

Investigation of the role of the basal ganglia in human learning and memory has provided some convergence with data from experimental animals. Patients with basal ganglia disorders exhibit impairments in a number of cognitive tasks. Cognitive impairment is a primary feature of Huntington's disease, which involves cell loss in the caudate and putamen. Patients with Parkinson's disease also exhibit a variety of cognitive deficits, though not typically as severe. In Parkinson's disease, cell death in the substantia nigra leads to a loss of dopaminergic input to the caudate and putamen. In both of these patient groups, motor deficits are the most obvious consequences of basal ganglia disease. However, as has been described in experimental animals, the demonstration of non-motor deficits following basal ganglia dysfunction also suggests a broader role for this brain region in human behavior; one that clearly includes learning and memory functions.[9]

It is believed that BG circuits are intimately involved in and necessary for new skill learning, but are of far less importance in the retention and recall of well-learned motor skills. Multiple lines of evidence indicate that the BG promotes new skill learning, but that other parts of the brain (cortex in particular) take over the storage and production of well-practiced skills.

4. Basal Ganglia–limbic System Interaction

Four main levels of interaction between the BG system and the limbic system are considered. (1) The BG receive direct afferents from several structures associated with the limbic system. Limbic cortical areas project to the striatum. (2) Projections from the amygdala to the nucleus accumbens and ventral caudate. (3) The striatum receives limbic information from cortical and subcortical limbic structures and project to the dopaminergic neurons of the substantia nigra (pars compacta), which in turn modulates their activity. (4) There is a significant overlap between the ventral portions of the BG, nucleus accumbens and ventral pallidum, and the ventral subcortical structures of the limbic system, extended amygdala and nucleus basalis.[12]

The cardinal manifestations of Parkinson's disease (PD) are tremor, rigidity, bradykinesia, and postural instability. Up to 70% of patients with PD exhibit psychiatric symptoms; hence they are a common occurrence. Depression is the most frequent mental disorder in patients with PD, being found in up to 50% of cases. Anxiety disorders (such as generalized anxiety, panic and phobic disorders) are found in up to 40% of patients with PD.[13]

References

1. Lanciego JL, Luquin N, Obeso JA. Functional neuroanatomy of the basal ganglia. *Cold Spring Harbor perspectives in medicine*, 2012; 2(12), a009621. doi: 10.1101/cshperspect.a009621

2. Knierim J. Basal Ganglia.neuroscience.uth. tmc.edu/s3/chapter04.html

3. Alexander GE, DeLong MR, Strick PL. Parallel organization of functionally segregated circuits linking basal ganglia and cortex. Annu Rev Neurosci 1986; 9: 357–81

4. Wichmann T, DeLong MR. Functional and pathophysiological models of the basal ganglia. Curr. Opin. Neurobiol. 1996; 6: 751–58.

5. DeLong MR, Wichmann T. Circuits and circuit disorders of the basal ganglia. Arch. Neurol. 2007; 64: 20–1016/S0959-4388(96)80024-9. PMID 9000030.

6. Galvan A, Devergnas A, Wichmann T. Alterations in neuronal activity in basal ganglia—thalamocortical circuits in the parkinsonian state. Front. Neuroanat., 05 February 2015 | https://doi.org/10.3389/fnana.2015.00005.

7. Parent A, Hazrati LN. Functional anatomy of the basal ganglia. I. The cortico-basal ganglia-thalamocortical loop. Brain Research Reviews. 1995; 20: 91–127.

8. Hikosaka O, Takikava Y, Kawagoe R. Role of the Basal Ganglia in the Control of Purposive Saccadic Eye Movements. Physiol Rev. 2000; 80: 953–78.

9. Packard MG, Knowlton BJ. Learning and memory functions of the basal ganglia. Annu. Rev. Neurosci. 2002; 25: 563–93.

10. Steiner H, Tseng KY. Handbook of Basal Ganglia Structure and Function. New York: Academic Press, 2010.

11. Leisman G, Braun-Benjamin O, Melillo R. Cognitive–motor interactions of the basal ganglia in development. Front. Syst. Neurosci., 13 February 2014 | https://doi.org/10.3389/fnsys.2014.00016

12. Buot A, Yelnik J. Functional anatomy of the basal ganglia: limbic aspects. Rev Neurol (Paris). 2012; 168: 569–75.

13. Wichmann T, DeLong M. The basal ganglia. In Principles of Neural Science. Fifth Edn. Editors: Kandel ER, et al. McGraw-Hill, New York, 2013.

Chapter 15

Thermoregulation

INTRODUCTION

Knowledge of the mechanisms of thermoregulation has progressed from the crude localization of areas within the central nervous system involved in thermoregulation to fuller, but still incomplete, understanding of one of the most important homeostatic mechanisms of the body. Ranson[1] used discrete local heating and lesioning to locate thermoregulatory structures within the hypothalamus. As research in this area has progressed over the last few decades, major advances in our understanding have been made, particularly in the central circuitry involved in thermoregulatory control and in the peripheral sensory mechanisms of temperature transduction. These advances have not filtered through to general textbooks that form the cornerstone of many medical-related and basic science courses.[2]

Thermometer is probably the most widely used instrument in clinical practice today. In 1868, Carl Wunderlich published temperature recordings from over 1 million readings in over 25,000 patients; made with a foot-long thermometer used in the axilla. He established the range of normal temperature from 36.3 to 37.5°C. The problem with this thermometer and all the earlier designs was not only their inconvenient size but also the fact that the reading had to be taken while the thermometer was still in the axilla. When taken out, it showed the temperature of the environment, not the patient. This problem was solved by Aitkin in 1852. He devised a mercury instrument with a bend in the capillary near the mercury reservoir, so that the mercury did not drop back even when it was no more in the axilla.[3] In 19th century, body temperature was measured by placing the thermometer in the mouth or axilla. In 1940s, the concept of body core and shell was introduced, and it was shown that the hypothalamic thermostat maintained homeostasis of only body core (measured as rectal temperature). Body shell temperature may show significant variations depending on the ambient temperature.

Concept of Body "Core" and "Shell" Temperature

Although human beings are homoeothermic, the temperature of the whole body is not kept constant around 37°C. Only the temperature of the body "core", that is, brain and thoracic and abdominal viscera (vital organs) is kept constant. On the other hand, temperature of the body "shell", that is, the skin and its underlying structures is allowed to vary with the change in the external temperature (Fig. 15.1). In cold weather the temperature of the shell may be several degrees lower than the core temperature. The lower temperature of

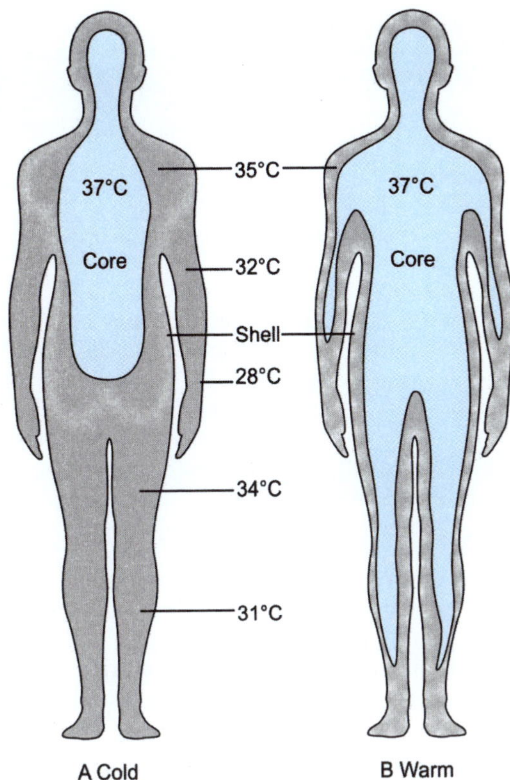

Fig. 15.1: Concept of body "core" and "shell"

the body shell decreases the loss of body heat to the environment by conduction and radiation. In hot environment, the shell temperature approaches the core temperature and this helps in heat loss by conduction and radiation. The surface area-to-mass ratio of shell areas is high (for example, the surface area-to-mass ratio of each hand is four to five times greater than that of the body), and it is known that the surface area-to-mass ratio is important to the transfer of thermal energy. With a high surface area-to-mass ratio, the hand will chill faster than the torso, which has a lower surface area-to-mass ratio. The shell temperature is influenced by blood flow to the skin.

Clinical Implication of Body Core and Shell Temperature

Oral temperature measurements are taken in most of the clinical settings. In several clinical trials, researchers have compared rectal and oral temperatures in both resting and exercising participants. The results were unfavorable for the use of oral temperature as an accurate measurement of core body temperature. In a systemic review of the subject, Mazerolle et al[4] reported that oral temperature assessment consistently provides inaccurate prediction of core body temperature during rest and exercise. These workers showed that oral temperature underestimated the core temperature measurement by 0.60°C ± 0.51°C. Most alarming, from the sports medicine perspective is, that the disparity between oral and core body temperatures increases with increasing hyperthermia. Many clinical situations (e.g. exertional heat stroke) warrant an immediate and accurate assessment of core body temperature for a proper diagnosis and for appropriate treatment.[4]

Fundamentals of body temperature regulation by hypothalamus was demonstrated first by Ranson in 1940.[1] By stimulation and lesions of specific areas of hypothalamus, it was concluded that a rise in core body temperature sets up *anterior hypothalamus*-mediated autonomic and somatic responses that lower the body temperature to normal set point. Similarly, a decrease in core body temperature sets up *posterior hypothalamus*-mediated autonomic and somatic responses that increase the body temperature to normal set point. During the next few decades, cutaneous thermal receptors and the physiologic autonomic and somatic responses to variations in core body temperature[2] were subsequently described (Table 15.1).

Recent researches on the subject have clarified the following aspects of thermoregulation.

a. Transduction of temperature has been clarified.

b. The afferent pathways involved in thermoregulation have been delineated.

c. The role of mediating both heat loss and heat gain mechanisms has been assigned to preoptic area of hypothalamus alone.

Table 15.1: Physiology of thermoregulation

Body temperature	Sensors	Control centre	Effectors	Responses
Increase	Peripheral and central thermo-receptors	Anterior hypothalamus	i. Cutaneous blood vessels ii. Sweat glands iii. Behavior	i. Vasodilation ii. Sweating iii. Reduced activity, loss of appetite
Decrease	Peripheral and central thermo-receptors	Posterior hypothalamus	i. Cutaneous blood vessels ii. Skeletal muscle iii. Arrector pili iv. Endocrines v. Behavior vi. Brown fat	i. Vasoconstriction ii. Shivering thermogenesis iii. Piloerection (cutaneous air trapping) iv. Increased adrenal medullary secretion: Chemical thermogenesis. Increased thyroxine secretion (long-term change) v. Increased activity, increased appetite vi. Non-shivering thermogenesis

d. Brown fat metabolism has been shown to have a significant role in thermoregulation even in adults.[2]

Transduction of Temperature

Major advances in the transduction processes in peripheral thermal sensation have been made since the discovery of the transient receptor potential (TRP) family of ion channels in the last decade and a half. TRP channels are cellular sensors for a wide spectrum of physical and chemical stimuli. TRP channels are a superfamily of proteins that can be expressed in cell membranes and in membranes of internal structures. Individual TRP channels have now been identified that may subserve these specific temperature discrimination roles.[2] Each has a relatively narrow band of temperature activation, yet overlapping of these sensitivities allows for a wide range of temperature discrimination. TRPV1 and TRPV2 were among the first to be identified with temperature sensitivity. They are activated by temperatures higher than 43°C and 52°C, respectively, and may mediate noxious hot sensation. TRPV4 and TRPV3 are activated by temperatures above 25°C and 31°C, mediating innocuous warm sensation.

TRPM8 is increasingly activated as temperature drops below 27°C and is likely to mediate innocuous cold sensation. TRPA1 is activated at temperatures below 17°C and may contribute to noxious cold sensation. The role of these TRP channels can now be effectively mapped onto textbook figures produced before the knowledge of TRP channels, describing discharge frequencies at different temperatures of neurons from these temperature receptors (Fig. 15.2). *However, this must be viewed cautiously at present as only few studies have identified these channels as the transduction mechanisms on primary thermosensitive afferents.*[2]

Afferent Pathways

Environmental temperature is sensed by thermoreceptors in the skin and the thermosensory information is transmitted to the spinal cord. The spinothalamocortical pathway reaches postcentral gyrus and mediates the perception of temperature. The thermosensory signaling from the spinal cord ascending through the lateral parabrachial nucleus (LPBN) (Fig.15.3) generates the emotion of thermal comfort and discomfort, which then likely drives thermoregulatory behavior.

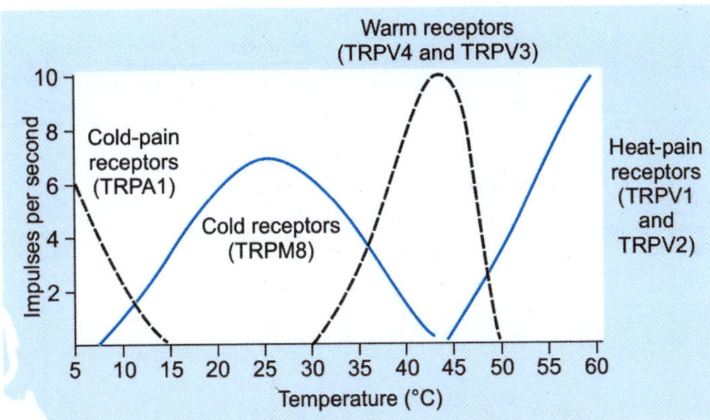

Fig. 15. 2: Various transient receptor potentials (TRPs) and their sensitivity range (after Tansey EA et al[2])

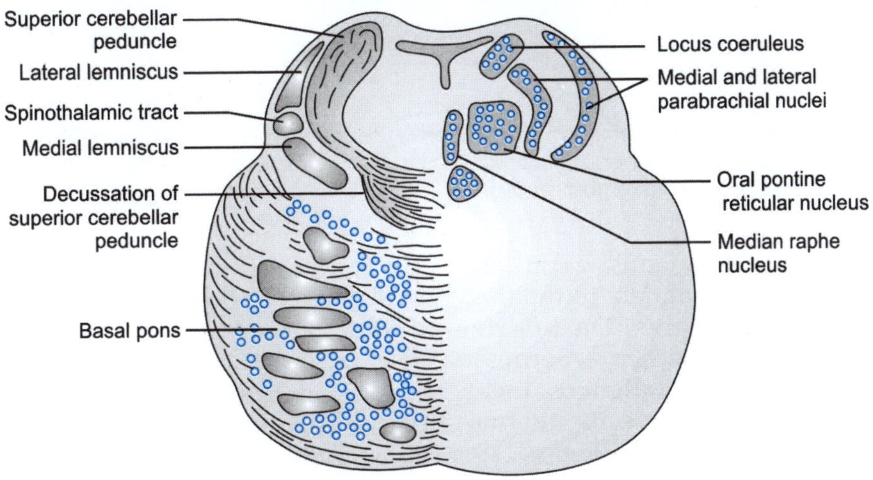

Fig. 15.3: TS pons (diagrammatic). Location of lateral parabrachial nucleus

The thermosensory signaling through the lateral parabrachial nucleus also feeds the information to the thermoregulatory center in the preoptic area to regulate autonomous thermoregulatory responses, such as shivering, sweating and skin vasoconstriction. Thus, pathway for thermoregulatory behavior is distinct from the spinothalamocortical pathway for perception of temperature (Fig. 15.4).

The concept given above is based on the work of Nakamura et al.[5–8] To confirm the contribution of the spinothalamocortical pathway, which is known to mediate thermosensory transmission for perception of skin temperature, these researchers lesioned thalamic regions mediating this pathway in rats. Thalamic-lesioned rats showed *compromised* electroencephalographic responses in the primary somatosensory cortex to changes in skin temperature, indicating lack of temperature perception. However, these lesioned rats displayed intact heat- and cold-avoidance thermoregulatory behaviors. Inactivation of neurons in the LPBN eliminated both heat- and cold-avoidance thermoregulatory behaviors. These results demonstrate that the LPBN, but not the thalamus, mediates the cutaneous thermosensory neural signaling required for behavioral thermoregulation. The spinal LPBN–POA (preoptic area)

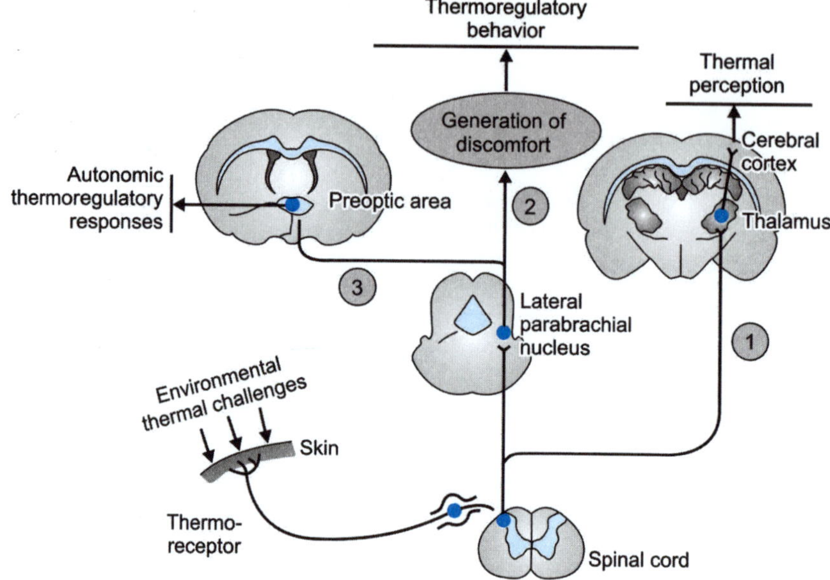

Fig. 15.4: Pathways for perception of thermal sensation (1), thermoregulatory behavior (2) and autonomic thermoregulatory responses (cutaneous vasodilation and sweating/cutaneous vasoconstriction, shivering, thermogenesis, etc.) (3), after Nakamura et al[9]

pathway transmits cutaneous warm- and cool-sensory signals separately to the thermoregulatory center in the POA to drive autonomous thermoregulatory responses to environmental thermal challenges, including shivering and non-shivering thermogenesis and cutaneous vasomotion.

To conclude, cool and warm temperature information stimulates separate populations of primary somatosensory afferents in the skin. These enter the spinal (or trigeminal) dorsal horn to synapse with second-order ascending neurons in the laminas.[1] There are subtypes of ascending spinothalamic fibers that relay temperature information from the skin to the central nervous system for perception of temperature sensation (Fig. 15.4). According to Morrison and Nakamura[6], a second subset of thermosensitive second-order neurons project to the LPBN where these second-order neurons synapse. *The external lateral subnucleus of LPBN relays spinal input from cold-sensitive neurons, whereas dorsal subnucleus of LPBN relays impulses from warm-sensitive neurons.* Third-order LPBN neurons arising from these regions project to the

median preoptic nucleus (MnPO) of the POA.[6,7] Activation of these pathways will initiate either heat gain or heat loss mechanisms (Fig. 15.4).

Afferents arising in the viscera travel to the central nervous system mainly in vagus and splanchnic nerves. The majority of this sensory information, including temperature, converges at the level of the nucleus tractus solitarius before passing to the LPBN, so the LPBN may integrate both skin and visceral thermal information.[8]

Central Control of Thermoregulatory Responses

Previously, temperature control was taught as involving integration of central and peripheral thermal signals coordinated in the hypothalamus and somehow involving a comparison of this integrated signal with a "set point" signal. If an error between the input temperatures and the set point occurs, then this would trigger appropriate heat gain or heat loss mechanisms. Although part of this model was based on the observation of thermal responses to heating and cooling of the skin and discrete regions of the hypothalamus, evidence of the central circuitry necessary for this model has been lacking. The application of several new techniques has allowed these concepts to be updated. What appears to be emerging is a model in which a single set point within the hypothalamus is no longer necessary? Evidence is accumulating that central and peripheral temperatures influence individual effector circuits independently.[6, 8] Thermosensitive neurons are activated when the appropriate temperature threshold for that neuron is reached, and action potentials ascend, via synaptic relays, to the preoptic area (POA) of the hypothalamus. These signals, along with thermosensitive signals arising within the POA, act on several effector outputs, and the influence of central and peripheral signals varies between different effectors.[8] Cold responses, in general, are more sensitive to skin temperature (potentially reflecting the preponderance of cold receptors), whereas warmth responses are more sensitive to core temperature, where warm receptors are more numerous.[8] Indeed, there are distinguishable variations in the sensitivity of specific *effector mechanisms* to skin and core temperature. For example, the shivering response and BAT activation are the most responsive to skin temperature compared with the cutaneous circulation being influenced more by core temperature changes.[9]

There is evidence that the individual effector circuits have distinctly different threshold temperatures. By carefully simultaneously recording the neural drive to individual effector organs, it has been possible to determine a threshold hierarchy for the activation of each effector.[10, 11] For example, activation of cutaneous sympathetic vasoconstriction occurs slightly before (higher core temperature) the activation of BAT in rats. Yet the contribution of all the relatively independent individual effector pathways collectively contributes to a steady core temperature of 37°C. In this way, the concept of the set point has been updated with a more

nuanced model, and the term "balance point" has been proposed as an alternative.[2]

Effector Systems

Cold-sensitive neurons within the median preoptic (MnPO) nucleus are activated by cutaneous/visceral afferent activity as well as direct stimulation of MnPO neurons themselves. These activated inhibitory neurons project to the medial preoptic (MPO) nucleus of hypothalamus. Disinhibition of MPO nucleus allows full expression of ongoing activity in the sympathetic vasoconstrictor outflow (via the spinal cord and preganglionic and postganglionic sympathetic nerves) and in outflow to skeletal muscle and BAT via dorsomedial nucleus and rostral raphe pallidus. These changes induce cutaneous vasoconstriction, shivering in skeletal muscle and increased metabolism in BAT (Fig. 15.5).

Warm-sensitive neurons within the MnPO nucleus are activated by cutaneous/visceral afferent activity as well as direct stimulation of MnPO nucleus neurons themselves. These

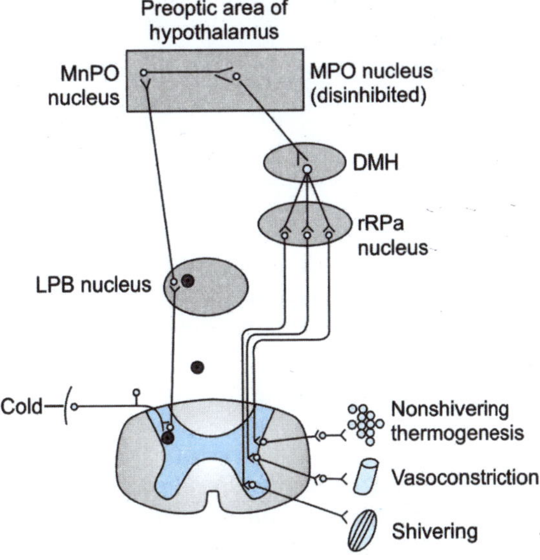

Fig. 15.5: Central and efferent mechanisms in response to stimulation of cold receptors (MnPO nucleus: Median preoptic nucleus; MPO nucleus: Medial preoptic nucleus; DMH: Dorsomedial hypothalamus; rRPa nucleus: Rostral raphe pallidus nucleus; LPB nucleus: Lateral parabrachial nucleus), after Nakamura et al[9]

neurons project to the medial preoptic hypothalamus, where they activate inhibitory output that travels to the rostral raphe pallidus via the dorsal medial hypothalamus. Finally, these inhibitory signals slow or stop ongoing activity arising in neurons that project to the spinal cord to control output to cutaneous blood vessels, BAT, and skeletal muscle and thereby result in cutaneous vasodilatation, sweating and reduced muscular activity (Fig. 15.6).

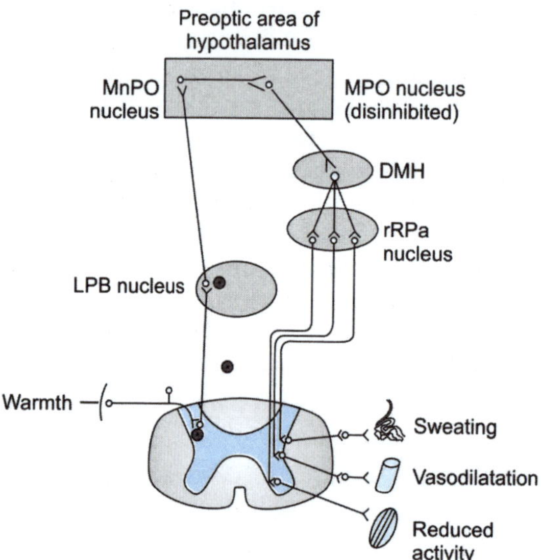

Fig. 15.6: Central and efferent mechanisms in response to stimulation of cold receptors (MnPO nucleus: Median preoptic nucleus; MPO nucleus: Medial preoptic nucleus; DMH: Dorsomedial hypothalamus; rRPa nucleus: Rostral raphe pallidus nucleus; LPB nucleus: Lateral parabrachial nucleus)

BROWN ADIPOSE TISSUE

Excess energy from food is stored in the form of triglycerides within *white adipocytes*. There are numerous *white adipose tissue* (WAT) depots throughout the body. White adipose tissue may constitute from 10% to more than 50% of total body weight. *Brown adipocytes* are another types of adipocytes with relatively restricted distribution. *These are equipped with abundant mitochondria, which uniquely express an uncoupling protein 1 (UCP-1). UCP-1 uncouples*

the process of oxidative phosphorylation in the mitochondria. This process, generates heat instead of adenosine triphosphate (ATP) generated by oxidation of triglycerides of white adipocytes. Thus, BAT is a specialized tissue that is responsible for nonshivering thermogenesis. In humans, the presence of *brown adipose tissue* (BAT) throughout life has been well documented for some decades. However, till recently, brown fat was thought to have a functional role only in human infants, defending newborns from hypothermia.[12]

White adipose tissue on gross appearance gives a whitish look. In histological slides, white adipocytes have rounded appearance and contain a large single lipid droplet, displacing the nucleus to the periphery (Figs 15.7 and 25.8). Brown adipose tissue has dark red (brown) gross appearance. In histological slides, brown adipocytes are characterized by their polygonal shape, smaller size, central nucleus, and numerous small lipid droplets, lending them a multilocular appearance. Brown adipocytes contain well-developed mitochondria filling most of the cytoplasm (Fig.15.9). In contrast, the mitochondrial content of white adipocytes is low. BAT is also endowed with a rich blood and nerve supply. The abundance of mitochondria containing respiratory chain cytochrome enzymes with iron as a cofactor, as well as the vasculature within BAT, gives rise to a darker red (brown) color, compared with the paler hue exhibited by WAT[12] (Fig. 15.7).

BAT Location in Adults

BAT in humans is located in the neck, thorax, and abdomen, juxtaposing deep viscera (heart, lungs, kidneys, adrenals, and intestines) and along the great vessels (carotids, aorta, pulmonary, and mesenteric vasculature). In rodents, BAT is most frequently and abundantly located in the interscapular region throughout life. In humans, the interscapular depot disappears rapidly after infancy. Despite the documentation that cervical, paravertebral, para-aortic, perirenal, and periadrenal BAT depots existed throughout

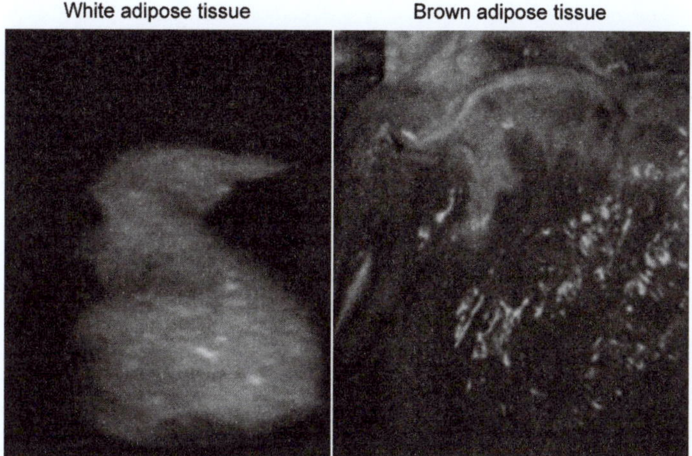

Fig. 15.7: Gross appearance of white and brown adipose tissues

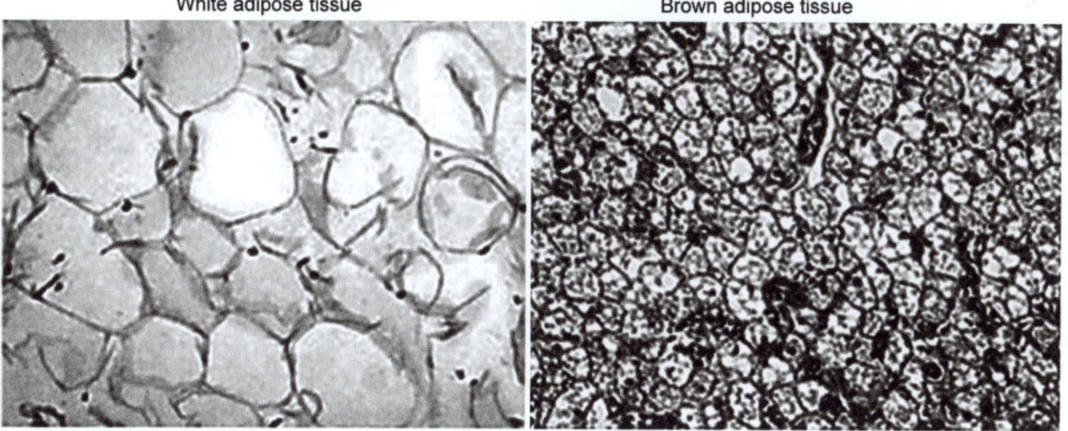

Fig. 15.8: White and brown adipose tissues (histological appearance)

life, the absence of interscapular BAT overshadowed these observations, propagating the belief that BAT disappeared in the adult human.[12,13]

The prevalence of BAT in humans can be estimated indirectly from PET-CT or directly from histological analysis of fat depots. PET imaging was introduced in the 1980s for the diagnosis and staging of cancers. PET is a form of metabolic imaging that uses tracers to determine physiological function. Hybrid computerized tomography (CT) became available in the early 2000s, which allowed for anatomical definition. The principle is that tumors have a higher metabolic activity than normal tissues so that labeled substrates such as glucose can be employed to discriminate between tissues that differ in metabolic activity. ^{18}F-fluorodeoxyglucose (^{18}FDG) imaging employing PET-CT technology is established for the diagnosis and staging of malignancies. ^{18}F tracer is constituted to deoxyglucose, which is taken up but not metabolized by cells. Based on this technique, the distribution of BAT is shown in Fig. 15.10.[14]

Effect of Cold Exposure on BAT

Studies in humans have established that acute cold exposure elevates brown fat thermogenesis[15] (Fig. 15.11). In addition, it has been

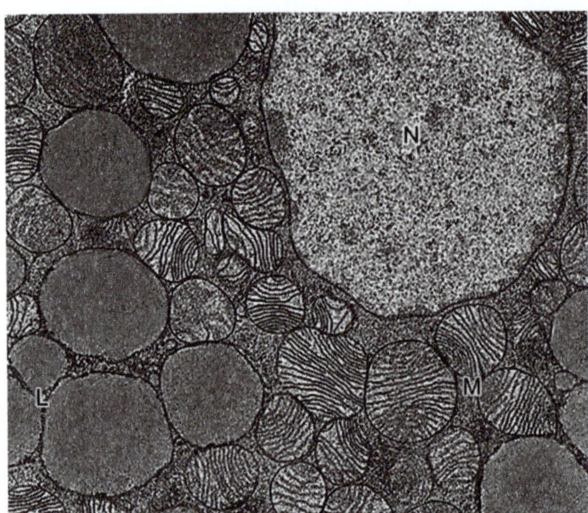

Fig. 15.9: Mitochondrial density in brown adipocyte (N: Nucleus; L: Lipid droplets; M: Mitochondria)

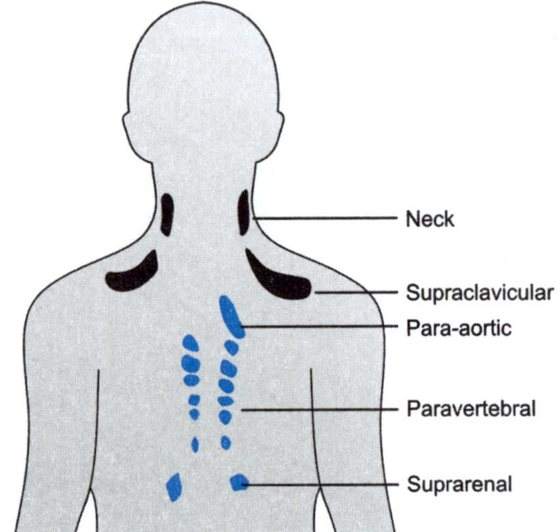

Fig. 15.10: Location of brown adipose tissue in adults, after Nedergaard et al[14]

observed that chronic cold exposure induces BAT hypertrophy and hyperplasia.[16] These reports carry important inferences of physiological significance for BAT in humans. Cold stimulation converts more than 80% of negative to positive PET-CT scans. Because there is 15- to 20-fold more BAT in PET-positive than in PET-negative fat, it is likely that cold exposure stimulates BAT activity by at least the same magnitude because proliferation of BAT is unlikely to occur after such a brief period of stimulation.[16] A distinct seasonal variation in the amount of BAT in humans has also been reported[17] (Fig. 15.12). Thus, a large regulatory capacity underlies BAT adaptation and function. Effect of cold on BAT seems to be mediated via sympathetic nervous system. BAT is densely innervated by sympathetic nerves. The SNS stimulates BAT activity. Sympathetic drive is augmented by cold exposure, which causes a near-doubling of circulating plasma noradrenaline levels.

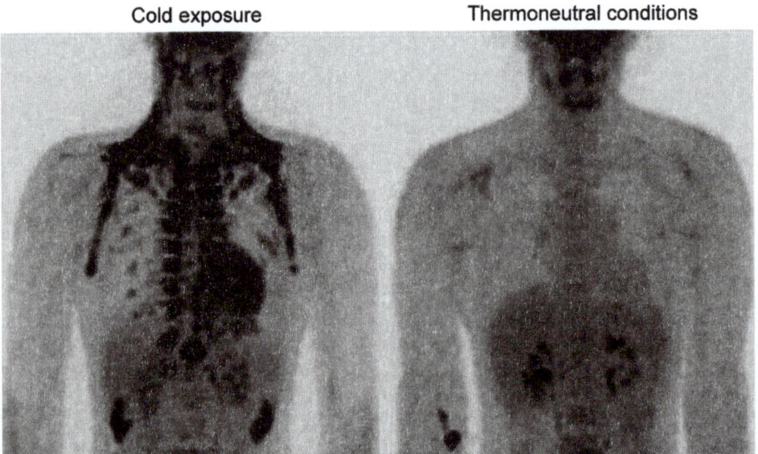

Fig. 15.11: Brown adipose tissue in a subject after acute cold exposure as compared to thermoneutral conditions, after van Marken Lichtenbelt et al[15]

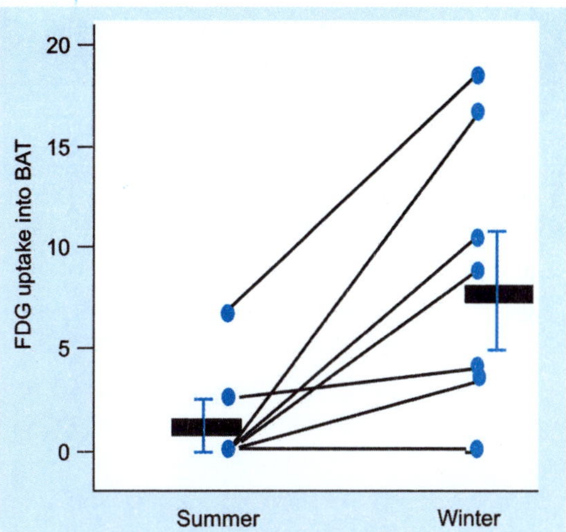

Fig. 15. 12: Seasonal variation in FDG intake into BAT, after Saito et al[17]

Effect of Catecholamines on BAT

Noradrenaline exerts powerful systemic effects on BAT in addition to its local effects as a neurotransmitter. Evidence for this comes from studies in rare patients with pheochromocytoma. Abdominal fat biopsies (perirenal, periadrenal, and omental) harvested during adrenalectomy revealed significantly higher abundance of BAT in patients with pheochromocytoma compared with healthy controls.[18] The high FDG uptake observed in BAT depots disappeared after cure of pheochromocytoma. In addition, BAT activity correlated with plasma catecholamine levels in patients with pheochromocytoma.[18]

Effect of Sex Steroids on BAT

In isolated brown adipocytes, 17β-estradiol stimulates, whereas testosterone inhibits mitochondrogenesis and UCP1 expression.[19] In humans, several PET-CT series have reported a higher prevalence of BAT in women than in men. Also, cold-stimulated PET studies reveal higher BAT mass and activity in women, who are 3 times more likely to exhibit active BAT than men.

References

1. Ranson SW. Regulation of body temperature. Assoc Res Nerv Ment Dis Proc. 1940; 20: 342–99.
2. Tansey EA, Johnson CD. Recent advances in thermoregulation. Advances in Physiology Education. 2015; 39: 139–48.
3. Marya RK, History of Medicine: Jaypee Publishers, New Delhi, 2009.
4. Mazerolle SM, Ganio MS, Casa DJ, et al. Is oral temperature an accurate measurement of deep body temperature? A systematic review. J Athl Train. 2011; 46: 566–73.
5. Yahiro T, Kataoka N, Nakamura Y, et al. Scientific Reports volume 7, Article number: 5031 (2017).
6. Morrisson SF, Nakamura K. Preoptic mechanism for cold-defensive responses to skin cooling. J Physiol. 2008; 586: 2611–20.
7. Morrisson SF, Nakamura KA. Thermosensory pathway mediating heat-defense responses. Proc Natl Acad Sci USA. 2010; 107: 8848–853.
8. Morrisson SF, Nakamura K. Central neural pathways for thermoregulation. Front Biosci. 2011; 16: 74–104.
9. Nakamura K. Central circuitries for body temperature regulation and fever. Am J Physiol Regul Integr Comp Physiol. 2011; 301: R1207–28.
10. McAllen RM, Tanaka M, Ootsuka Y, et al. Multiple thermoregulatory effectors with independent central controls. Eur J Appl Physiol. 2010; 109: 17–33.
11. Heaton JM. The distribution of brown adipose tissue in the human. J Anat. 1972;112: 35–39.
12. Michael PMM, Ho SKY. Brown adipose tissue in adult humans: A metabolic renaissance. Endocrine Reviews. 2013; 34: 413–38.
13. Blaza S. Brown adipose tissue in man: a review. J R Soc Med. 1983; 76: 213–216.
14. Nedergaard J, Bengtsson T, Cannon B. Unexpected evidence for active brown adipose tissue in adult human. Am J Physiol Endocrinol Metab. 2007;293: E444–52.
15. van Marken Lichtenbelt WD, Vanhommerig JW, Smulders NM, et al. Cold-activated brown adipose tissue in healthy men. N Engl J Med. 2009; 360: 1500–08.
16. Cannon B, Nedergaard J. Brown adipose tissue: function and physiological significance. Physiol Rev. 2004; 84: 277–359.
17. Saito M, Okamatsu-Ogura Y, Matsushita M. High incidence of metabolically active brown adipose tissue in healthy adult humans. Diabetes. 2009; 58: 1526–31.
18. Lean, ME, James, WP, Jennings G, et al. Brown adipose tissue in patients with phaeochromocytoma. Int J Obes. 1986; 10: 219–27.
19. Rodriguez-Cuenca S, Monjo M, Gianotti M, et al. Expression of mitochondrial biogenesis-signaling factors in brown adipocytes is influenced specifically by 17β-estradiol, testosterone, and progesterone. Am J Physiol Endocrinol Metab. 2007; 292: E340–46.

Chapter 16

Appetite Regulation and Weight Control

INTRODUCTION

Despite daily variation in food intake and energy expenditure, for most individuals, body weight remains remarkably stable over long periods of time. For this, food intake and energy expenditure must be constantly modulated and balanced. The hypothalamus is essential for the regulation of appetite and energy balance. Focus on the hypothalamic region as a possible regulator of food intake began when Hetherington and Ranson in 1940 were able to induce obesity in rats by bilateral hypothalamic lesions in the region of the ventromedial nuclei (Fig. 16.1).

Anand and Brobeck in 1951, demonstrated that, while bilateral lesions in rats in the ventromedial regions produced "hyperphagia", lesions in lateral hypothalamus area (LHA) led to complete 'aphagia' and death of the animal due to starvation. Further, they found that medial hypothalamic lesions produced hyperphagia only when the LHA was intact. On stimulation of lateral hypothalamic area in unanesthetized animals, through chronically implanted electrodes, the animal would start eating immediately, irrespective of whether it had been fasted or previously fed. Similarly, stimulation of the medial hypothalamic area stopped eating even when the animals were hungry and eating. All these studies have provided evidence that there are two opposing mechanisms in the hypothalamus that regulate food intake, namely a mechanism in the lateral region of the hypothalamus which initiates feeding and is therefore designated as the "feeding center", and one in the medial part of the hypothalamus which brings about satiety after a meal has been taken and is thus termed the "satiety center".[1] These observations formed the basis of **dual center hypothesis** of regulation of food intake (Fig. 16.2).

Later, repetitions of the earlier lesion studies showed that in fact animals with VMN lesions would not eat continually without limit nor would the animals with LHA lesions

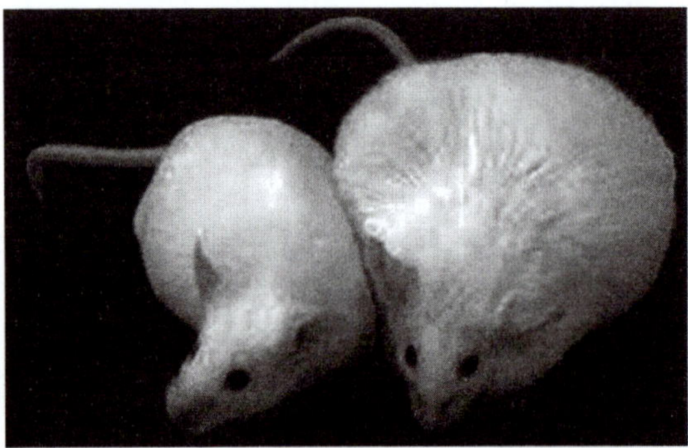

Fig.16.1: A bilateral ventromedial nucleus-lesioned rat (right) and a normal rat of same age (left)

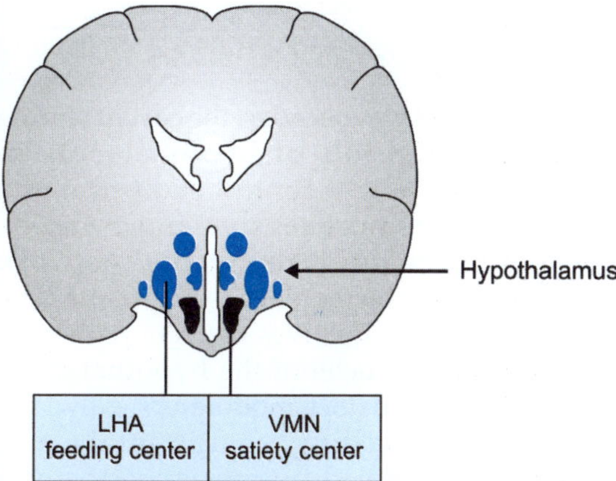

Fig. 16.2: The "dual center" hypothesis of regulation of food intake (LHA: Lateral hypothalamic area; VMN: Ventromedial nucleus)

necessarily starve to death. Rather, the VMN lesioned animals would eat excessively until they reached some new higher body weight at which point they would then reduce their food consumption to maintain this new set point. Likewise, once the LHA lesioned animals had dropped weight to some new level, they would again eat normally to maintain the new lower body weight set point.

GLUCOSTATIC HYPOTHESIS

The glucostatic theory of food intake control postulated that reduced glucose utilization in critical brain regions leads to perception and expression of hunger, and increased glucose utilization in these same glucosensitive sites leads to decreased hunger and cessation of eating. Mayer proposed that decreased glucose utilization or 'metabolic hypoglycemia', the point at which the peripheral arteriovenous difference in blood glucose becomes negligible and glucose is no longer entering 'metabolizing cells', was the signal for meal initiation (Fig. 16.3). This hypothesis was based on a number of observations: When gold thioglucose is administered to animals, it is taken up into the brain and destroys areas of the ventromedial hypothalamus with subsequent development of obesity. Other data which are consistent with the idea of Mayer include the following: (1) Injections of insulin which lower glucose can stimulate feeding; (2) administration of 2-deoxy-D-glucose, an analogue of glucose which blocks glucose metabolism will stimulate feeding.[2]

Glucostatic hypothesis received further support with the ability of techniques to monitor blood glucose continuously in freely behaving animals in contrast to discrete blood sampling at fixed intervals. In 1980, it was observed that a fall in blood glucose was correlated with meal initiation in the rat. Similar results were obtained in humans who were blinded as to the time of the day. Indeed, the initiation of meals and perception of hunger were synchronized with transient and dynamic blood glucose declines.[3] As a refinement of glucostatic hypothesis, the role of glucose in regulation of food intake has shifted from a satiety signal to an initiation signal. Moreover, its role has shifted from a static one aimed at maintaining constant levels to a dynamic one in which the brain or liver are responding to dynamic changes in glucose concentration to initiate food seeking. Moreover, the use of the word static for the "glucostatic" theory is unfortunate, since it implies a static or unchanging system. Glucose oscillates in the plasma under normal

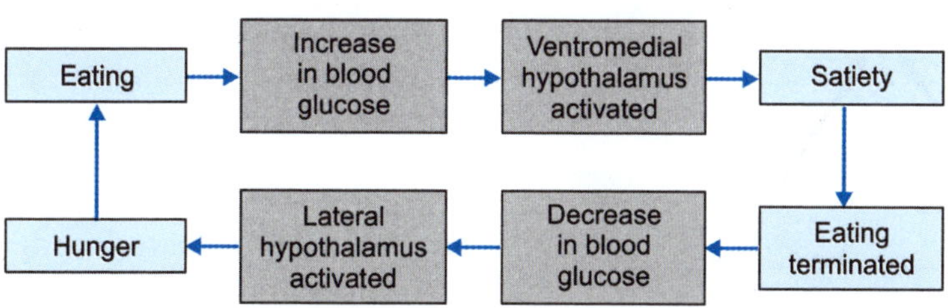

Fig. 16.3: Glucostatic hypothesis of control of food intake

circumstances and responds to a variety of external circumstances. The word "dynamic" would be more appropriate in light of the way in which food intake begins after the nadir in glucose suggesting that it is the pattern of dynamic change in glucose which is sensed rather than its level. It would thus be preferable to refer to this as the glucodynamic theory of food intake rather than the glucostatic theory.[3]

LIPOSTATIC HYPOTHESIS

In 1953, Kennedy[4] presented a hypothesis that it was not the on/off switch for eating behavior that was being disturbed by hypothalamic lesions, but rather the regulation of total body fat. Indeed, Kennedy coined the term "lipostasis" to refer to the process whereby body fat was regulated through the control of food intake and energy expenditure.

Kennedy's concept of the regulation of body fat shifted attention from short-term circulating metabolites that might signal the stimulation and termination of eating behavior to the search for signals that enable the regulation of adiposity. It was only in 1990s that leptin was discovered. Leptin is a hormone that is produced by adipocytes that is released into the circulation in proportion to the adipose tissue mass, and that stimulates receptors in the hypothalamus (and other brain areas) that control energy intake and expenditure (Fig. 16.4).

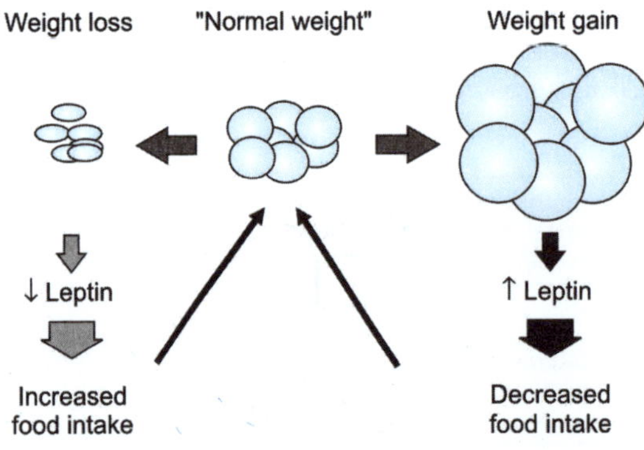

Fig. 16.4: The lipostatic hypothesis of food intake

HYPOTHALAMIC ARCUATE NUCLEUS: KEY PLAYER IN ENERGY HOMEOSTASIS

During the last two decades, the earlier simple concept of the role of hypothalamus in appetite and weight control evolved into a much more comprehensive and complex concept of the integrated neural network responsible for the regulation of appetite. The new information involves discrete pathways within specific nuclei of the hypothalamus, and various regulatory modulators. Now, the regulation of feeding, energy intake and expenditure, and body weight is considered a homeostatic process.[5] *Information regarding long-term weight control is communicated predominantly via long-term humoral signals, whereas meal initiation and termination are believed to be regulated via short-term signals, such as neural signals from the brain and humoral signals from the gut. The receipt and integration of these signals occur mainly in the hypothalamus, and are largely regulated by the hypothalamic arcuate nucleus (ARC), located at the base of the region* (Fig. 16.5). Within the ARC, there exists two distinct populations of neurons responsible for appetite regulation:

i. *Anorexigenic (appetite-inhibiting)* pro-opiomelanocortin (POMC) and cocain- and amphetamine-regulated transcript (CART) neurons.

ii. *Orexogenic (appetite stimulating)* neuropeptide Y (NPY) and agouti-related peptide (AgRP) neurons.

Gut hormone receptors are located on neuronal populations within the ARC, which is accessible to circulating appetite modulators due to its incomplete isolation from the blood–brain barrier (BBB). Signals from the periphery result in changes in the relative activity of these two neuronal sub-populations and the release of their respective neuropeptides, subsequently influencing feeding behavior and energy expenditure[6,7] (Fig. 16.5). Overall central regulation of appetite is discussed in detail later in this chapter. Peripheral signals involved in appetite control are discussed first.

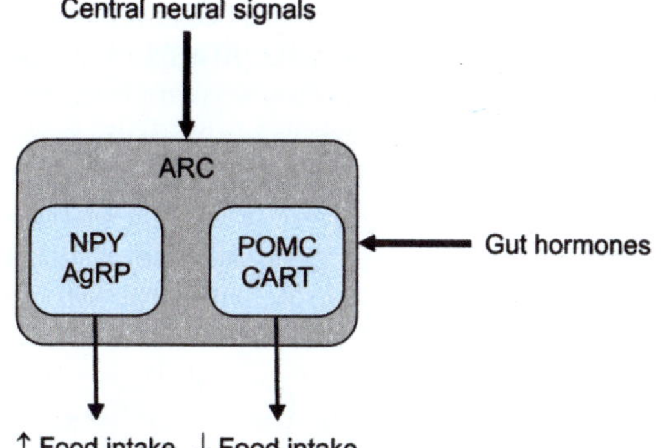

Fig.16.5: Role of hypothalamic arcuate nucleus (ARC) in the control of food intake (NPY: Neuropeptide Y; AgRP: Agouti-related peptide; POMC: Pro-opiomelanocortin; CART: Cocain- and amphetamine-regulated transcript)

I. PERIPHERAL SIGNALS CONTROLLING APPETITE (Fig. 16.6)

A. Adipose Tissue

i. Leptin

Although originally thought to be inert tissue solely for the storage of energy, it has now become clear that adipose tissue is an active endocrine organ. One of its most important

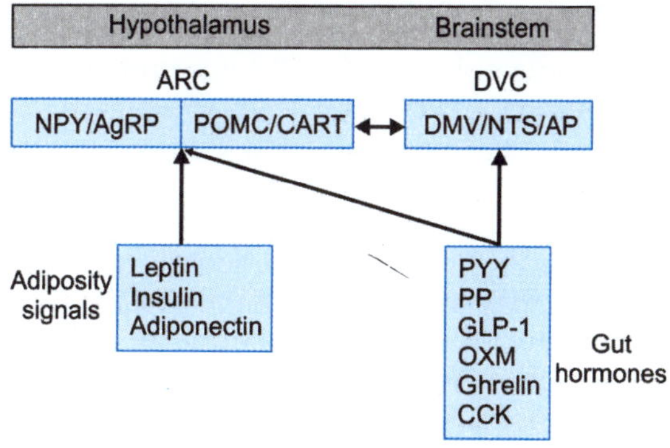

Fig. 16.6: Peripheral signals regulating food intake and role of hypothalamus and brainstem in the regulation of food intake. (ARC: Arcuate nucleus, NPY: Neuropeptide Y; AgRP: Agouti-related peptide; POMC: Pro-opiomelanocortin; CART: Cocain- and amphetamine-regulated transcript; DVC: Dorsal vagal complex; DMV: Dorsal motor nucleus of vagus; NTS: Nucleus of tractus solitarius; AP: Area postrema; GLP-1: Glucagon-like peptide-1; CCK: Cholecystokinin; PP: Pancreatic polypeptide; PYY: Peptide YY; OXM: Oxyntomodulin)

hormones produced by adipose tissue is leptin. Leptin is a peptide hormone with numerous actions, including influences on energy homeostasis and neuroendocrine and immune function. Leptin is the product of the ob gene (ob for obesity) expressed predominantly in adipocytes. Circulating leptin levels reflect both energy stores and acute energy balance. Plasma leptin levels are highly correlated with adipose tissue mass, but food restriction results in suppression of circulating leptin[8], which can be reversed by refeeding or insulin administration. Exogenous leptin administration, both centrally and peripherally, reduces spontaneous and fasting-induced hyperphagia[8] whilst chronic peripheral administration reduces food intake resulting in loss of fat mass and body weight.[9]

Circulating leptin crosses the blood–brain barrier. Leptin receptors are found in many places within the hypothalamus but are found particularly in the arcuate nucleus (ARC). In the ARC, it inhibits the neurons which release orexigenic (appetite increasing) neuropeptides, neuropeptide Y (NPY) and agouti-related peptide (AgRP).[10] On the other hand, leptin activates anorectic (appetite decreasing) pro-opiomelanocortin (POMC) and cocaine- and amphetamine-regulated transcript (CART) releasing neurons in the arcuate nucleus.[11] Besides ARC, leptin receptors are also found in the many other nuclei of hypothalamus, e.g. ventromedial hypothalamus (VMH), dorsomedial hypothalamus (DMH), lateral hypothalamic area (LHA) and medial preoptic area (MPOA). Leptin receptors are also found in appetite-modulating pathways in the brainstem.

The absence of leptin has profound effects on body weight. Lack of circulating leptin, due to a mutation in the ob gene, leads to hyperphagia, obesity, as well as neuroendocrine and immune disturbance in the leptin deficient ob/ob mouse, which can be normalized by leptin administration.[12] Similarly, leptin deficiency in humans, though rare, causes severe obesity and hypogonadism, which can be ameliorated by recombinant leptin therapy.[13] In addition to its effects on food intake, leptin also modulates energy expenditure in rodents (though not in humans); the hypothalamopituitary control of the gonadal, adrenal and thyroid axes; and the immune response.[14]

ii. Adiponectin

Adiponectin is a 244-amino acid protein secreted from adipose tissue. Its circulating levels are up to 1,000-fold higher than other circulating hormones such as leptin and insulin. Function(s) of adiponectin is largely unknown but is postulated to regulate energy homeostasis. The plasma concentration of adiponectin is inversely correlated with adiposity in rodents, primates, and humans. Peripheral administration of adiponectin to rodents has been shown to attenuate body weight gain, by increased oxygen consumption (metabolism), without affecting food intake.[15]

iii. Resistin

Resistin ('resistance to insulin') is also produced by adipose tissue and appears to increase insulin resistance. Circulating resistin is increased in obese rodents and falls after weight loss. Resistin, originally described as an adipocyte-specific hormone, has been suggested to be an important link between obesity, insulin resistance and diabetes. Increasing evidence indicates that resistin plays important regulatory roles apart from its role in insulin resistance and diabetes in a variety of biological processes: atherosclerosis and cardiovascular disease.[16]

B. Pancreatic Hormones

i. Insulin

The pancreatic hormone insulin was one of the first adiposity signals to be described as regulator of energy balance acting via hypothalamus.[17] And, like leptin, it is positively correlated with long-term energy balance.

However, unlike leptin levels, which are relatively insensitive to acute food intake, insulin secretion increases rapidly after a meal. There is considerable evidence that insulin acts as an *anorectic signal* within the central nervous system (CNS). Centrally administered insulin decreases food intake and body weight. Insulin infusion into the cerebral ventricle in rodents or primates dose-dependently decreases food intake resulting in weight loss over a period of weeks.[18] Intrahypothalamic (PVN) insulin injection also decreases food intake and weight gain in rats.[19] Conversely, antibodies to insulin injected into the VMH of rats stimulate food intake.[20]

Insulin receptors are widely distributed in the brain, particularly in hypothalamic nuclei involved in food intake (ARC, DMH, PVN, suprachiasmatic and periventricular regions). The pathways mediating insulin's effects on food intake remain to be fully elucidated. It is likely that both the NPY and melanocortin systems are important downstream mediators of insulin's actions on food intake and body weight.[21]

ii. Pancreatic Polypeptide

Pancreatic polypeptide (PP) is secreted from PP cells in the pancreatic islets of Langerhans. PP appears to reduce food intake directly through the Y4 receptors in the brainstem and hypothalamus. The anorectic effects of PP are abolished by vagotomy in rodents, suggesting that PP may also act via the vagus nerve to reduce food intake. Circulating PP concentrations rise after a meal in proportion to the calorific load. The anorectic effects of PP have been demonstrated in a number of experimental models. In mice, acute and chronic peripheral administration of PP reduces food intake. In leptin-deficient ob/ob mice, repeated intraperitoneal injection of PP decreases body weight gain and ameliorates insulin resistance and hyperlipidemia. In normal-weight human subjects, intravenous infusion of PP results in a 25% reduction in 24-hour food intake.[22]

C. Gut Hormones

i. Peptide YY (PYY)

The role of PYY in regulation of body weight is less clear.[21] PYY is released by L cells of the distal gut. Circulating PYY concentrations are low in fasted state and rise rapidly following a meal with a peak at 1–2 hours and remain elevated for several hours. Interestingly, the increase in plasma PYY concentrations is seen rapidly after food intake, well before nutrients are in contact with the L cells of the distal intestine.[21] Ingestion of fat results in greater release of PYY than observed with ingestion of carbohydrate or protein meals with a similar calorie content. Peripheral PYY administration shows a decrease in food intake and body weight gain in rats. The anorectic effects of PYY may be mediated centrally via a direct action in the ARC, through an indirect action involving the vagus and brainstem, or by a combination of both pathways.[22]

ii Ghrelin

Ghrelin is the only known orexigenic gut hormone. It is produced and released primarily by the gastric oxyntic cells; however, total gastrectomy only reduces plasma ghrelin by 50–60%. The remaining circulating ghrelin is released by the duodenum, ileum, cecum, and colon. Levels of circulating ghrelin increase in preprandial stage and fall rapidly in the postprandial period.[21] Both central and peripheral administrations of ghrelin increase food intake and body weight with a reduction in fat utilization in rodents. In human, negative correlations between circulating ghrelin levels and body mass index are found. Fasting plasma levels of ghrelin are high in patients with anorexia nervosa[23] and in subjects with diet-induced weight loss. Evidence suggests that ghrelin mediates its orexigenic action via stimulation of NPY/AgRP co-expressing neurons within the ARC of hypothalamus. Brainstem and vagus nerve may also contribute to the effects of ghrelin on food intake.[22]

iii. Glucagon-like Peptide-1(GLP-1)

GLP-1 is released by the L cells of the small intestine following nutrient ingestion, and circulating GLP-1 levels are inversely correlated with body mass. Acute GLP-1 injection into the third or fourth ventricle or into the PVN reduces calorie intake. GLP-1 may influence energy homeostasis via the brainstem also. In humans, GLP-1 dose-dependently decreases food intake. However, when the infusions mimic postprandial concentrations, the effect is small.[24]

iv. Oxyntomodulin (OXM)

OXM is released from L cells of the intestine in response to ingested food and in proportion to caloric intake. Administration of OXM reduces food intake and increases energy expenditure in both rodents and humans. Both central and peripheral OXM administration acutely reduce food intake in rodents; repeated administration reduces body weight gain and adiposity. OXM also reduces hunger and food intake (by 19.3%) in healthy human volunteers, an effect which continues for 12 hours post-infusion.[25, 26] Thus, OXM may play a physiological role in regulation of energy balance.

v. Cholecystokinin (CCK)

In animals, repeated CCK administration does not alter body weight, for although meal size is reduced, meal frequency increases and there is no overall change in food intake. Similarly, continuous CCK infusion is not effective after the first 24 hours. The evidence for a role of CCK in long-term body weight regulation, and hence as a potential therapy for obesity, remains contradictory.

II. CENTRAL REGULATORS OF APPETITE

A. Hypothalamic Structure and Neuronal Pathways Regulating Appetite

Researches by Hetherington and Ranson in 1940 and by Anand and Brobeck in 1951 led to a simple "dual center" hypothesis of regulation of food intake, which involved the role of hypothalamic LHA and VMN.[1] However, new researches have assigned central role to arcuate nucleus (ARC) of hypothalamus in the regulation of energy homeostasis. Besides ARC, many other hypothalamic nuclei as well as structures in the brainstem seem to be involved in regulation of appetite and weight homeostasis (Figs 16.5 and 16.7).

1. ARC

The arcuate nucleus (ARC) of hypothalamus is thought to play a pivotal role in the integration of signals regulating appetite. The ARC lies immediately above the median eminence. The ARC-median eminence area is one of the circumventricular organs where the blood–brain barrier (BBB) is specially modified to allow access of peripheral proteins and peptides, such as insulin and leptin, both of which are considered to be signals of fat mass. Two primary neuronal populations within the ARC integrate signals of nutritional status and influence energy

homeostasis. A subpopulation of neurons in the medial ARC express the orexigenic neuropeptides NPY and AgRP. These neurons project primarily to the ipsilateral PVN[25] but also locally within the ARC. A subpopulation of ARC NPY neurons release GABA locally to inhibit the adjacent POMC neurons. More laterally lies a second subpopulation that inhibits food intake via the expression of CART and POMC, which is processed to α-melanocyte stimulating hormone (α-MSH). This subpopulation projects much more widely within the CNS, to hypothalamic nuclei such as the DMH, LHA, and perifornical area (PFA) as well as the PVN.[21] Schwartz et al[26] proposed a model of appetite regulation whereby arcuate neurons act as the primary hypothalamic site of action of peripheral hormones, such as insulin and leptin. These modulate activity of arcuate neurons, which in turn project to secondary hypothalamic nuclei, e.g. the PVN or LHA. Here, the release of further anorectic or orexigenic peptides is modulated to adjust energy intake and expenditure to maintain a stable body weight.

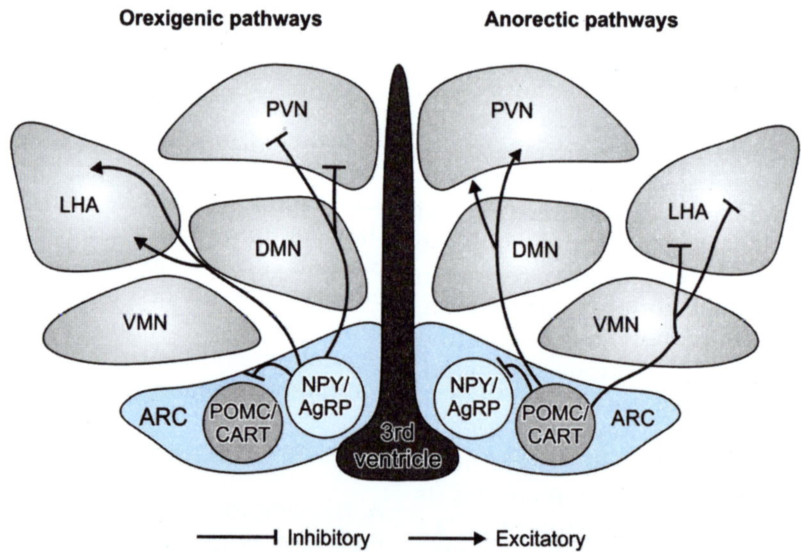

Fig. 16.7: Projections of arcuate nucleus (ARC) of hypothalamus (VMN: Ventromedial nucleus; PVN: Paraventricular nucleus; DMN: Dorsomedial nucleus; LHA: Lateral hypothalamic area; NPY: Neuropeptide Y; AgRP: Agouti-related peptide; POMC: Pro-opiomelanocortin; CART: Cocain and amphetamine-regulated transcript)

2. PVN

The PVN lies adjacent to the superior part of the third ventricle in the anterior hypothalamus. The PVN can integrate signals from neuronal pathways that are important to the control of energy intake, such as POMC/CART and NPY/AgRP neurons in the ARC and orexin neurons in the LH. *Microinjection into the PVN of almost all known orexigenic and anorectic signals alters appetite.* Food intake can be affected by infusion of diverse substances into or near the PVN, whereas lesions of the PVN cause an increase in food intake (hyperphagia) and a decrease in energy expenditure and obesity.[27] The PVN can also interact with the brainstem regions that are critical for appetite regulation. Neurons in the PVN project to the nucleus tractus solitaries.

3. DMH

Dorsomedial hypothalamus (DMH) also has a role in the modulation of energy intake. Destruction of the DMN results in hyperphagia and obesity, although less dramatically than VMN lesion. Injection of orexigenic peptides, NPY, galanin, and GABA into the DMN increases food intake.[22] The DMH has extensive connections with other hypothalamic nuclei. It receives AgRP/NPY neurons from the ARC.

4. LHA/PFA

The lateral hypothalamic area and perifornical area (LHA/PFA) are involved in downstream signaling. Indeed, the PFA is one of the most sensitive areas for NPY-induced feeding, more so than the PVN. The LHA/PFA contains melanin-concentrating hormone (MCH) expressing neurons. MCH appears to have a powerful role in appetite regulation. Repeated intracerebroventricular injection of MCH increases food intake and adiposity in rats. Conversely, MCH-1 receptor antagonists inhibit feeding, and chronic administration leads to a sustained reduction in body weight gain.[28] This suggests MCH may be a downstream mediator of leptin's and POMC's effects on feeding.

5. VMH

The ventromedial hypothalamus has been known to play a role in energy homeostasis for many years, since the finding that bilateral VMH lesions induce hyperphagia and obesity. The VMH receives NPY, AgRP, and α-MSH secreting projections from arcuate neurons and, in turn, VMH neurons project onto both hypothalamic nuclei (e.g. DMH) and brainstem regions (e.g. NTS). Brain-derived neurotrophic factor (BDNF) is highly expressed in the VMN, and its expression is regulated both by food deprivation and melanocortin agonists.[29] Mice with reduced BDNF receptor expression have increased food intake and body weight. Therefore, BDNF neurons in the VMH may act as an additional downstream pathway through which nutritional status and the melanocortin system modulate energy homeostasis.[30]

B. Hypothalamic Appetite-regulating Mediators

a. Neuropeptide Y (NPY)

ARC is the major hypothalamic site of NPY release. Hypothalamic levels of NPY reflect the body's nutritional status. Hypothalamic NPY release increases with fasting and decreases after refeeding. NPY is the most potent orexigen known, and repeated third ventricle or PVN injection of NPY causes marked hyperphagia and obesity. In the rat, central administration of NPY also inhibits brown fat thermogenesis, suppresses sympathetic nerve activity, and inhibits the thyroid axis, thereby reduces energy expenditure.[21]

b. Agouti-related Peptide (AgRP)

Agouti-related peptide is expressed in the CNS, primarily in the medial part of the arcuate nucleus. AgRP mRNA expression is increased by fasting. Central administration of AgRP is able to block α-MSH-induced anorexia and increase nocturnal food intake in rat. AgRP and NPY are co-localized (to occur together in the same cell) in 90% of ARC neurons.[30]

Activation of ARC NPY/AgRP neurons potently stimulates feeding via a number of pathways: The orexigenic effect of NPY released in the PVN, AgRP antagonism of melanocortin receptors in the PVN, and local release of NPY and GABA within the ARC to inhibit the arcuate POMC neurons.

c. Pro-opiomelanocortin (POMC)

Pro-opiomelanocortin release by hypothalamus is regulated by nutritional status. Low POMC levels are recorded in fasting rat that are restored by exogenous leptin administration or 6 hours after refeeding. Human POMC gene mutations or abnormal POMC peptide processing result in early-onset obesity.

d. Cocaine- and Amphetamine-regulated Transcript (CART)

Cocaine- and amphetamine-regulated transcript is the third most abundantly expressed mRNAs within the hypothalamus and is expressed in the ARC (with POMC), LHA, and PVN. Food deprivation reduces ARC expression of CART.[30]

C. Brainstem Regulators of Appetite

It is now becoming evident that the brainstem can integrate much the same signals as have been shown to modulate hypothalamic activity. Leptin receptors are expressed at several strategically located brainstem sites, and selective stimulation of these receptors suppresses food intake at doses comparable with those used in forebrain injections.

Extensive reciprocal connections exist between the hypothalamus and brainstem, particularly the nucleus of tractus solitarius (NTS).[31] The brainstem plays an important role in the regulation of energy balance. The NTS is in close anatomical proximity to the area postrema, a circumventricular organ with an incomplete BBB. Like the ARC, the NTS is therefore in an ideal position to respond to peripheral circulating signals but in addition also receives vagal afferents from the gastrointestinal tract and afferents from the glossopharyngeal nerves.[21]

Classical concept of regulation of feeding describes the two systems balancing each other: The hypothalamus, monitoring the periphery for signals alerting central circuits to diminishing energy stores, and the brainstem, receiving oral and gastrointestinal information as an online signal of the amounts and qualities of the food that was being ingested. This arrangement would allow the hypothalamus to function as a long-term control orchestrating *meal initiation* and the brainstem served as a short-term control for *meal termination*[33] (Fig. 16.6).

Brainstem Appetite-regulating Mediators

i. GLP-1

GLP-1 forms the major brainstem circuit regulating energy homeostasis. In the CNS, GLP-1 is synthesized exclusively in the caudal NTS, and these preproglucagon neurons also express leptin receptors. GLP-1 secreting neurons project widely, but particularly to the PVN and DMN, with fewer projections to the ARC. GLP-1 receptors are found both within the hypothalamus (PVN, DMH, and supraoptic nucleus) and in the brainstem (subfornical organ, organum vasculosum laminae terminalis, and area postrema). Central administration of GLP-1, either into the third or fourth ventricle, potently reduces fasting and NPY-induced food intake Blockade of endogenous GLP-1 with the GLP-1 receptor antagonist increased food intake.[35]

ii. Others

NPY neurons from the brainstem project forward to the PVN[32], and extracellular NPY levels within the NTS are modulated by feeding. There is also evidence for a separate melanocortin system in the NTS. POMC-derived peptides are synthesized in the NTS of the rat and caudal medulla of humans.[20]

III. REWARD SYSTEM AND REGULATION OF APPETITE

Even in the absence of an energy deficit, the hedonic nature (hedonic—giving a sensation of pleasure) of food may act as a stimulus to feeding. However, there is interaction between nutritional status and the sensation of reward, as the subjective palatability of food differs between the fed and fasting states. Signals of energy status, such as leptin, are able to modulate reward or hedonic pathways.

The reward circuitry is complex, involving interactions between several signaling systems, including opioid and dopaminergic systems. Opioids play an important role. Mice lacking either enkephalin or β-endorphin lose the reinforcing property of food, regardless of the palatability of the food tested. However, the reinforcing effect is regained in fasted animals; thus homeostatic mechanisms can override the hedonistic pathways. The nucleus accumbens (NAc) forms an important part of the reward circuit. Microinjection of opioid agonists into the NAc stimulates the preferential consumption of highly palatable sucrose and fat. Conversely, opioid antagonists administered into the NAc reduce sucrose ingestion rather than less palatable substances.[35]

Conclusion

As has been outlined above, organismal energy balance is regulated by many factors through complex and multi-level integration processes that involve multiple neuronal circuits. The homeostatic system is basically influenced by long-term (leptin and insulin) and short-term (GI hormones and vagal inputs) signals that act in concert to engage specific neuronal circuits in the hypothalamus and brainstem aimed at fulfilling whole-body metabolic needs. In addition to this homeostatic module, hedonic, and emotional stimuli are involved in a non-homeostatic manner. Therefore, these complex interactions between the homeostatic and non-homeostatic systems culminate in coordinated appetite and energy balance regulation through the modulation of endocrine, autonomic, and behavioral outputs.[36]

References

1. Mallick HN. Review of prof. BK Anand's scientific study following his discovery of feeding center. Indian J Physiol Pharmacol 2001; 45: 269–95.
2. Campfield LA, Smith FJ. Transient declines in blood glucose signal meal initiation. Int J Obes. 1990; 14: 15–31.
3. Bray GA. What's in a name: The glucostatic or glucodynamic hypothesis for regulation of food intake? Obes Res. 1996; 4: 489–492.
4. Kennedy GC. The role of depot fat in the hypothalamic control of food intake in the rat. Proc R Soc B Biol Sci. 1953; 140: 578–92.
5. Perry B, Wang Y. Appetite regulation and weight control: the role of gut hormones. Nutrition and Diabetes. 20122; e26.doi:10.1038/nutd.2011.21
6. Williams G, Bing C, Cai XJ, et al. The hypothalamus and the control of energy homeostasis: different circuits, different purposes. Physiol Behav. 2001; 74: 683–701.
7. Cone RD, Cowley MA, Butler AA, et al. The arcuate nucleus as a conduit for diverse signals relevant to energy homeostasis. Int J Obes Relat Metab Disord. 2001; 25: S63–67.
8. Maffei M, Halaas J, Revussin E, et al. Leptin levels in human and rodent: measurement of plasma leptin and ob RNA in obese and weight-reduced subjects. Nat Med. 1995; 1: 1155–61.
9. Halaas JL, Gajiwala KS, Maffei M, et al. Weight-reducing effects of the plasma protein encoded by the obese gene. Science. 1995; 269: 543–46.
10. Elias CF, Aschkenasi C, Lee C, et al. Leptin differentially regulates NPY and POMC neurons projecting to the lateral hypothalamic area. Neuron. 1999; 23: 775–86.
11. Cheung CC, Clifton DK, Steiner RA, et al. Pro-opiomelanocortin neurons are direct targets for leptin in the hypothalamus. Endocrinology.1997; 138: 4489–92.
12. Halaas JL, Gajiwala KS, Cohen SL, et al. Weight-reducing effects of the plasma protein encoded by the obese gene. Science. 1995; 269: 543–46.
13. Strobel A, Issad T, Camoin L, et al. Aleptin missense mutation associated with hypogonadism and morbid obesity. Nat Genet. 1998; 18: 213–15.
14. Farooqi IS, O'Rahilly S. 20 years of leptin: Human disorders of leptin action. J Endocrinol. 2014; 223: T63–70.
15. Fruebis J, Tsao TS, Javorschi S, et al. Proteolytic cleavage product of 30-kDa adipocyte comple-

ment-related protein increases fatty acid oxidation in muscle and causes weight loss in mice. Proc Natl Acad Sci USA. 2001; 98: 2005–10.

16. Jamaluddin MS, Weakley SM, Yao Q, et al. Functional roles and therapeutic considerations for cardiovascular disease. Br J Pharmacol. 2012; 165: 622–32.

17. Schwartz MW, Figlewicz DP, Baskin DG, et al. Insulin in the brain: a hormonal regulator of energy balance. Endocr Rev. 1992; 13: 387–414.

18. Woods SC, Lotter EC, McKay LD, et al. *Chronic intracerebroventricular infusion of insulin reduces food intake and body weight of baboons.* Nature. 1979; 282: 503–5.

19. Menendez JA, Atrens DM. Insulin and the paraventricular hypothalamus: modulation of energy balance. Brain Res.1991; 555: 193–201.

20. Strubbe JH, Mein CG. *Increased feeding in response to bilateral injection of insulin antibodies in the VMH.* Physiol Behav. 1977; 19: 309–13.

21. Stanley S, Wynne K, McGowan B, et al. Hormonal Regulation of Food Intake. Physiol Rev. 2005; 85: 1131–58.

22. Suzuki K, Jayasena CN, Bloom SR. The gut hormones in appetite regulation. J Obesity. 2011, Article ID 528401,http://dx.doi.org/10.1155/2011/528401

23. Nakazato M, Murakami N, Date Y, et al. A role for ghrelin in the central regulation of feeding. Nature. 2001; 409: 194–98.

24. Flint A, Raben A, Ersboll AK, et al. The effect of physiological levels of glucagon-like peptide-1 on appetite, gastric emptying, energy and substrate metabolism in obesity. Int J Obesity Related Metab Disorders. 2001; 25: 781–92.

25. Cohen MA, Ellis SM, Le Roux CW, et al. Oxyntomodulin suppresses appetite and reduces food intake in humans. J Clin Endocrinol Metab. 2003; 88: 4696–4701.

26. Schwartz MW, Woods SC, Porte D Jr, et al. Central nervous system control of food intake. Nature. 2000; 404: 661–71.

27. Shor-Posner G, Azar AP, Insinga S, Leibowitz SF. Deficits in the control of food intake after hypothalamic paraventricular nucleus lesions. Physiol and Behav. 1985; 35: 883–90.

28. Borowsky B, Durkin MM, Ogozalek K, et al. Antidepressant, anxiolytic and anorectic effects of a melanin-concentrating hormone-1 receptor antagonist. Nat Med. 2002; 8: 825–830.

29. Xu B, Goulding EH, Zang K, et al. Brain-derived neurotrophic factor regulates energy balance downstream of melanocortin-4 receptor. Nat Neurosci. 2003; 6: 736–42.

30. Hahn TM, Breininger JF, Baskin DG, et al. Coexpression of Agrp and NPY in fasting-activated hypothalamic neurons. Nat Neurosci. 1998; 1: 271–72.

31. Kristensen P, Judge ME, Thim L, et al. Hypothalamic CART is a new anorectic peptide regulated by leptin. Nature. 1998; 393: 72–76.

32. Ricardo JA, Koh ET. Anatomical evidence of direct projections from the nucleus of the solitary tract to the hypothalamus, amygdala, and other forebrain structures in the rat. Brain Res. 1978; 153: 1–26.

33. Broberger G. Brain regulation of food intake and appetite: molecules and networks. Jour Intern Med. 2005; 258: 301– 27.

34. Turton MD, O'Shea D, Gunn I, et al. A role for glucagon-like peptide-1 in the central regulation of feeding. Nature. 1996, 379: 69–72.

35. Katherine A. Simpson, Bloom SR, et al. Hypothalamic regulation of appetite. Expert Rev Endocrinol Metab. 2008; 3: 577–92.

36. Ahima RS, Antwi DA. Brain regulation of appetite and satiety. Endocrinol Metab Clin North Am. 2008; 37: 811–23.

Chapter 17

Sleep-Wakeful Cycles

INTRODUCTION

Sleep is a naturally recurring state of mind and body, characterized by altered consciousness, relatively inhibited sensory activity, inhibition of nearly all voluntary muscles, and reduced interactions with surroundings. It is distinguished from wakefulness by a decreased ability to react to stimuli, but is more easily reversed than the state of being comatose. The study of sleep physiology was initiated in 1916, when von Economo, a Viennese neurologist, began to see patients with a new type of encephalitis, eventually called encephalitis lethargica. Patients suffering from the disorder seemed to fall asleep whilst eating or working. The disease swept through Europe and North America. Although the virus that caused it was never identified, von Economo was able to identify the lesions in brainstem. On the basis of neuropathological studies, in 1929, von Economo postulated a sleep center located in the border zone of midbrain and diencephalon, with distinct parts for wakefulness and sleep. After the World War II, experiments revealed that lesion of the hypothalamus induce sleep. Giuseppe Moruzzi (1910–1986) and Horace Magoun (1907–1991) identified the existence of reticular formation, which regulates the awaking/sleeping cycle on the level of the forebrain. Only during the past few decades, the accuracy of von Economo's[1] observations come to be appreciated, as the key components of the sleep-wake regulatory system.

In 1929, Berger[2] recorded the electrical activity of the brain as series of waves, which he called electroencephalogram. This discovery of the electroencephalography (EEG), gave further impetus to the studies on sleep. In 1935, Bremer[3] demonstrated that a section at the level of upper border of midbrain results in EEG change from awake (high frequency, low voltage) pattern to EEG pattern of deep sleep (low frequency high voltage) (Fig. 17.1).

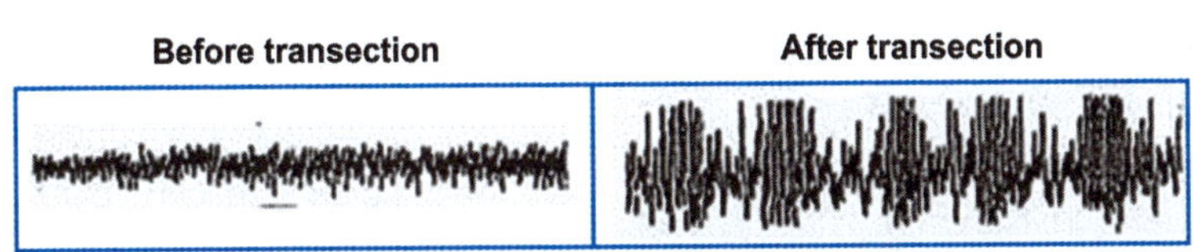

Fig. 17.1: Effect of a transection at the level of upper border of midbrain on the EEG pattern

The stages of sleep were first described in 1937 by Loomis and his coworkers, who separated the different electroencephalographic features of sleep into five levels (A to E), representing the spectrum from wakefulness to deep sleep.[4] REM type of sleep was discovered in 1953.[5] The stages of sleep described by Loomis et al came to be known as non-REM (NREM) sleep.

NREM AND REM SLEEP CYCLES

A sleep episode begins with a short period of NREM stage 1, progressing through stage 2, followed by stages 3 and 4 and finally to REM. However, individuals do not remain in REM sleep the remainder of the night but, rather, cycle between stages of NREM and REM throughout the night. NREM sleep constitutes about 75 to 80% of total time spent in sleep, and REM sleep constitutes the remaining 20 to 25%. The average length of the first NREM-REM sleep cycle is 70 to 100 minutes. The second, and later, cycles are longer lasting—approximately 90 to 120 minutes. In normal adults, REM sleep increases as the night progresses and is longest in the last one-third of the sleep episode. As the sleep episode progresses, stage 2 begins to account for the majority of NREM sleep, and stages 3 and 4 may sometimes altogether disappear (Fig. 17.2).

Stages of NREM Sleep

NREM sleep was divided into four stages in the Rechtschaffen and Kales (R&K) standardization of 1968.[6] That has been reduced to three in the 2007 update by The American Academy of Sleep Medicine (AASM)[7] (Fig. 17.3).

Stage N1: Alpha waves are associated with this wakeful relaxation state and are characterized by a frequency of 8 to 13 cycles per second (alpha waves). NREM stage 1 sleep serves a transitional role in sleep-stage cycling. This stage usually lasts 1 to 7 minutes in the initial cycle and is easily interrupted by a disruptive noise. Brain activity on the EEG in stage N1 is a transitions from wakefulness (marked by rhythmic **alpha waves**) to low-voltage, mixed-frequency **theta waves**. The sleeper may be aware of sounds and conversations, but feels unwilling, rather than unable, to respond to them. A person awakened during this period will often believe he has never slept at all. Typically, this stage represents only about 5% of the total sleep time.

Stage N2 is the first unequivocal stage of sleep, during which muscle activity decreases still further and conscious awareness of the outside world begins to fade completely. Stage 2 sleep lasts approximately 10 to 25 minutes in the initial cycle and lengthens with each successive cycle, eventually consti-

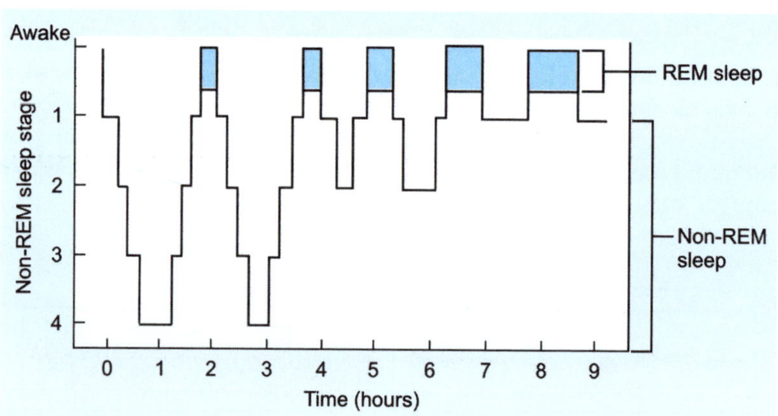

Fig. 17.2: A typical night of sleep in a young human adult showing cycles of non-REM and REM sleep

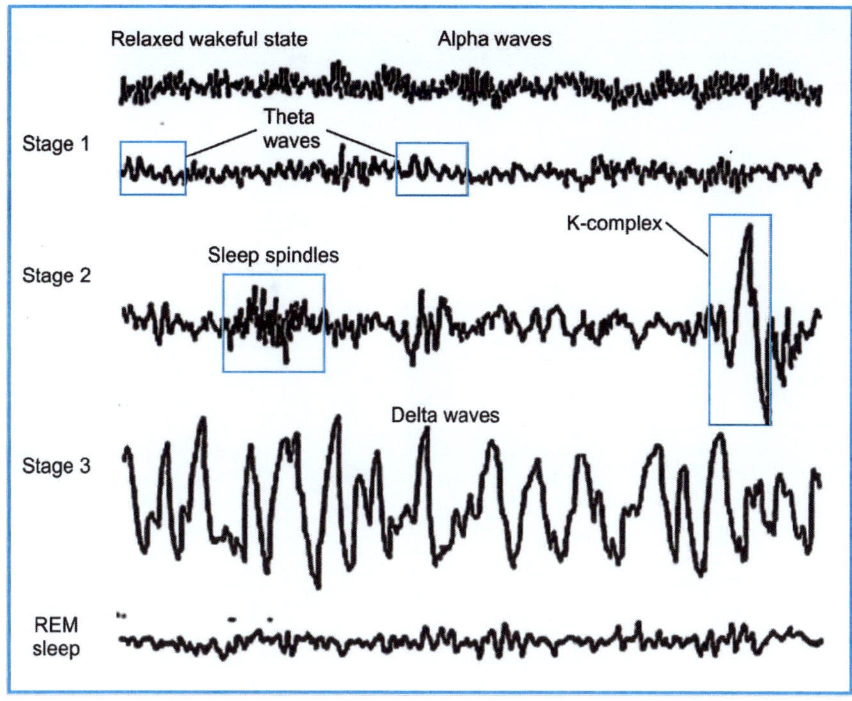

Fig. 17.3: EEG record during wakefulness, NREM and REM sleep

tuting between 45% and 55% of the total sleep episode. An individual in stage 2 sleep requires more intense stimuli than in stage 1 to awaken. Brain activity on an EEG shows relatively low-voltage, mixed-frequency activity **theta waves** characterized by the presence of **sleep spindles** (short bursts of brain activity in the region of 12–14 Hz) and **K-complexes** (short negative high voltage peaks, followed by a slower positive complex, and then a final negative peak).

Stage N3 (sleep stages 3 and 4 in earlier classification) is referred to as slow-wave sleep (SWS), most of which occurs during the first third of the night. The EEG shows high-voltage, slow-wave **delta activity**. The arousal threshold is highest of all the NREM stages. This stage of NREM sleep lasts approximately 20 to 40 minutes in the first cycle and gradually decreases during the subsequent cycles. It makes up about 10 to 15% of total sleep. During this stage, the subject is even less responsive to outside environment, essen-

tially unaware of any sounds or other stimuli. Other physiological indicators, such as body temperature, breathing rate, heart rate and blood pressure, are all at their lowest levels. Dreaming is more common during this stage than in the other non-REM sleep stages, although not as common (nor as vivid and memorable) as during REM sleep.

REM Sleep

REM sleep is defined by the presence of desynchronized (low-voltage, mixed-frequency) brain wave activity, muscle atonia, and bursts of rapid eye movements. **Saw-tooth waveforms, theta activity** (3 to 7 cycles per second), and **slow alpha activity** also characterize REM sleep. During the initial cycle, the REM period may last only 1 to 5 minutes; however, it becomes progressively prolonged as the sleep episode progresses. There are numerous differences in the body functions between NREM and REM sleep[8] (Table 17.1).

Table 17.1: Differences between NREM and REM sleep

Characteristic	NREM sleep	REM sleep
Proportion of sleep cycle	75–80%	20–25%
Ability to arouse	Easy	Difficult
Heart rate	Decreased	Irregular especially during dreaming
Blood pressure	Decreased, steady	Irregular, high values during dreaming
Respiratory rate	Decreased	Increased
Airway resistance	Greater than awake	Greater than awake
Cough reflex	Suppressed	Suppressed
Sympathetic activity	Decreased	Increased
Muscle tone	Reduced	Markedly reduced
Babinski sign	+/–	+
BMR	Reduced	Similar to awake state
Growth hormone secretion	Increased	Decreased
Pupils	Constricted	Constricted
Eyeballs	Rolled up, steady	Rapid eyeball movement
Dreaming	Occurs, not recallable	Occurs, recallable
Sexual arousal (penile erection, clitoral engorgement)	Absent	Present
Body temperature regulation	Maintained at slightly lower level	Impaired

MECHANISMS OF WAKEFULNESS AND SLEEP

I. Wakefulness

Wakefulness is produced by a complex interaction between multiple neurotransmitter systems arising in the brainstem and ascending through the midbrain, hypothalamus, thalamus and basal forebrain. The posterior hypothalamus plays a key role in the maintenance of the cortical activation that underlies wakefulness. Several systems originating in this part of the brain control the shift from wakefulness into sleep and sleep into wakefulness.

Wakefulness and EEG desynchronization require excitatory input into the forebrain. Being awake depends on the discharge activity in several, apparently redundant, parallel ascending neurotransmitter pathways, which include glutamate, acetylcholine (ACh), and the monoamines (serotonin and norepinephrine). None of these systems is absolutely necessary for the expression of wakefulness, but all appear to contribute.[9] Many of the corresponding cell bodies for these systems are located in the brainstem. Two studies were critical for the discovery that wakefulness entails forebrain activation by the brainstem. Bremer in 1935, reported that a complete transection above the brainstem produces a preparation that remains in a state resembling normal sleep.[3] Subsequently, Moruzzi and Magoun in 1949 demonstrated that electrical stimulation of the reticular core of the upper brainstem produced immediate and long-lasting EEG desynchronization in a previously sleeping preparation.[10] The latter study thus established upper brainstem as the location of an *ascending reticular activating system* (ARAS) (Fig. 17.4) that maintained wakefulness although the neurons that were responsible remained unknown for many decades.

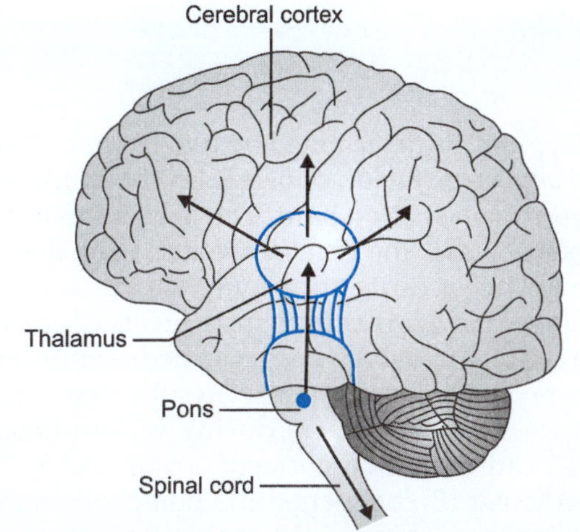

Fig. 17.4: The concept of ascending reticular activating system

The neurotransmitter systems that comprise the ARAS were subsequently identified. However, when these transmitter systems were described individually, the term ARAS became less accurate, especially because some of the cell bodies were found to lie outside the core of the reticular formation. The *ascending activating system* (AAS) is a general term that is now more typically used.

A major component of the AAS was determined by Steriade et al[11], who recorded the activity of neurons in chronic, freely moving cats. They identified neurons projecting the pons-midbrain junction to thalamus, whose discharge rates increased before the first sign of the change to an EEG desynchronized state. Later studies established that many of these cells which project to the thalamus (Th) were *cholinergic* and localized them to the laterodorsal tegmentum/pedunculopontine tegmental (LDT/PPT) region. These neurons remain active whenever the EEG is desynchronized during wakefulness and REM sleep. Conversely these neurons have low discharge rates during NREM sleep.[11]

Besides the LDT/PPT cholinergic system, many other significant brainstem reticular neuronal projections to the thalamus, utilizing monoaminergic projections that contain *norepinephrine* from the locus coeruleus *(LC) and serotonin* from the dorsal raphe (DR) also play a role and are part of the AAS[12] (Fig. 17.5). The *monoaminergic cells* send widespread projections throughout the forebrain to the cortex as well as the thalamus. Importantly, they *demonstrate significant state-related changes in discharge rates, being highest during wakefulness, decreasing during NREM sleep and becoming virtually quiescent during REM sleep.* A similar state-related change in discharge rate is also shown by cells that contain another monoamine, *histamine.* These cells are located uniquely in the *tuberomammillary nucleus (TMN)* of the caudolateral hypothalamus. *Histaminergic cells,* like the other monoaminergic components of the AAS, have been found to be *a substrate of wakefulness* on the basis of pharmacological, electrophysiological, lesion and stimulation studies.[13] A *second ascending cholinergic projection,* originating in the nucleus basalis of Meynert in the basal forebrain (BF), sends widespread projections to the cortex and thalamus[14] (Fig. 17.5). This cholinergic system, like that from the LDT/PPT, is also important for EEG desynchronization, receives brainstem input.

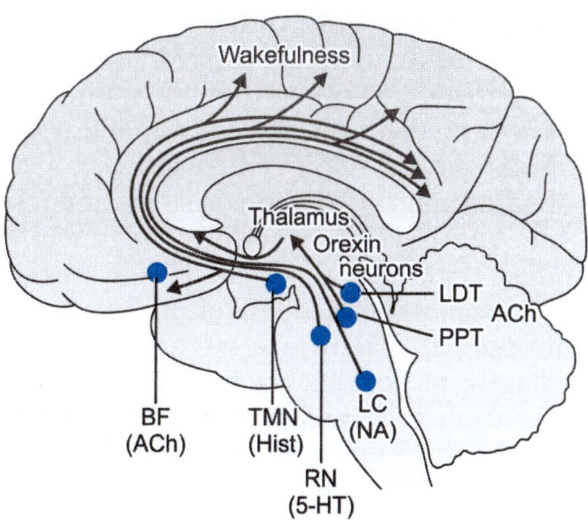

Fig. 17.5: Pathways and their neurotransmitters of ascending activating system (AAS) involved in wakefulness (BF: Basal forebrain; TMN: Tuberomammillary nucleus; RN: Raphe nuclei; LC: Locus coeruleus; LDT: Lateral dorsal tegmental nucleus; PPT: Pedunculopontine tegmental nucleus)

Thus, there are two major relay pathways for brainstem activation of cortex, one through the thalamus and one through the BF and nearby hypothalamus, as well as the direct projections of LC and DR to cortex.

Another relatively recent addition to the list of neurotransmitters that contribute to wakefulness is the *neuropeptide, orexin.* Orexin-containing cell bodies are located uniquely in the lateral hypothalamus and send widespread projections throughout the neuraxis.[15] The importance of orexin to the control of sleep and wakefulness was discovered serendipitously when narcolepsy was found to be caused by the absence of an orexin signal, either through the loss of the receptors, the neurons or the neuropeptide. Orexin supports wakefulness through excitatory projections to other components of the AAS, though its role in the regulation of sleep and wakefulness is likely to be more critical than the enhancement of general arousal and EEG desynchronization. *Orexin is important for controlling the gate from one state to another (arousal to sleep and vice versa).*

Studies involving *stimulation* of the brain areas and neurotransmitter systems comprising the AAS consistently report EEG activation and wakefulness as a result. These studies include both older techniques of electrical stimulation and infusion of specific inactivating pharmacological agents. In contrast to the stimulation experiments, studies where *localized inactivation* of individual neurotransmitter systems or nuclei of the AAS have been performed, generally produce relatively minor changes in cortical EEG or the amount of wakefulness in a 24-hour period. There are several possible explanations for this dichotomy between stimulation and inactivation experiments. First, the AAS systems are strongly interconnected, mutually excitatory to each and converge onto common effector systems at the level of thalamic and cortical neurons. Second, there is considerable redundancy in the system, and inactivation of any individual component of the system is compensated for by the other systems.[16]

II. Sleep

i. The Neurobiology of NREM Sleep

Sleepiness is determined by two factors. One, a *homeostatic* factor, depends on the duration since the last sleep bout. The longer the time spent awake, the greater the drive for sleep. Evidence points to adenosine, a neurotransmitter with extracellular levels coupled to cellular metabolism as one mediator of the homeostatic sleep response. Results show that *adenosine* accumulates during wakefulness and can inhibit components of the AAS, in particular BF cholinergic and non-cholinergic neurons.[17] This would make it a *potential homeostatic sleep factor.* The second factor is *circadian,* which varies with a 24-hour periodicity and is independent of the amount of preceding sleep or wakefulness. The circadian influence is modulated through the suprachiasmatic nucleus (SCN) with inhibitory projections to components of the AAS. The discharge activity of cells in the SCN depends on the circadian cycle. Together, the homeostatic and circadian factors modulate the need for sleep and influence the balance between alertness and sleepiness.

A sleep-promoting area: The existence of a *sleep-promoting area* in the brain was first implied from observations of the neuropathology of encephalitis by von Economo, who published a correlation between damage to the anterior hypothalamus and profound insomnia in 1930.[18] The importance of this region was later confirmed by stimulation, electrophysiological and histological studies, which identified the critical area as being within the hypothalamic preoptic area of the thalamus. In particular, a group of cells that are concentrated in the *ventrolateral preoptic area (VLPO)* has been shown to be active specifically during sleep and to contain the inhibitory neurotransmitters, γ-*aminobutyric acid (GABA)* and *galanin* (an inhibitory, hyperpolarizing neuropeptide) (Fig. 17.6). VLPO cells show their first signs of activity during drowsiness and continue to discharge selectively throughout NREM sleep. They are

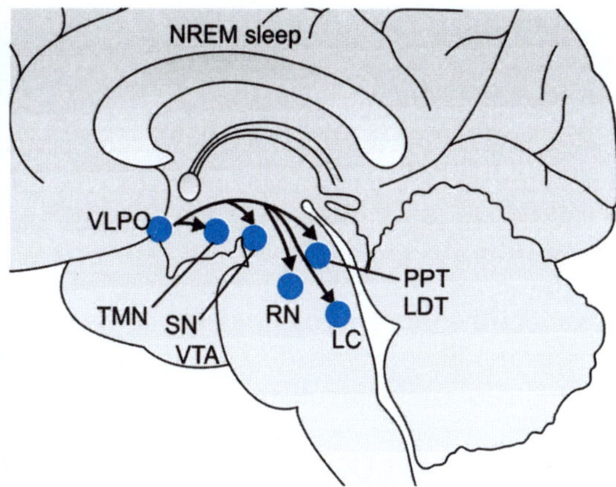

Fig. 17.6: Mechanism of sleep.VLPO sends inhibitory impulses to brainstem centres concerned with arousal (VLPO: Ventrolateral preoptic nucleus; TMN: Tuberomammillary nucleus; RN: Raphe nuclei; LC: Locus coeruleus; LDT: Lateral dorsal tegmental nucleus; PPT: Pedunculopontine tegmental nucleus; VTA: Ventral tegmental area; SN: Substantia nigra)

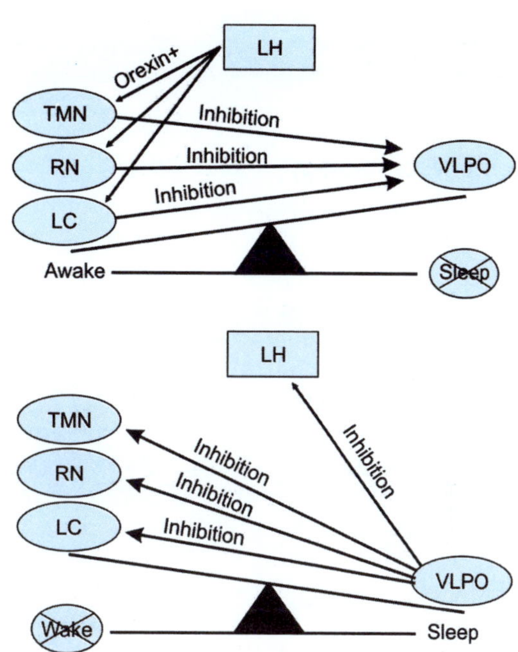

Fig. 17.7: The flip-flop model of sleep and wakefulness (VLPO: Ventrolateral preoptic nucleus; TMN: Tuberomammillary nucleus; RN: Raphe nuclei; LC: Locus coeruleus; LH: Lateral hypothalamus)

inhibited by the neurotransmitters of the AAS and also send direct inhibitory projections to all components of the AAS.

Most sleep-active neurons in the VLPO are believed to be active during both NREM and REM sleep. Many of the VLPO neurons are activated by sleep-inducing factors including adenosine and prostaglandin D_2. These neurons are sensitive to warmth and heating this area of the brain increases their activity. VPLO neurons inhibit tuberomammillary nucleus (TMN), dorsal raphe nucleus (DRN) and locus coeruleus (LC) as well as inter-neurons of the lateral dorsal tegmental/pedunculopontine tegmental (LDT/PPT) region (Fig. 17.6). The neurons in the VLPO receive inhibitory projections from the RN, LC and TMN. Destruction of the VLPO impairs sleep.[11]

A "Flip-flop switch" between sleep and wakefulness (Fig. 17.7): The ability to remain in a stable period of sleep or wakefulness is a result of what scientists call "mutual inhi-bition" between the wake-promoting neurons and the sleep-promoting neurons. So, for example, the areas of the brain that maintain wakefulness by activating the cortex also inhibit VLPO neurons. Conversely, when VLPO neurons fire rapidly and induce sleep, they also inhibit activity in the arousal centers such as RN and LC.[19]

Transitions between the stable states of wakefulness and sleep occur relatively quickly, often in just seconds. Some rese-archers have compared the neurological mechanism that controls these rapid transitions to the "flip-flop switch" in an electrical circuit. In the brain, the mechanism that maintains stability through mutual inhibition is triggered by changes in factors such as the body's drive for sleep or the circadian alerting signal. When one of these forces becomes strong enough, it drives the transition to the opposite state.[19] The same "flip-flop switch" analogy also describes the brain mechanisms involved in switching between REM and non-REM sleep. However, different neurotransmitters and different groups of neurons in the brainstem are involved in the transitions between REM and NREM sleep.[19]

ii. The Neurobiology of REM Sleep

The neurobiology of REM sleep is yet to be fully elucidated. After a series of transection studies, Jouvet and collaborators localized the neurons required for the generation of REM sleep to the brainstem.[20] A transection made just above the junction of the pons and midbrain produced a state in which the periodic occurrence of REM sleep signs could be found in the isolated brainstem while, in contrast, recordings in the isolated forebrain showed no sign of REM sleep. Having established the importance of the brainstem in REM sleep, subsequent work has provided a more detailed picture of the circuits controlling the REM sleep rhythm. As in humans, the cardinal signs of REM sleep in all species in which the state occurs include EEG desynchronization, rapid eye movements and muscle atonia. Importantly, these components can be dissociated, indicating that each is under the mechanistic control of a different groups of neurons, which have been found to be especially concentrated in the pontine reticular formation (RF). These cell groups have been termed REM sleep effector neurons.[17]

REM sleep effector neurons: Intracellular recordings of effector neurons have shown that they remain in a relatively hyperpolarized state and generate almost no action potentials during NREM sleep, but that they begin to depolarize even before the occurrence of the first sign of EEG desynchronization, that signals the approach of REM sleep. Then, as depolarization of the pontine RF neurons proceeds and the threshold for action potential production is reached, they begin to discharge. A high rate of discharge is maintained throughout the REM sleep episode, due to continuing depolarization.

These different groups of effector neurons, that control the components of REM sleep, are centered in the pontine RF. This is an anatomically complex region and hence detailed separation and demarcation between the groups is only approximate.[11]

- Eye movements are controlled by a group of medial pontine RF neurons.
- Muscle atonia by a group of neurons in the lateral pontine RF, including the ventrolateral and dorsolateral pontine RF. These send inhibitory projections to the motor neurons in spinal cord.
- EEG desynchronization by neurons in an extensive pontomesencephalic field, which includes the LDT/PPT cholinergic neurons. Additionally, BF cholinergic neurons affect EEG desynchronization.

REM sleep modulatory circuit in the hypothalamus: Although the pons is critical for the expression of REM sleep, recent research has suggested that melanin-concentrating hormone/GABAergic cells in the lateral hypothalamus "gate" REM sleep.[21] Lateral hypothalamus melanin-concentrating hormone (MCH)-expressing neurons are active during REM sleep. Acute activation of MCH neurons at the onset of REM sleep extended the duration of REM, but not non-REM, sleep episodes. It is now believed that activation of MCH neurons trigger REM sleep.[22]

References

1. von Economo C. Encepahlitis lethargica. *Wiener Klinische Wochenschrift*. 1917; 30: 581–85.
2. Berger H. Ueber das Elektrenkephalogramm des, Menschen," in ArchivfürPsychiatrie. 1929; 87: 527–70;
3. Bremer F. Cerveau "isole" etphysiologie du sommeil. C Roy Soc Biol. 1935: 118: 1235–41.
4. Loomis AL, Harvey EN, Hobart GA. Cerebral states during sleep as studied by human brain potentials. J Exp Psychol. 1937; 21: 127–44.
5. Aserinsky E, Kleitman N. Regularly occurring periods of eye motility and concomitant phenomenon during sleep. Science. 1953; 118: 273–74.
6. Rechtschaffen A, Kales A, editors. Los Angeles: Brain Information Service/Brain Research Institute, University of California. A manual of standardized terminology, techniques and scoring system of sleep stages in human subjects 1968.

7. Schulz H. Rethinking sleep analysis. Comment on the AASM Manual for the Scoring of Sleep and Associated Events. J Clin Sleep Med. American Academy of Sleep Medicine. 2008; 4: 99–103

8. Virk JS, Kotecha B. Otorhinolaryngological aspects of sleep-related breathing disorders. J Thorac Dis. 2016; 8: 213–23.

9. Jones BE. From waking to sleeping: neuronal and chemical substrates. Trends Pharmacol Sci. 2005; 26: 578–86.

10. Moruzzi G, Magoun HW. Brain stem reticular formation and activation of the EEG. Electroencephalogr. Clin. Neuro. 1949; 1: 455–73.

11. McCarley RW, Sinton CM. Neurobiology of sleep and wakefulness (2008). Scholarpedia, doi: 10.4249/scholarpedia.3313

12. McCormick DA. Cholinergic and noradrenergic modulation of thalamocortical processing. Trends Neurosci. 1989; 12: 215–21.

13. Monti JM. Involvement of histamine in the control of the waking state. Life Sci. 1993; 53: 1331–38.

14. Metherate R, Cox CL, Ashe JH. Cellular bases of neocortical activation: modulation of neural oscillations by the nucleus basalis and endogenous acetylcholine. J Neurosci. 1992; 12: 4701–11.

15. Ebrahim O, Howard RS, Kopelman MD, et al. The hypocretin/orexin system. J R Soc Med. 2002; 95: 227–30.

16. Brown RE, Basheer R, McKenna JT, et al. Control of Sleep and Wakefulness. Physiol Rev. 2012; 92: 1087–1187.

17. McCarley RW. Neurobiology of REM and NREM sleep. Sleep Med. 2007; 8: 302–30.

18. van Economo C. Sleep as a problem of localization. J Nerv Ment Disease. 1930; 71: 249–59.

19. Saper CB, Chou TC, Scammell TE. The sleep switch: hypothalamic control of sleep and wakefulness. Trends Neurosci. 2001; 24: 726–31.

20. Jouvet M. What does a cat dream about? Trends Neurosci. 1979; 2: 280–282.

21. Siegel JM. The neurobiology of sleep. Semin Neurol. 2009; 29: 277–96.

22. Jego S, Glasgow SD, Herrera CZ, et al. Optogenetic identification of a rapid eye movement sleep modulatory circuitin of the hypothalamus. Nature Neuroscience. 2013; 16: 1637–43.

Chapter 18

Circadian Rhythms

INTRODUCTION

Patrick and Gilbert in 1896, observed that during a prolonged period of sleep deprivation, sleepiness increases and decreases with a period of approximately 24 hours. In 1918, Szymanski showed that animals are capable of maintaining circadian rhythms or 24-hour activity patterns in the absence of external cues such as light and changes in environmental temperature. The circadian rhythm is an internal cycle in the body that occurs within a period of approximately 24 hours. In 1977, the International Committee on Nomenclature of the International Society for Chronobiology formally adopted the definition, which states: Circadian—relating to biologic variations or rhythms with a frequency of 1 cycle in 24 ± 4 hours [*circa* (about, approximately) and *dies* (day or 24 hours)].

A. CIRCADIAN RHYTHMS

Nearly all human bodily functions show significant daily variations including arousal, psychophysical performance, food and water consumption, metabolism, body temperature, heart rate, blood pressure, hormone production, etc. These variations may merely reflect different patterns of behavior imposed by the cyclical environment. A critical finding, however, is that when humans or experimental animals are held in temporal isolation, deprived of direct contact with the solar or social world, their daily cycles do not become disorganized or peter out. Rather, they continue with a period of about 1 day, hence circadian (Fig.18.1) and these rhythms can persist for weeks, even years, free-running with a high amplitude and exquisite precision.[1] Under natural conditions, the circadian clocks are synchronized (entrained) to

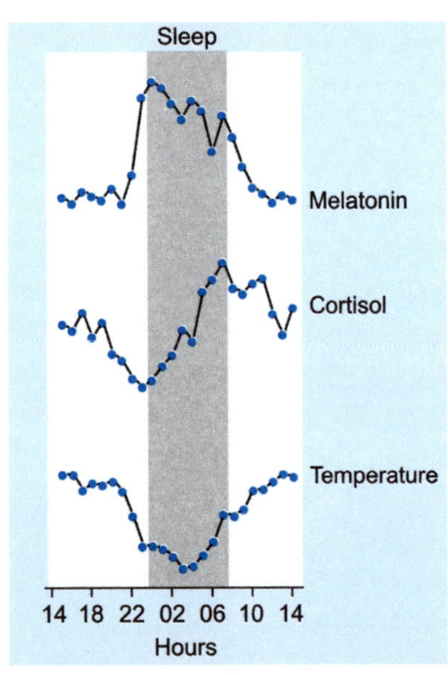

Fig. 18.1: Circadian rhythms in plasma melatonin, plasma cortisol and core body temperature

the external world, principally via light–dark cycles.[2] This ensures that they run with a period of exactly 24 hours, and so provide an internal representation of solar time phase locked to dawn and dusk. The biological advantage of such entrained clocks is that they enable the individual to anticipate and thereby prepare for the challenges and opportunities of day and night and also seasonal changes. This pre-adaptation makes our bodies more efficient biological machines.[3]

A. The Central Circadian Oscillator: Suprachiasmatic Nucleus

The principal circadian clock in the brain of mammals consists of the paired supra-chiasmatic nucleus (SCN) at the base of the hypothalamus, above the optic chiasm on either side of the third ventricle. SCN neurons are subdivided into a *ventral 'core'* region, which also express vasoactive intestinal polypeptide (VIP) and receive direct inner-vation from the retina and areas of brainstem that are responsible for entrainment, and a *dorsal 'shell'* region containing arginine vasopressin positive cells, which appear to be the primary pacemaker, its output driving behavioral and other rhythms.[4,5] Damage to the SCN can render experimental animals arrhythmic and cause sleep disorders in pa-tients, whilst intracerebral grafts of perinatal SCN can restore behavioral circadian rhythms of SCN-ablated rodents.[5] Electrical firing rates within the SCN exhibit spontaneous circadian rhythms both *in vivo* and in isolated culture. The suprachiasmatic nuclei control endocrine cycles in particular and wider metabolic rhythms in general by two general means. First, via their anatomical connections to centers controlling sleep and wakefulness, the SCN determines the timing of sleep and thereby the timing of sleep-dependent events, e.g. the nocturnal secretion of prolactin and growth hormone.[1] Secondly, through their connections to the neuroendocrine and autonomic systems, the SCN clocks can drive hormonal and other rhythms independently of sleep; hence these rhythms, e.g. melatonin

and cortisol, continue under constant routines where subjects are prevented from sleeping. Many circadian rhythms exhibit intermediate effects, e.g. core body temperature cycles have greater amplitude when subjects sleep but are nevertheless clearly expressed even during sleep deprivation.[6] To be called circadian, a biological rhythm must meet these three general criteria.[6]

i. **The rhythm has an endogenous free-running period that lasts approximately 24 hours:** The rhythm persists in constant conditions (i.e. constant darkness) with a period of about 24 hours. The rationale for this criterion is to distinguish circadian rhythms from simple responses to daily external cues. A rhythm cannot be said to be *endogenous* unless it has been tested and persists in conditions without external periodic input.

ii. **The rhythms are entrainable:** The rhythm can be reset by exposure to external stimuli (such as light and heat), a process called entrainment. The external stimulus used to entrain a rhythm is called the Zeitgeber, or "time giver". Travel across time zones illustrates the ability of the human biolo-gical clock to adjust to the local time; a person will usually experience jet lag before entrainment of their circadian clock has brought it into sync with local time.

iii. **The rhythms exhibit temperature com-pensation.** In other words, they maintain circadian periodicity over a range of physiological temperatures. Many orga-nisms live at a broad range of tempera-tures, and differences in thermal energy will affect the kinetics of all molecular processes in their cells. In order to keep track of time, the organism's circadian clock must maintain roughly a 24-hour periodicity despite the changing kinetics, a property known as temperature com-pensation.

B. Peripheral Circadian Oscillators

A number of mammalian peripheral tissues appear to contain the molecular machinery necessary for circadian oscillations. For example, *in vitro*, fibroblast cell division exhibits a circadian pattern similar to the animal from which the cells have been harvested. Moreover, peripheral tissues exhibit damped circadian oscillations in the absence of the SCN. Such observations have led to the concept that besides SCN, peripheral tissues have their own oscillators, called peripheral circadian oscillators. The suprachiasmatic nucleus (SCN), reacting to the external environment, provides timing to the brain and to the whole organism. Its rhythmic signal to mainly hypothalamic structures results in a synchronized hormonal and autonomic output to the body that coordinates behavior and physiology. Consequently, the expression of clock genes in all organs having a rhythm, is dictated by the SCN. Together with these clock genes, a number of cellular processes follow a similar rhythm. That is why, it has been proposed that these events are driven at least, in part, by clock genes. Thus, the peripheral tissues are said to be driven by peripheral circadian oscillators (Fig. 18.2). Together, this forms a multiple oscillating system that interacts and under normal conditions is synchronized by the SCN. The autonomic and hormonal outputs from the SCN are examples of messages that are clearly targeted; the behaviors driven by the SCN are examples of

messages that may have more diffuse targets. For example, food intake and locomotor activity, which are normally driven by the SCN, have the capacity to drive the rhythm of clock genes in cells of the liver.[7]

C. Circadian Photoreception

Some new advances have been made regarding circadian photoreception. Besides image forming photopigments (rhodopsin and photopsin) present in the rods and cones, the retina contains another photopigment known as *melanopsin*. Melanopsin, a non-image forming photopigment, is present in a small percentage of ganglion cells. These light-sensitive cells are known as the *intrinsically photosensitive retinal ganglion cells* (ipRGCs). These cells are the primarily concerned with light input into circadian photoentrainment. Whereas other ganglion cells project to lateral geniculate body, the ipRGCs project directly to the SCN, where they help in the entrainment (synchronization) of the master circadian clock. From ipRGCs, the signals follow a pathway called the retinohypothalamic tract, leading to the SCN (Fig. 18.3). The SCN takes the information on the lengths of the day and night from the retina, interprets it, and passes it onto the pineal gland. In response, the pineal secretes the hormone melatonin. Secretion of melatonin peaks at night and ebbs during the day and its presence provides information about night-length.[8] The SCN-pineal gland pathway is described later in this chapter.

D. Biological Markers of Circadian Rhythms[9]

The classic phase markers for measuring the timing of a mammal's circadian rhythm are:

- Melatonin secretion by the pineal gland

- Core body temperature minimum

- Plasma cortisol

Melatonin is absent from the system or undetectably low during daytime. Its onset in dim light approximately at 21:00 can be

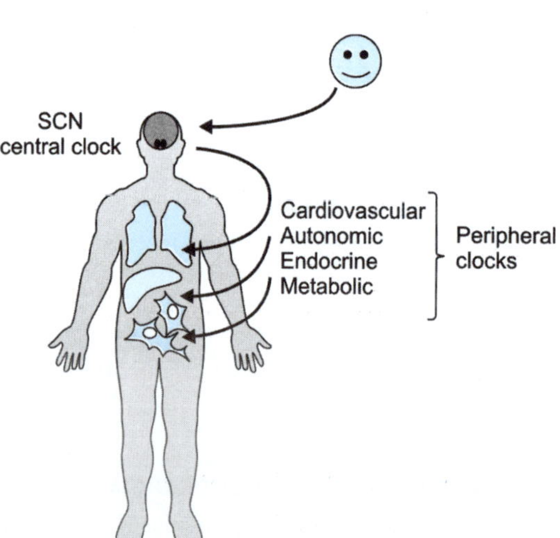

Fig. 18.2: Central and peripheral circadian oscillators

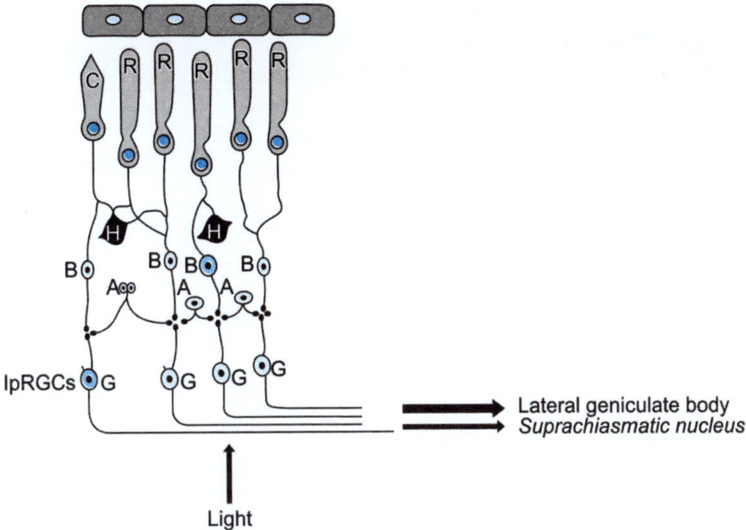

Fig. 18.3: The intrinsically photosensitive retinal ganglion cells (ipRGCs) and other ganglion cells (G) of the retina and their central projection. R: Rod; C: Cone; B: Bipolar cells; A: Amacrine cell; H: Horizontal cell

measured in the blood or the saliva. Peak plasma melatonin level is recorded at approximately 02:00 hours. The average human adult's temperature reaches its minimum at about 05:00 hours, about two hours before habitual wake time. Maximum core temperature is recorded in the evening. The circadian peak in glucocorticoid release is locked to the activity phase of the animal: It occurs in the early *morning in diurnal* and in the *early night in nocturnal animals*.

B. THE PINEAL GLAND: MELATONIN

i. Structure

The pineal gland develops from the roof of the diencephalon, behind the third ventricle. It measures approximately 8 mm × 4 mm in adults, and weighs approximately 0.10 to 0.18 g. The pineal gland has rich blood supply, and is the second organ (other than kidney) with the highest blood flow in relation to tissue mass. The endothelium of the pineal capillaries is fenestrated. The blood supply is greater during the night, probably related to increased production of melatonin. Removal of the superior cervical ganglion concomitantly decreases the metabolic activity of the gland and the blood flow decreases to 2/3 of normal.[10] It is pertinent to remember that the pineal gland is one of the circumventricular organs and lacks blood–brain barrier.

Histologically, two types of cells are found in the pineal gland: *Pinealocytes*, constituting about 95% of the cells, are large, light-staining round nuclei) and glial *astrocytes* having darkly-staining, elongated nuclei). Pinealocytes produce the hormone melatonin. Aside from the cells the pineal gland also contains "sand" (Fig. 18.4). These are calcium-containing concretions in the pineal parenchyma, which increase in size and number with age. These concretions are radiopaque, and, since the pineal is located in the midline of the brain, they provide a *good midline-marker in radiograms of the skull*. They have no known physiological function.

ii. Innervation

Impulses from retinal ipRGCs transmit information about light/darkness to SCN through retinohypothalamic tract. From SCN, information reaches cells of paraventricular nucleus, which in turn, make synapses with autonomic neurons in the intermediolateral column of the upper thoracic spine. From

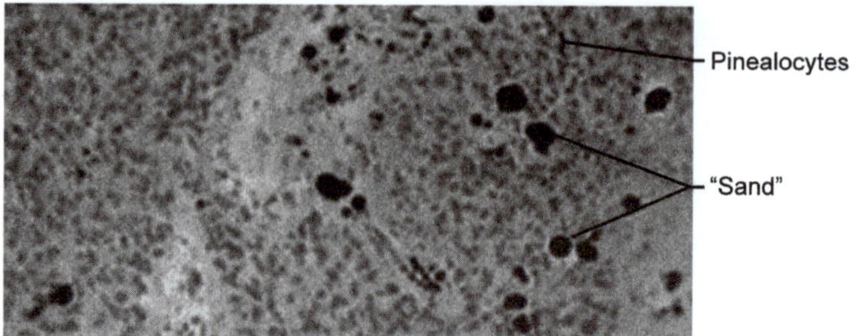

Fig. 18.4: Histological structure of pineal gland

there, the preganglionic axons leave the spinal cord, making synapses with the superior cervical ganglion neurons, where the postganglionic neurons arise and finally end in the pineal gland[11] (Fig. 18.5).

It is via this neural pathway that external light regulates pineal activity. Typically, during the day, light inhibits the production of melatonin. Conversely, the production of melatonin is stimulated at night. Sectioning of the sympathetic innervation or the use of α-adrenergic blockers inhibits the metabolic activity of the pineal cells. Postganglionic sympathetic nerve endings release norepinephrine in the pineal gland in accordance with a circadian rhythm, i.e. increasing secretion of the neurotransmitter during the dark phase.

iii. Melatonin Secretion

In humans, melatonin secretion increases soon after the onset of darkness, with peaks between two and four in the morning, and gradually falls during the second half of the night (Fig. 18.6). There are daily and seasonal modulations of a number of physiological processes which are related to melatonin. *Serum* concentrations of melatonin also vary considerably according to age. It is maximal in the first years of life, falling immediately preceding puberty and becoming minimal with old age. Thus, it is postulated that melatonin also has an important role in determining the physiological changes associated with life cycle (growth, maturation and aging).[11]

The melatonin levels can be interpreted as a signal of night-time to the organism; its increase indicates the onset of dusk and its decrease signals the onset of dawn. From this cyclic increase in plasma melatonin, information on the 24-hour light-dark cycle can be determined by the organism. Furthermore, as the melatonin onset and offset times vary

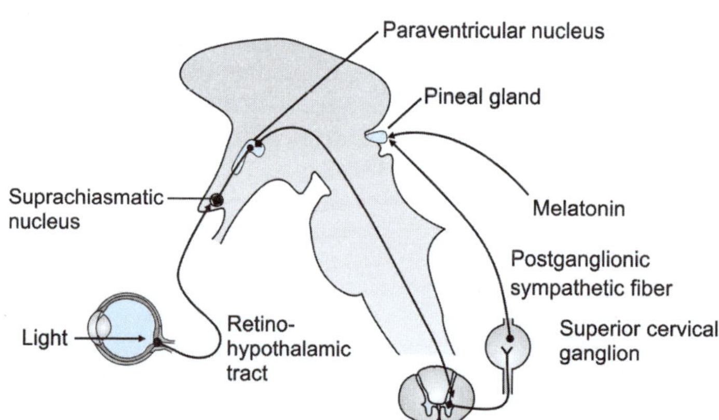

Fig. 18.5: Pathway through which melatonin is secreted in response to retinal exposure to light/darkness

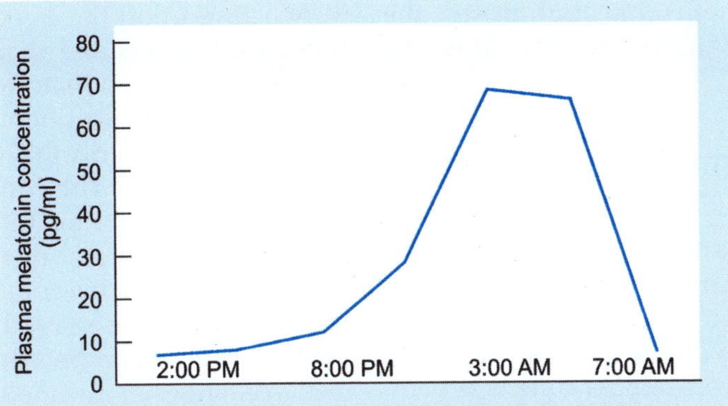

Fig. 18.6: Circadian rhythm of melatonin secretion

throughout the year, the melatonin signal may also reflect seasonal changes. Thus, from an evolutionary perspective, melatonin can been seen to promote adaptive restorative behavior.[12]

iv. Functions of Melatonin

1. **Sleep-wake cycles:** The circadian rhythm of sleep-wake state is determined by night-time release of melatonin by pineal gland. Therapeutically melatonin is used for the management of sleep disorders and jet lag.

2. **Role of melatonin on the immune system:** Most authors agree on an immunostimulating effect of melatonin. Earlier studies reporting that pinealectomized rats displayed a structurally modified thymus and that melatonin administration or pineal grafting prevented thymic involution.[13] *In vivo*, high exogenous doses of melatonin result in a general stimulation of the immune system. It increases T cell activity, lymphocyte growth, humoral responses, and may inhibit thymus involution with age. *In vitro*, melatonin also increases T helper and NK cell activities, the production of interleukin 2 and interferon gamma, and the expression of interleukin 1 mRNA in human monocytes.[14]

3. **Antioxidant/antiaging property of melatonin:** The publication of a revitalizing effect of melatonin or of youth pineal transplants to old mice created a general interest for melatonin as an antiaging/antioxidant molecule.[15] The use of oxygen in cell metabolism leads to the production of cytotoxic by-products that are reactive free radical species, which destroy macromolecules like DNA, lipids, and proteins leading to cell death via apoptosis. High doses of melatonin (in the micromolar range) are reported to neutralize most of these cytotoxic molecules, but especially the hydroxyl radical. In addition, melatonin is reported to stimulate the activity of various antioxidant enzymes, like superoxide dismutase or glutathione peroxidase.[16]

In view of the facts that (i) melatonin could be a very powerful antioxidant molecule, (ii) the production of melatonin decreases with age, and (iii) the free radical effects are involved in the processes of aging and cancer, it has been suggested that maintaining melatonin at a high level could slow age- and cancer-related alterations.[17] The anticarcinogenic effect of melatonin is best described *in vivo* and *in vitro* on the estrogen-responsive mammary tumors.[18] It may be stressed that most of these effects necessitate pharmacological doses of melatonin (in the micromolar range), whereas the physiological plasma melatonin concentrations are in the picomolar range. Recent studies, however, suggest that melatonin could display antioxidant properties even at physiological levels. Even if beneficial at high doses, the therapeutic effect of melatonin should not be ignored.[19]

C. CIRCADIAN RHYTHM IN CORE BODY TEMPERATURE

The body temperature of homoeothermic organisms including humans is regulated within narrow limits by a complex feedback system. Obviously, a "set point" exists, a value at which the core temperature is kept by a thermoregulatory center located within the hypothalamus. This thermoregulatory center consists of two groups of neurons. One of them activate the heat loss mechanisms when core temperature starts to rise; the other activates heat gain mechanisms when body temperature starts to fall. Mechanisms of heat production include metabolic heat production, physical activity or non-shivering thermogenesis. The gradient between the body core and skin temperature allows heat to be exchanged with the environment by means of convection and radiation. As a consequence, the core body temperature is maintained at a rather stable level, in spite of a wide range of ambient temperatures or during exercise.

The circadian changes in core temperature are probably due to a rhythmic input from the SCN acting upon the hypothalamic thermoregulatory centers and altering the thresholds for cutaneous vasodilatation and sweating.[20] Under resting conditions, in thermoneutral environments, mainly the mechanisms of heat loss are involved. The contribution of metabolic heat production to the observed circadian rhythm at rest has been found to be much less. It has been estimated that only about 25% of the observed thermal circadian rhythm at rest can be attributed to the circadian changes in metabolic heat production. In humans, there is a circadian rhythm of heat loss from the distal limbs, with rhythms of skin temperature and blood flow in these regions showing peaks in the late evening and minimal in the morning.[21] In subjects living a conventional lifestyle, circadian changes in core temperature are phased such that there is a plateau of temperature at about 14:00 to 20:00 hours, and a minimum at about 05:00 hours (Fig. 18.7). This rhythm reflects the combined effects of the body clock, sleep, and physical and mental activity. If sleep is prevented, but physical and mental activities and the timing and composition of meals are all maintained constant, even then a circadian rhythm of core temperature continues, with an amplitude that is reduced by about half. This reduction in amplitude indicates the effects of the sleep-wake cycle.[22]

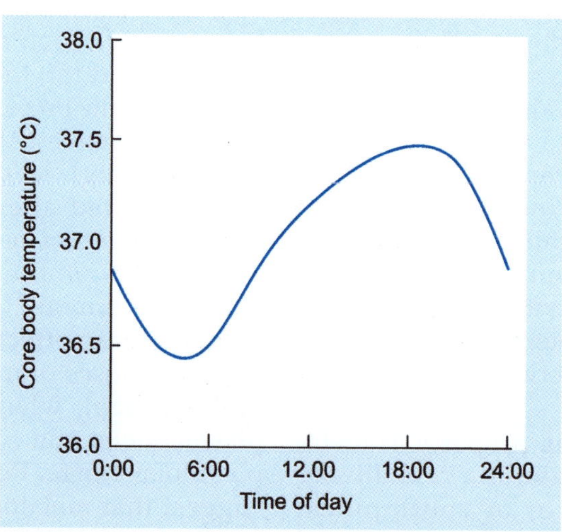

Fig. 18.7: Circadian rhythm of core body temperature

The circadian rhythm of core temperature is associated with widespread effects throughout the body, and it is believed to be one of the main determinants of the difference between the active "ergotropic" phase when awake and the "trophotropic" phase when asleep. The term ergotropic denotes a state of nervous system that favors activity and expenditure of energy, in contrast to trophotropic phase that favors reconstitution of energy stores. The rhythm of core temperature is correlated with rhythms of sleep propensity, physical performance, and mental performance. Also, it must, like melatonin, be considered to be an internal Zeitgeber for the organism, synchronizing different circadian functions.

D. CIRCADIAN RHYTHM IN CORTISOL SECRETION

Glucocorticoid hormones fulfil many different functions in body homeostasis and stress responses. Their secretion shows pronounced temporal regulation, with circadian rhythmicity.[23] The circadian peak in glucocorticoid release is locked to the activity phase of the animal: It occurs in the early morning in diurnal animals (active in daytime, e.g. birds and most mammals including humans) (Fig. 18.8), and in the early night in nocturnal animals (active at night, e.g. rats, owls). These rhythms are paralleled by similar rhythms in the adrenocorticotropic hormone (ACTH),

albeit less robust and of lower amplitude. The circadian aspects of the rhythmic release of glucocorticoids and their tropic hormone ACTH are under control of the SCN, since SCN lesions abolish the rhythms.[24] Although the rhythmic glucocorticoid secretion has been appreciated since many decades, the specific physiological relevance of the rhythm remains to be further elucidated.

E. CIRCADIAN REGULATION OF BLOOD PRESSURE

Day-night variations in blood pressure (BP) and heart rate (HR) are among the best known circadian rhythms of physiology. In humans, there is a 24-hour variation in BP with a sharp rise before awakening, the highest BP value is around midmorning[25] (Fig. 18.9). Concomitantly, many cardiovascular events, such as sudden cardiac death, myocardial infarction and stroke, display diurnal variations with an increased incidence in the morning. These events, as well as kidney albuminuria and progression to end-stage renal diseases, are relatively common in patients whose blood pressure fails to decline during the night, so-called non-dippers. Inverse dippers—BP rises instead of decreases at night—showed even higher cardiovascular mortality. These time-dependent effects are not just consequences of the sleep/wakefulness cycle or the rhythms in neuroendocrine constituents. They are also attributable to the intrinsic properties of the hearts and blood vessels whose functions show significant fluctuations during the course of the day.[26]

F. CIRCADIAN RHYTHMS AND HUMAN PERFORMANCE

It is generally accepted that human performance declines at night, when the body and mind desire rest.[26] A 'normal sleep period' is when the body and mind restore from the days exertions and prepare for the next period of wakefulness. The normal sleep period contains several sleep cycles of about 90 minutes, with four to five cycles in an 8-hour sleep period. Research indicates that several short

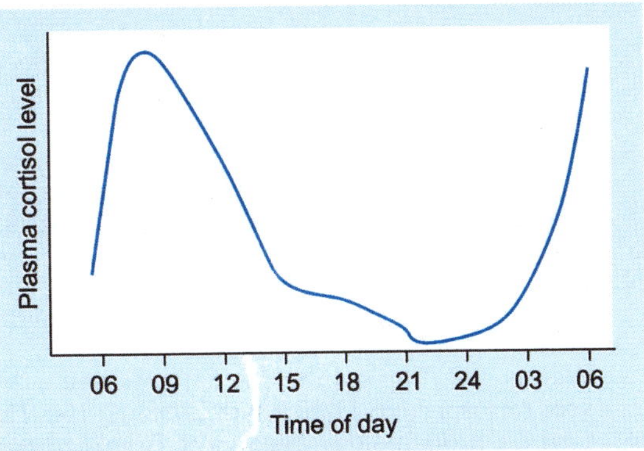

Fig. 18.8: Circadian rhythm in plasma cortisol level in humans

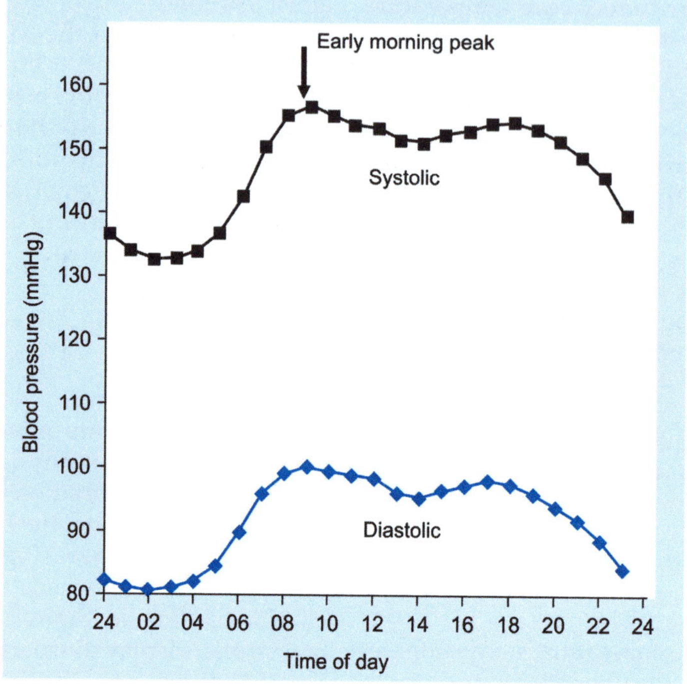

Fig. 18.9: Circadian rhythm in blood pressure

periods of sleep do not have the same mental restorative properties as one long period. In 1995, a study[27] reported that the quality of sleep experienced by night shift workers is significantly worse than a normal night sleep period. Post-night-shift sleep periods are reduced in length, since the wake up occurs as the body temperature starts to rise near midday. Furthermore, night shift workers often have two shorter periods of sleep, one after and one before the night shift. The amount of rapid eye movement (REM) sleep, a person experiences during a sleep cycle increases with the length of the sleep period. Research indicates that several short periods of sleep do not have the same mental restorative properties of one long period, due to people not getting as much cognitively-restoring REM sleep.

The NASA Ames Fatigue Countermeasures Program has underlined the fact that sleep loss, and circadian disruption lead to significant decrements in alertness and performance. A 1995 report on the effects of sleep deprivation[27] highlights human performance decrements include problems with vigilance, alertness, mental agility, ability to multi-task, decision-making and perceptive skills. These observations are especially relevant to commercial pilots and night-shift workers.

References

1. Czeisler CA, Kleiman EB. Circadian and sleep dependent regulation of hormone release in humans. Rec Prog Horm Res. 1999; 54: 97–130.
2. Pittenrigh CS. Temporal organization: reflections on Sarwanian clock-watchers. Ann Rev Physiol. 1993; 55: 16–54.
3. Woelfle JK, Dijk DJ, Ritz CA, et al. The adaptive value circadian clocks: an experimental assessment in cyanobacteria. Curr Biol. 2004; 14: 1481–86.
4. Hastings MH, Herzog ED. Clock genes, oscillators and cellular networks in suprachiasmatic nuclei. J Biol Rhyth. 2004; 19: 400–13.
5. Weiver DR. The suprachiasmatic nuclei: a 25 year retrospective. J Biol Rhyth. 1998; 13: 100–12.
6. Carl J. Chronobiology: Biological Timekeeping. Sunderland, Massachusetts, USA: Sinauer Associates, Inc. 2004; pp. 67–105.

7. Buijs R, Salgado R, Sabath E, et al. Peripheral circadian oscillators: time and food. Prog Mol BiolTransl Sci. 2013; 119: 82–103.

8. Skene DJ, Arendt J. Human circadian rhythms and their therapeutic relevance of light and melatonin. Ann Clin Biochem. 2006; 43: 344–53.

9. Benloucif S, Guico MJ, Reid KJ, et al. Stability of melatonin and temperature as circadian phase markers and their relation to sleep times in humans. J Biol Rhyth. 2005; 20: 178–88.

10. Tenorio FC, Simoes MJ, Teixera VW, et al. Effects of melatonin and prolactin in reproduction: review of literature. Rev. Assoc. Med. Bras. vol. 61 no. 3 São Paulo May/June 2015.

11. Yanovski J, WitcherJ, Adler N, et al. Stimulation of the paraventricular nucleus area of the hypothalamus elevates urinary 6-sulphatoxy melaton in during daytime. Brain Res Bull. 1987; 19: 129–33.

12. Tordjman S, Chokoron S, Delmore R, et al. Melatonin: Pharmacology, functions and therapeutic benefits. Curr Neuropharmacol. 2017; 15: 434–43.

13. Liebmann PM, Wolfler A, Felsner P, et al. Melatonin and the immune system. Int Arch Allergy Immunol. 1997; 112: 203–211.

14. Simonneaux V, Ribelayga C. Generation of the melatonin endocrine message in mammals: A review of the complex regulation of melatonin synthesis by norepinephrine, peptides, and other pineal transmitters. Pharmacol Rev. 2003; 55: 325–95.

15. Pierpaoli W, Regelson W. Pineal control of aging: effect of melatonin and pineal grafting on aging mice. Proc Natl Acad Sci USA. 1994; 91: 787–91.

16. Reiter RJ, Tan DX, Osuna C, et al. Actions of melatonin in the reduction of oxidative stress. A Review. J Biomed Sci. 2000b; 7: 444–58.

17. Kiefer T, Ram PT, Yuan L, et al. Melatonin inhibits estrogen receptor transactivation and cAMP levels in breast cancer cells. Breast Cancer Res Treat. 2002; 71: 37–45.

18. Benot S, Goberna R, Reiter RJ, et al. Physiological levels of melatonin contribute to the antioxidant capacity of human serum. J Pineal Res. *1999*; 27: 59–64.

19. Armstrong SM, Redman JR. Melatonin: a chronobiotic with antiaging properties. Med Hypothese. 1991; 34: 300–309.

20. Smolander J, Harma M, Lindqvist A, et al. Circadian variation in peripheral blood flow in relation to core temperature at rest. Eur J Appl Physiol Occup Physiol. 1993; 67: 192–96.

21. Waterhouse J, Drust B, Weinert D, et al. The circadian rhythm of core temperature: origin and some implications for exercise performance. Chronobiol Int. 2005, 22: 207–25.

22. Lightman SL. The neuroendocrinology of stress: a never ending story. J Neuroendocrinol. 2008; 20: 880–84.

23. Moore RY, Eichler VB. Loss of a circadian adrenal corticosterone rhythm following suprachiasmatic lesions in the rat. Brain Research. 1972; 42: 201–06.

24. Millar-Craig MW, Bishop CN, Raftery EB. Circadian variation of blood pressure. Lancet. 1978; 1: 795–97.

25. Chen L, Yang G. Recent advances in circadian rhythms in cardiovascular system. Front. Pharmacol. 01 April 2015.doi: 10.3389/fphar.2015.00071

26. Monk TH. Subjective ratings of sleepiness: The underlying circadian mechanisms. Sleep 1987; 10: 343–53.

27. Sallinen M, Sihvola M, Puttonen S, Ketola K, et al. Sleep, alertness and alertness management among commercial airline pilots on short-haul and long-haul flights. Accid Anal Prev. 2017; 98: 320–29.

Chapter 19

The Blood–Brain Barrier

INTRODUCTION

To maintain normal brain function, the neural environment must remain in a narrow homeostatic range. This requires tight regulation of transport of cells, molecules and ions into the neural tissues. Such regulation is achieved by unique anatomical and physiological barriers between blood and brain tissues.

The initial studies that led to the concept of the blood–brain barrier were performed by Ehrlich at the beginning of 20th century. Ehrlich observed that many intravenously injected dyes stained the tissues of practically the whole body but not the brain and spinal cord (Fig. 19.1). However, when the dye was injection in the subarachnoid space, the brain and spinal cord took up the stain, but not rest of the body. Later, Lewandowsky also showed that the Prussian blue reagents did not pass from blood to brain, and he formulated the concept of the BBB. Goldmann's experiments with trypan blue demonstrated very distinctly the existence of this BBB.[1]

Interrelation between cerebral intracellular fluid compartment with cerebral interstitial compartment and CSF compartment is shown in Fig. 19.2. The interstitial fluid in the central nervous system and CSF in the intraventricular and subarachnoid space are in

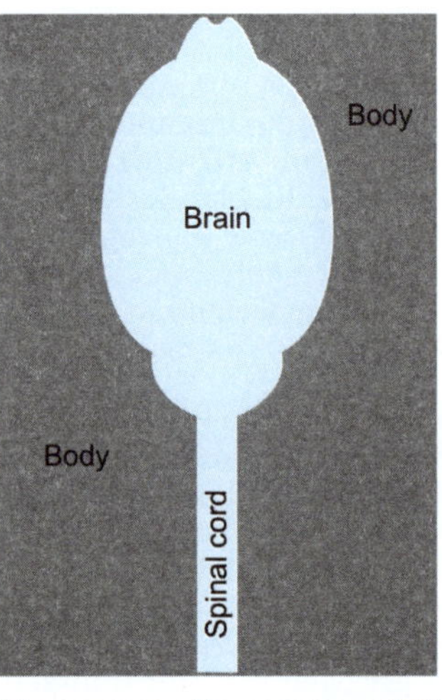

Fig. 19.1: Diagrammatic representation of the experiment by Ehrlich. Intravenous injection of a dye stains all tissues of the body except central nervous system

separate compartments. Homeostasis in these fluids and ultimately intracellular compartments of brain is regulated by blood–brain and blood–CSF barriers. However, there is a free exchange between interstitial compartment of brain and CSF[2] (Fig. 19.2).

Three barrier layers separate the blood from neural tissues:

1. A highly specialized *endothelial cell layer* constitutes the blood–brain barrier (BBB) and separates the blood and the brain

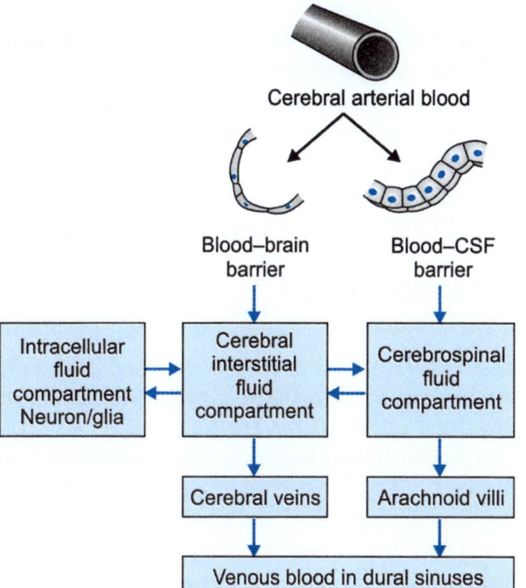

Fig. 19.2: Interrelation between cerebral intracellular fluid compartment with cerebral interstitial compartment and CSF compartment (Latewrra and Goldstein)

interstitial fluid. The BBB is situated along 99% of the brain's capillary endothelium.

2. The blood–CSF barrier constituted by choroidal plexus epithelium which secretes the cerebrospinal fluid into the four cerebral ventricles.

3. The *arachnoid epithelium* which separates dural sinus blood from subarachnoid CSF.

1. THE BLOOD–BRAIN BARRIER (BBB)

A. Anatomy of BBB

Most of the systemic capillaries are so-called continuous type. In this type, capillary wall is a continuous membrane of endothelial cells, about 1 µm thick except at the site of the nucleus, where it is thicker. The endothelial cells rest on a basement membrane. Actually, the endothelial cells forming the capillary walls are not continuous with each other. A small cleft of 10 nm size always separates the two adjacent endothelial cells. It is believed that most of the water-soluble ions and molecules pass across the capillary through these clefts (or slit-pores) (Fig. 19.3A). The endothelial lining is enveloped by pericytes. Both pericytes and endothelial cells share a common basement membrane. These two types of cells can be easily distinguished from one another based on the presence of the prominent round nucleus of the pericyte compared to the flat elongated nucleus of the endothelial cells. Pericytes also project finger-like extensions that wrap around the capillary wall, allowing the cells to regulate capillary blood flow.[3,4]

In *cerebral capillaries, intercellular clefts are non-existent;* instead there are tight

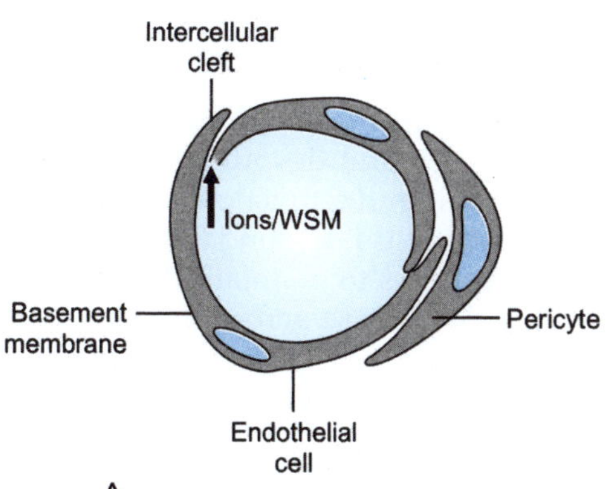

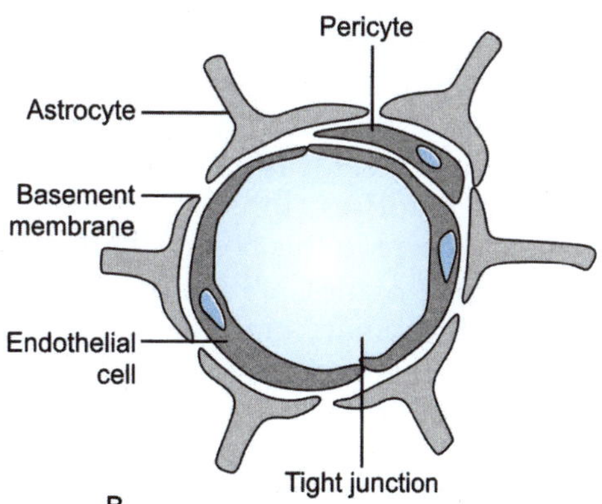

Fig. 19.3: Structure of a general systemic capillary (A) and a cerebral capillary (B) (WSM: Water-soluble molecules)

junctions connecting the adjacent endothelial cells (Fig. 19.3B). Therefore, transcapillary exchange can only be transcellular. Only lipid-soluble solutes can diffuse freely through the capillary endothelial membrane. In capillaries of other parts of the body, such transcellular exchange is overshadowed by other nonspecific exchanges through the clefts. Despite the estimated total length of 650 km and total surface area of 12 m^2 of capillaries in human brain, this barrier is very efficient and makes the brain practically inaccessible for lipid-insoluble compounds such as polar molecules and small ions.[5] Moreover, the pericyte/endothelial cells ratio is high in the brain compared to the vasculature of other organs, e.g. 1:3 in brain *vs.* 1:100 in skeletal muscle.[6] Besides pericytes, in the cerebral capillaries, footplates of astrocytes are closely applied to the outer surface of the endothelial tubes. In fact, the pericytes and footplates of astrocytes constitute a continuous layer adjacent to the basement membrane. Thus, three layers separate blood from cerebral interstitium, (i) endothelial cells, (ii) basement membrane and (iii) a layer constituted by pericytes and footplates of astrocytes. However, the *blood–brain barrier function is primarily served by tight junctions among the adjacent endothelial cells only.* Pericytes seem to play a critical role during formation and maturation of BBB. In addition, pericytes have a regulatory control over capillary circulation. In animal models, destruction of pericytes in aged animals was shown to result in loss of BB integrity. Astrocytes seem to play important role in the maintenance of BBB and in homeostasis of extracellular concentration of neurotransmitters, metabolites, ions and water, explained elsewhere in this book (Chapter 13).

Some regions of the CNS do not express the classical BBB capillary endothelial cells, but have microvessels similar to those of the periphery. These areas are adjacent to the ventricles of the brain and are termed the circumventricular organs (CVOs). The CVOs include the choroid plexus, the median eminence and neurohypophysis, the pineal gland, organum vasculosum of the lamina terminalis, subfornical organ, subcommissural organ and the area postrema. The choroid plexus may be of importance when considering the transport of peptide drugs, because it is the major site of cerebrospinal fluid (CSF) production, and both the CSF and brain ECF freely exchange[6,7] (Fig. 19.2).

B. Physiology of BBB

In the cerebral capillaries, the tight junctions between the adjacent endothelial cells preclude any paracellular transport. All the transport has to be transcellular. Consequently, only lipid-soluble substances such as O_2 and CO_2, anesthetic agents, caffeine, nicotine and alcohol can easily diffuse across the BBB. All other non-polar solutes and charged ions can cross the BBB either by *carrier-mediated diffusion, active transport* or by *endocytosis.*

i. Water Transport

Large water fluxes continuously take place between the different compartments of the brain as well as between the brain parenchyma and the blood or cerebrospinal fluid. The molecular pathways by which water molecules cross the blood–brain barrier are not well-understood, although the discovery of aquaporin 4 (AQP4) in the brain improved the understanding of some of these transport processes, particularly under pathological conditions.[6–8]

ii. Glucose Transport

How the brain meets its continuous high metabolic demand despite varying plasma glucose levels and a functional BBB is poorly understood. GLUT-1, found in high density at the BBB appears to maintain the continuous shuttling of glucose across the blood–brain barrier irrespective of the plasma concentration.[9] It is now generally accepted that glucose enters the brain via a carrier-mediated, facilitated diffusion process which is not energy dependent. Glucose transport across BBB is Na$^+$- and insulin-independent;

and displays saturation kinetics and transport competition among hexoses.

iii. Amino Acid Transport

a. Essential amino acids, and also the precursor of dopamine, L-dopa, enter the brain as rapidly as glucose. These amino acids are transported into the brain by the leucine-preferring or the L-type transport proteins, by facilitated diffusion. These amino acids compete with each other for entry into the brain. Therefore, an elevation of plasma level of one will inhibit uptake of the others. This competition may be important for certain metabolic diseases such as phenylketonuria (PKU), where high levels of phenylalanine in plasma reduce brain uptake of other essential amino acids.

b. Small neutral amino acids, such as alanine, glycine, proline and GABA (gamma-aminobutyric acid), are markedly restricted in their entry into the brain. These non-essential amino acids are transported by alanine-preferring or A-type transport protein. These transporters are sodium-dependent active transporters. The A-type transporters are present on the external (antiluminal) rather than luminal surface of blood–brain barrier. Therefore, these small neutral amino acids appear to be transported out of the brain across the blood–brain barrier into blood capillaries.[10]

iv. Transport of Ions

Due to control of K^+ at BBB, potassium concentration in brain interstitium is maintained at 2.5–2.9 mEq/L despite higher concentration of potassium in plasma (3.5–5.0 mEq/L). BBB is largely impermeable to ions such as Ca^{2+} and Mg^{2+}. pH is actively regulated at BBB. Neuronal firing and synaptic transmission are associated with influx of Na^+ and Ca^{2+} and increased extracellular concentration of K^+ and neurotransmitters. Astrocytes have a key role in the homeostatic mechanisms maintaining extracellular environment of the brain within a narrow normal limits (*see* Chapter 13).

v. Large Macromolecules: Proteins and Peptides

Larger polypeptides, proteins and lipoproteins also enter the brain endothelial cells by receptor-mediated endocytosis, and to some extent can reach the brain side using a transcytosis mechanism. Receptor-mediated endocytosis and possibly transcytosis pathways have been largely described and discussed in particular for insulin receptor, transferrin receptor, low density lipoprotein receptor, and more recently low density lipoprotein receptor-related protein 1 (LRP1). LRP1 involvement in the influx of insulin-like growth factor I into the brain has been shown *in vitro*. These receptor-mediated endo-/trans-cytosis pathways are currently used in strategies aiming at improving macromolecules (drugs) delivery to the brain.[11]

vi. "Enzymatic Blood–Brain Barrier"

Metabolic processes within the brain capillary endothelial cells are important to blood–brain function. Most neurotransmitters present in the blood do not enter the brain because of their low lipid solubility and lack of specific transport carriers in the luminal membrane of the capillary endothelial cell. In contrast, L-dopa, the precursor for dopamine, has an affinity for the L-type transporter. Therefore, it enters the brain more easily from the blood than would be predicted based on its lipid solubility. Patients with Parkinson's disease are treated with L-dopa rather than with dopamine because of this fact. However, the penetration of L-dopa into the brain is limited by the presence of enzymes L-dopa decarboxylase and monoamine oxidase within the capillary endothelial cells. This "enzymatic blood–brain barrier" limits passage of L-dopa into the brain and explains the need for large doses of L-dopa in the treatment of Parkinson's disease. Therapy is currently enhanced by concurrent treatment with an inhibitor of L-dopa decarboxylase.

Endothelial monoamine oxidase may also play a role in the inactivation of neurotransmitters released by neuronal activity.

Monoamines show very little uptake when presented from the luminal side. The uptake systems for monoamines are present in the antiluminal surface of the brain capillary endothelial cells, i.e. the mechanism helps to destroy neurotransmitter present in interstitium, rather than protecting the brain from monoamines present in the blood. The brain endothelial capillary also contains a variety of other neurotransmitter-metabolizing enzymes such as cholinesterases, GABA transaminases, aminopeptidase and endopeptidases. In addition, several drug and toxin metabolizing enzymes are also found in the brain capillaries. Thus, the "enzymatic blood–brain barrier" protects the brain not only from circulating neurotransmitters but also from many toxins.[10]

2. THE BLOOD–CEREBROSPINAL FLUID (BLOOD–CSF) BARRIER

The observations of human choroid plexus seen at autopsy has given wrong impression about its structure during life. At autopsy, when all the CSF is drained out, choroid plexus is seen to extend as a finger-like projection along the floor of the lateral ventricles, hang-down from the roof of third ventricle and overly the roof of fourth ventricle (Fig. 19.4). Recent advances in technology have enabled the insertion of micro-video probe into the third ventricle. The result shows that the choroid plexus with its villi fills the entire ventricle, pulsating with each heartbeat[12] (Fig. 19.5).

Structure of choroid plexus is similar in all the ventricles. There is a vascular network surrounded by a single layer of cuboidal cells joined together by tight junctions. The endothelial cells of capillaries in choroid plexus, unlike those at BBB, are fenestrated and highly porous, which facilitates the movement of fluid out of the capillaries. The epithelial cells of choroid plexus have a large number of mitochondria (12–15% of cell volume), consistent with high energy demands for the secretory functions of the cells.[12]

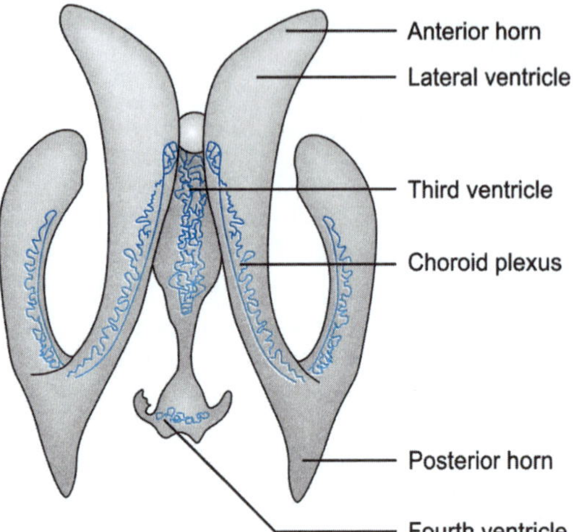

Fig.19.4: Choroid plexus as observed postmortem (retracted to the side of ventricles)

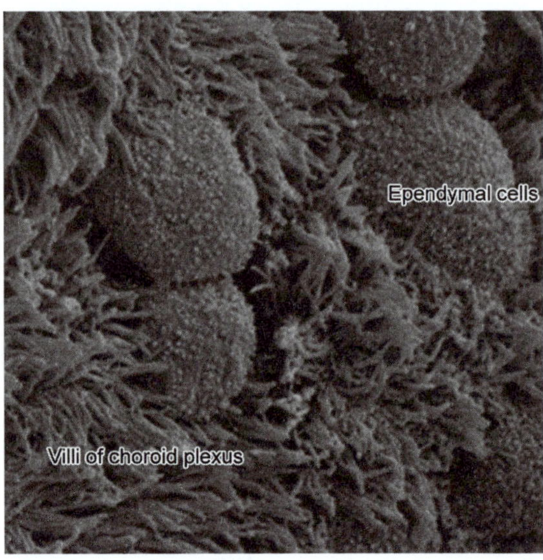

Fig. 19.5: Choroid plexus and ependymal cells during life (Steve Gschmeissner/Science Photo Library/Getty Images)

In humans, the rate of CSF secretion is 0.3 to 0.4 ml/min, about 500 ml/24 hours. The total volume of CSF is estimated to be 100 to 150 ml in normal adults, so that CSF is totally replaced three or four times each day. Several constituents are maintained at concentrations in CSF different from those in plasma, indicating that CSF is not simply a protein-free ultrafiltrate of plasma. This was clearly demonstrated in experiments showing that the concentrations of some ions in the CSF are very carefully regulated, and more importantly are independent of variations in the plasma concentrations of these ions, e.g. K^+, HCO_3^- and Ca^{2+}. The composition of an ultrafiltrate could not be regulated in this manner. Cerebrospinal fluid production by the choroid plexus is driven by active or carrier-mediated ion transport systems that results in a net secretion of Na^+ and Cl^-, the main ionic constituents of CSF (Fig. 19.6). The exact mechanisms involved have yet to be determined fully.[13]

The movement of substances from the blood into the CSF is, in many ways, analogous to that from the blood into the brain, with many of the same transporters present in both tissues. Quantitatively, however, there are major differences between transport by the choroid plexus and across the blood–brain barrier. In terms of O_2, CO_2, glucose and amino acid entry into the brain, transport through the blood–brain barrier predominates. For some other compounds, however, the choroid plexus is the major site of entry. This is the case for Ca^{2+}, where the CSF influx rate constant is tenfold greater than across the blood–brain barrier value. This reflects the role of the choroid plexus in not only CSF formation but also brain Ca^{2+} homeostasis.[13]

Ion Transport Across Blood–CSF Barrier

Of all substances crossing the blood–CSF barrier, ion transport has been extensively studied. The choroid plexuses secrete Na^+ and Cl^-, and HCO_3^-. Furthermore, the whole secretory process is highly dependent on carbonic anhydrase-mediated HCO_3^- secretion. Drugs which inhibit carbonic anhydrase reduce cerebrospinal fluid (CSF) flow.[14] The epithelium of the choroid plexus also mediates the net absorption of *K^+ from the CSF to the blood*. The transport of K^+ is important in maintaining the K^+ homeostasis in the CSF.

In the choroid plexus, activity of the Na^+-K^+ ATPase pump also is closely associated with the secretion of CSF. Inhibitors of the pump, e.g. the cardiac glycoside ouabain, have been

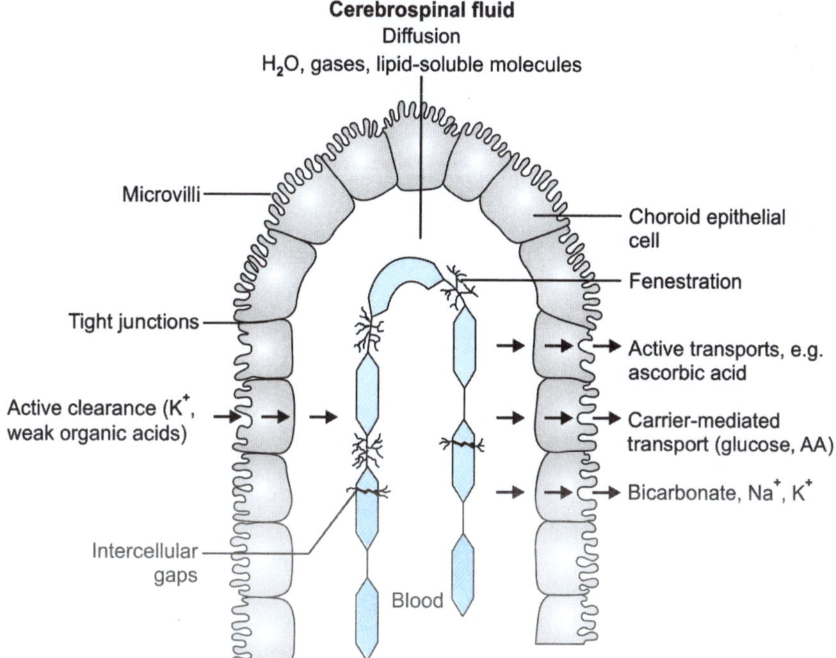

Fig. 19.6: Diagrammatic representation of choroid plexus and proposed mechanism of production of CSF

shown to reduce CSF secretion, and the movement of Na^+ into the CSF. The sodium potassium ATPase pump is situated on the apical brush border membrane in the choroid plexus (Fig. 19.7). This is in marked contrast to most epithelial cells where the Na^+-K^+ ATPase pump is expressed only in the basolateral membrane. The apical localization of the pump in the choroid plexus is vital because the *pump is the major pathway for Na^+ secretion into the CSF.* Furthermore, the Na^+-K^+ ATPase is also the most important mechanism for K^+ *uptake from the CSF into the epithelial cells.*[15]

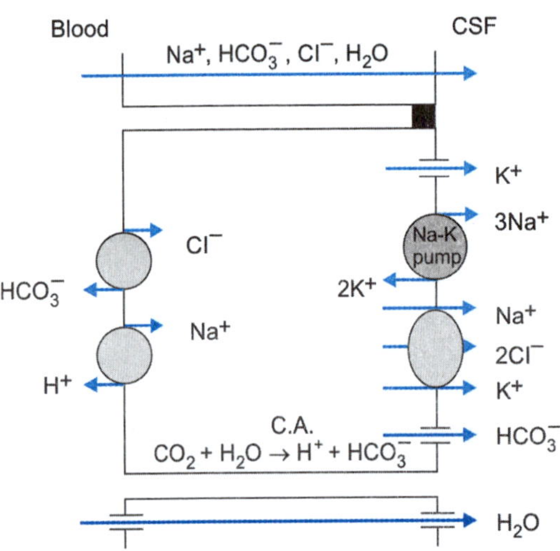

Fig. 19.7: Ion transport across blood–CSF barrier

In brief, major processes involved in cerebrospinal fluid (CSF) secretion are (Fig. 19.7):

1. Carbonic anhydrase catalyzes the production of bicarbonate ions and protons from carbon dioxide.
2. Bicarbonate is exchanged for chloride across the basolateral membrane.
3. Bicarbonate and chloride flow into the CSF through anion channels, down an electrochemical gradient, and through sodium–potassium–chloride cotransport.
4. Sodium is exchanged for protons across the basolateral membrane and for potassium across the apical membrane.
5. Water follows the osmotic gradient created by the combined secretion of sodium, chloride, and bicarbonate into the CSF.[15]

3. THE ARACHNOID BLOOD–CSF BARRIER

The CSF is absorbed through the arachnoid granulations and villi. Arachnoid granulations consist of clusters of villi formed by herniation of arachnoid membrane through the dura into the lumen of superior sagittal sinus and other venous structures (Fig. 19.8) The arachnoid cells are joined together by tight junctions and do not show selective permeability. The arachnoids appear to act as valves that allow one-way flow of CSF from subarachnoid spaces into the venous blood. This one-way flow of CSF is sometimes called bulk flow, since all constituents of CSF leave with the fluid including small molecules, proteins, microorganisms or even red blood cells.

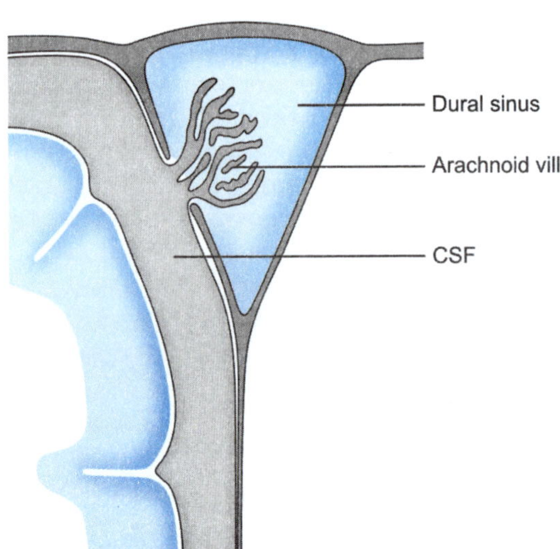

Fig. 19.8: CSF drainage through arachnoid villi

Recently, a plausible mechanism of one-way CSF transport at arachnoid villi has been elucidated. Cells of the villus membrane have been observed to contain giant vacuoles that transport fluid from CSF side of the cells to the side toward blood. The vacuoles are formed by pinocytosis of CSF by arachnoid cell. The pinocytic vacuole gradually moves to the border of arachnoid cells facing blood and releases CSF droplet into blood sinus by exocytosis[2] (Fig. 19.9).

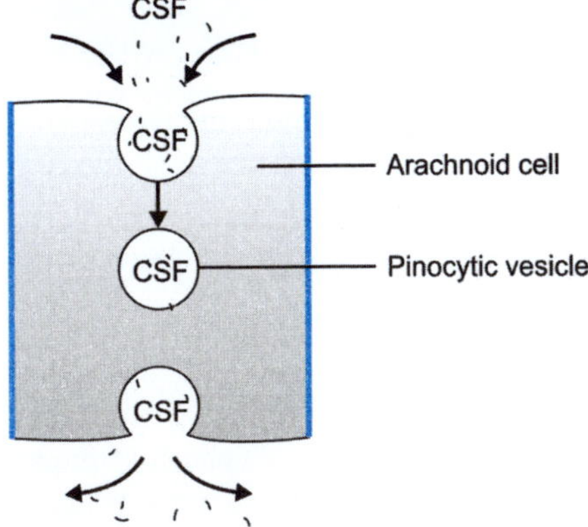

Fig.19.9: Mechanism of one-way transport of CSF into dural sinus blood

Functions of Blood-Brain and Blood-CSF Barriers

1. **Protection against variations in electrolyte composition of plasma:** In view of the fact that neuronal function is very sensitive to electrolyte composition of extracellular fluid, cerebral tissue needs better protection than any other tissue. Otherwise, even minor and transient changes in the electrolyte composition of blood would have disastrous consequences.

2. **Protection against circulating hormones and neurotransmitters of non-cerebral origin:** Epinephrine, norepinephrine, dopamine, histamine, serotonin, glucagons and somatostatin are present in the blood as hormones. In the CNS, these agents also function as neurotransmitters. Therefore, if these agents could cross the BBB, they would play havoc with the neuronal function.

3. **Protection against circulating toxins of exogenous and endogenous origin:** The blood–brain barrier acts very effectively to protect the brain from many common infections. Thus, infections of the brain are very rare. However, since antibodies are too large to cross the blood–brain barrier, infections of the brain which do occur are often very serious and difficult to treat.

Clinical Implications of Altered Blood-Brain (and blood-CSF) Barriers

1. In the fetal and neonatal life, BBB does not seem to be fully developed. As a result, if a neonate develops physiological jaundice or hemolytic disease of the newborn, the unconjugated bilirubin crosses the BBB. If the plasma bilirubin level exceeds 18 mg%, bilirubin deposition in certain regions of the brain, especially the basal ganglia, results in permanent motor disabilities (kernicterus). Similar degree of jaundice in adults does not produce kernicterus.

2. Most of the **antibiotics** cannot cross the BBB. Therefore, in a patient with a cerebral infection only an antibiotic which crosses the BBB would be beneficial.

3. Blood–brain barrier may be **disrupted** by hypertension, plasma hyperosmolality, exposure to radiations, trauma, ischemia and inflammations. In all such conditions, brain becomes far more vulnerable to exogenous or endogenous toxins.

4. In the treatment of certain brain tumors, BBB is *deliberately disrupted* by injection of hypertonic mannitol into the carotid and vertebral arteries. Thereby, anti-cancer drugs are able to cross the BBB and attack the tumor tissue.[6]

References

1. Ehrlich P. Farbenanalytischestudie. Berlin. Das sauerstubudurfnis des organismus, 1885.
2. Latewrra J, Goldstein GW. The blood–brain barrier, choroid plexus and cerebrospinal fluid. In Principles of Neural Science, Ed. Kandel ER, Schwartz, JH, et al. 6th Edition, McGraw-Hill, New York, 2013.
3. Abbott NJ, Ronnback L, Hansson E. Astrocyte–endothelial interactions at the blood–brain barrier. Nat Rev Neurosci. 2006; 7: 41–53.
4. Liddellow SA. Fluids and barriers of the CNS: a historical viewpoint. Fluid Barriers CNS. 2011; 8: 2. Published online 2011 Jan 18. doi: 10.1186/2045-8118-8-2
5. Davson H, Segal MB. Physiology of the CSF and blood–brain barriers. CRC Press, Florida, USA, 1995.
6. Misra A, Ganesh S, Shahiwala A, et al. Drug delivery to the central nervous system: a review. J Pharm Pharmaceut Sci. 2003; 6: 252–73.

7. Sherpo D, Morel NM. Pericyte physiology. FASEB J. 1993; 7: 1031–38.

8. Francesca B, Rezzani R. Aquaporin and Blood Brain Barrier. Curr Neuropharmacol. 2010; 8: 92–96.

9. McAllister MS, Krizanac-Bengez L. Mechanisms of glucose transport at the blood–brain barrier: an *in vitro* study. Brain Res. 2001; 904:20–30.

10. nba.uth.tmc.edu/neuroscience/s4/chapter 11.htm

11. Strazielle N, Ghersi-Egat JF. Physiology of blood–brain interfaces in relation to brain disposition of small compounds and macromolecules. Mol. Pharmaceutics, 2013; 10: 1473–491.

12. Spector R, Keep RF, Robert S, et al. A balanced view of choroid plexus structure and function: Focus on adult humans. Exper Neurol. 2015; 267: 78–86.

13. Laterra, John; Keep R, Betz LA, et al, Blood–cerebrospinal fluid barrier. In Basic Neurochemistry: Molecular, Cellular and Medical Aspects. 6th edition, 1999.

14. Vogh BP. The relation of choroid plexus carbonic anhydrase activity to cerebrospinal fluid formation: study of three inhibitors in cat with extrapolation to man. J Pharmacol Exp Ther. 1980; 213: 321–31.

15. Brown PD, Davies SL, Speake T, et al. Molecular mechanisms of cerebrospinal fluid production. Neuroscience. 2004; 129: 957–70.

Chapter 20

The Executive Functions of Prefrontal Cortex

INTRODUCTION

Until recently, the prefrontal areas of human brain were referred to as the "silent areas" or "uncommitted cortex", because injury to these regions was not accompanied by any sensorimotor deficit. This concept persisted despite the result seen in 19th century in a patient, Phineas Gage, in whom left frontal lobe was destroyed when a large iron rod was driven through his head in an 1848 accident. Although Gage retained normal memory, speech and motor skills, his personality changed radically: He became irritable, quick-tempered, and impatient—characteristics he did not previously display—so that friends described him as "no longer Gage". Whereas, he had previously been a capable and efficient worker, afterward he was unable to complete tasks.Gage's animal propensities (basic needs for survival—eating, sleeping, sex life and defense) were now sharply pronounced while his intellectual abilities suffered; garrulous or obscene jokes replaced his once quick wit.[1] He was fitful, irreverent, indulging at times in the grossest profanity, manifesting a little deference for his fellows, impatient of restraint of advice when it conflicts with his desires, at times pertinaciously obstinate. Subsequently, in 1940s and 1950s, prefrontal lobotomies were performed in over 50,000 patients worldwide in patients suffering from psy-

chiatric illness. Although the patients got rid of suicidal tendencies or aggressive behavior, their state of mind was worse than death. The patients showed no emotions in the face of most tragic events. There was complete absence of the sense of social or moral behavior. In short, the patients were no better than zombies. The procedure was declared "contrary to the principles of humanity" and "through lobotomy an insane person is changed into an idiot." By the 1970s, numerous countries including USA and USSR had banned the procedure.[2]

In spite of the clinical observations on the effects of damage of frontal lobes described above, the role of frontal lobes remained a mystery. As late as 1975, Penfield[3], who pioneered the mapping of the cortex in conscious humans, called the PFC as "uncommitted cortex". He observed that unlike sensory (which included visual and auditory) and motor convolutions, which are committed to function at birth, "some of the convolutions that are used eventually for what may be called psychical functions are *uncommitted to their exact function at the time of birth.*"During the last couple of decades the advent of modern electrophysiological and imaging (functional magnetic resonance imaging, proton emission tomography) techniques have provided a wealth of insight into its functions.

EVOLUTIONARY CONSIDERATIONS

According to Brodmann[4] the prefrontal cortex constitutes 3.5% of the total cerebral cortex in the cat, 11.5% in the dog, 12.5% in the rhesus monkey, 17% in the chimpanzee, and 29% in the human (Fig. 20.1). The disproportionate evolutionary growth of the prefrontal cortex parallels that of the associative cortex of temporal and parietal regions. It is a legitimate inference, that the evolutionary expansion of the cortex of association, both posterior and prefrontal, is closely related to the evolution of cognitive functions.[5]

ANATOMY OF PREFRONTAL CORTEX

The **prefrontal cortex** (**PFC**) is the cerebral cortex which covers the front part of the frontal lobe. The prefrontal cortex consists of Brodmann areas 8, 9, 11, 12, 13, 14, 24, 25, 32, 44, 45, 46 and 47 (Fig. 20.2).

Connections of PFC

Athough there are no absolute boundaries between the different regions of the prefrontal cortex on the basis of their structure, connectivity and dominant functions, it is best to divide it into three components, e.g. lateral, medial and orbitofrontal (Fig. 20.3).

Prefrontal cortex, as studies in the monkey demonstrate, is arguably the best connected of all cortical structures. The three prefrontal regions, **medial, lateral, and orbitofrontal,** are reciprocally connected with one another and with the nuclei of the anterior and dorsal *thalamus*. The medial and orbital regions, in addition, are connected with the *hypothalamus* and *other limbic structures*; some of these connections are indirect, through the thalamus. The lateral region sends connections to the *basal ganglia*; in addition, it is profusely connected with *the association cortex of occipital, temporal, and parietal* regions. The precise functional role of the connections of the prefrontal cortex is not entirely known, but can be inferred from the functional role of the structures with which it is connected. In general terms, the *prefrontal-limbic connections are involved in the control of emotional behavior, whereas the prefrontal-striatal connections are involved in the coordination of motor behavior. Of special importance for the cognitive aspects of all forms of behavior are the reciprocal connections of the lateral prefrontal cortex with the hippocampus and with the posterior association cortices. There are well-demonstrated reciprocal connections between the hippocampus and the prefrontal cortex, especially its lateral region, although their exact path has not been completely clarified.* Given the proven role of the hippocampus in the acquisition of memory, it appears very likely that those connections participate in the formation of networks of motor or executive memory in the prefrontal cortex.[6] In generalized terms, the afferent input and efferent output of PFC are shown in Fig. 20.4.

Executive Functions of Prefrontal Cortex

1. **Awareness:** Age-appropriate insight of one's strengths and weaknesses.

2. **Planning:** Spontaneous use of planning behaviors in novel tasks; anticipate future events; grasp main idea; prioritize, reduce/cluster information, re-arrange material or information (Fig. 20.5).

3. **Goal setting:** Set intermediate and long-term goals appropriate to abilities.

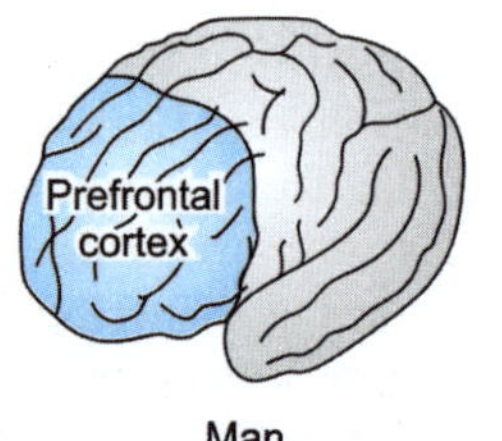

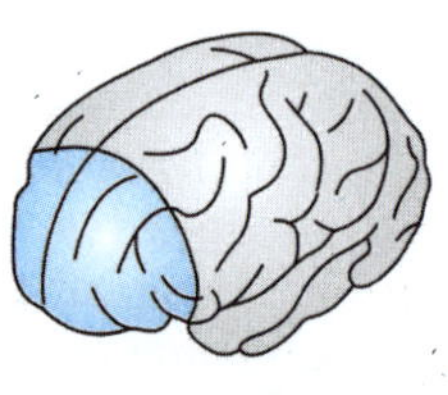

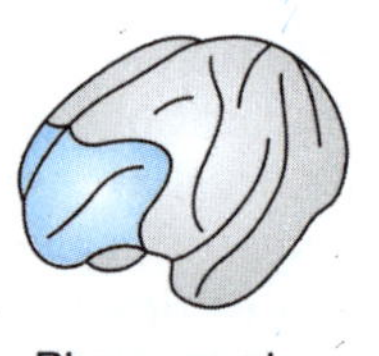

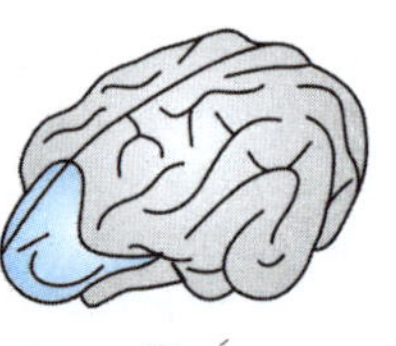

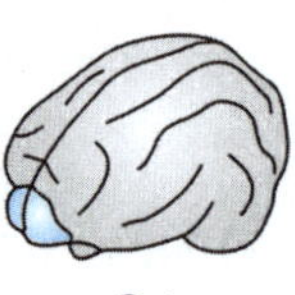

Fig. 20.1: Evolutionary development of prefrontal cortex

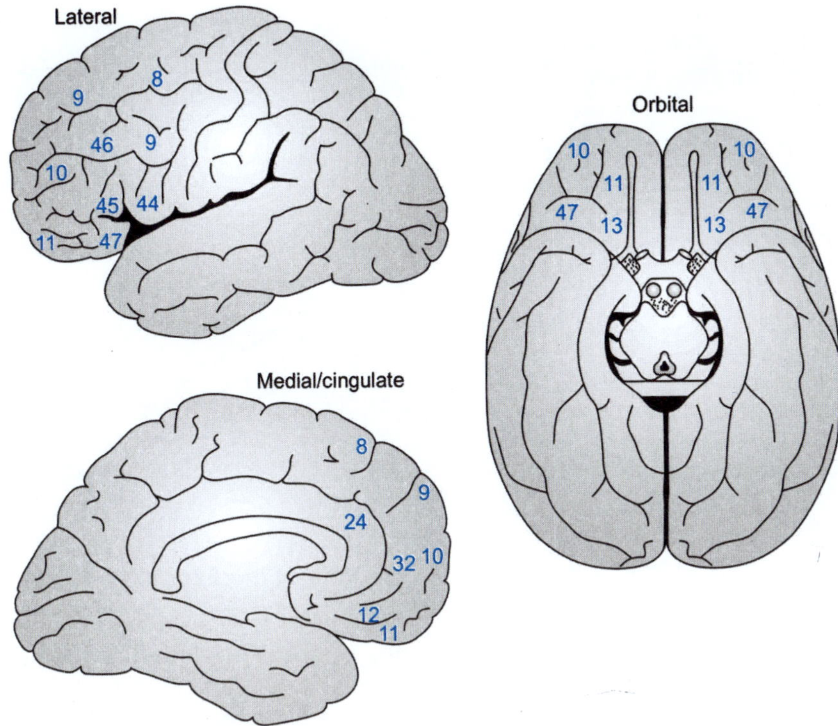

Fig. 20.2: The Brodmann areas included on prefrontal cortex (*see* text)

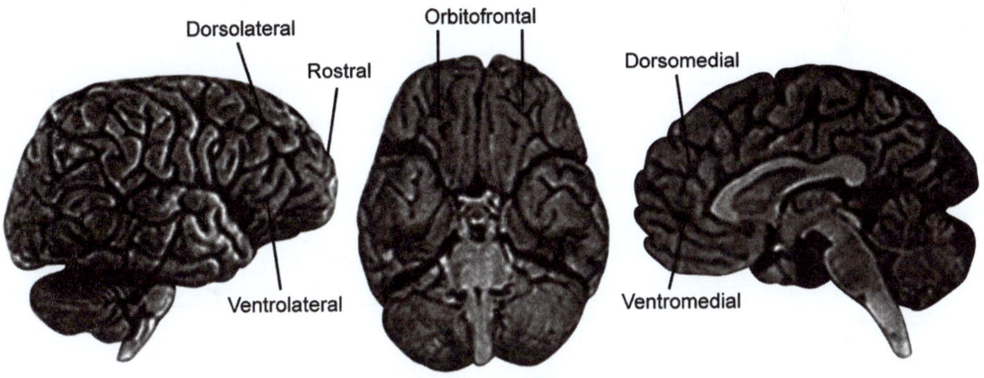

Fig. 20.3: Components of prefrontal cortex

4. **Self-initiation:** Independently initiate new activities; spontaneous in conversation; begins activity without procrastination, seek and search for information, generate ideas, persist, complete all parts of an activity.

5. **Self-monitoring:** Independently assess behaviors/responses and make changes as needed.

6. **Self-inhibiting:** Behavior is appropriate to changing situations; control impulses, think before acting, pace actions; follow or select according to rules or criteria; manage extraneous distractions, irrelevant information or interference (Fig. 20.6).

7. **Ability to change:** Demonstrate appropriate variations in behaviors; independently consider a variety of solutions in

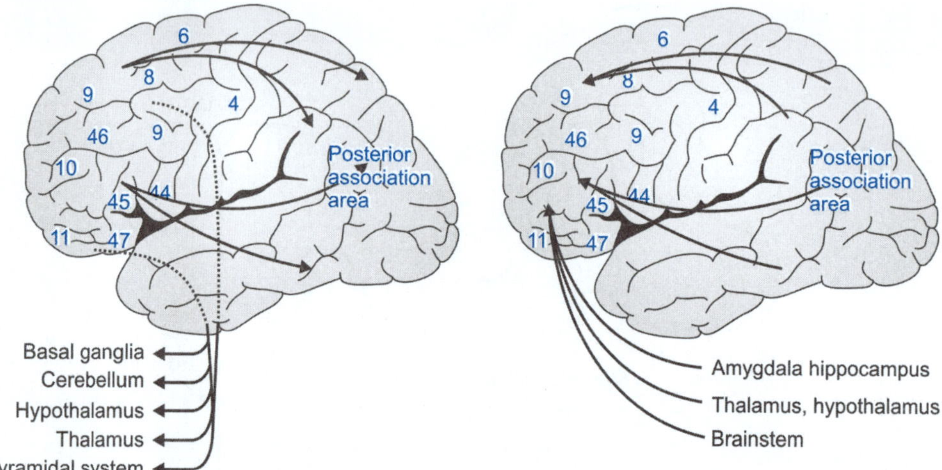

Fig. 20.4: Afferent input (right) and efferent output from prefrontal cortex

Fig. 20.5: Some of the executive functions

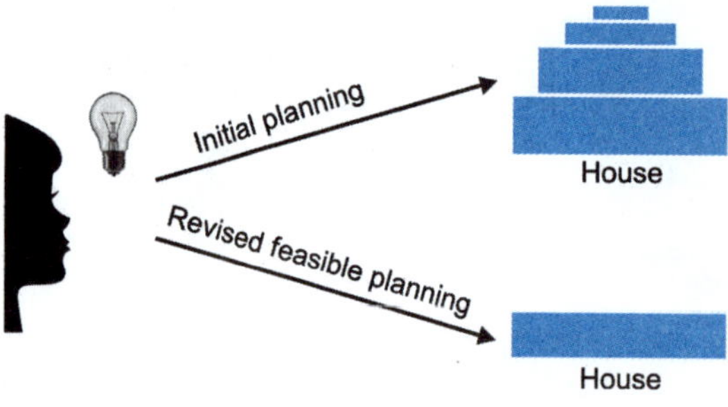

Fig. 20.6: Self-inhibition

problem solving; revise plans, move freely from one step/activity/situation to another.

8. **Strategic behavior:** Create useful strategies for functional use.

9. **Working memory:** Hold information in mind for the purpose of completing a task.

The Role of the Executive System (Fig. 20.7)

The role of the executive system is to handle novel situations outside of the domain of some of our more automatic psychological processes. There are five types of situations in which routine activation of behavior would not be sufficient for optimal performance, and where executive functions must kick in.

1. Situations that involve planning or decision making.

2. Situations that involve error correction or troubleshooting.

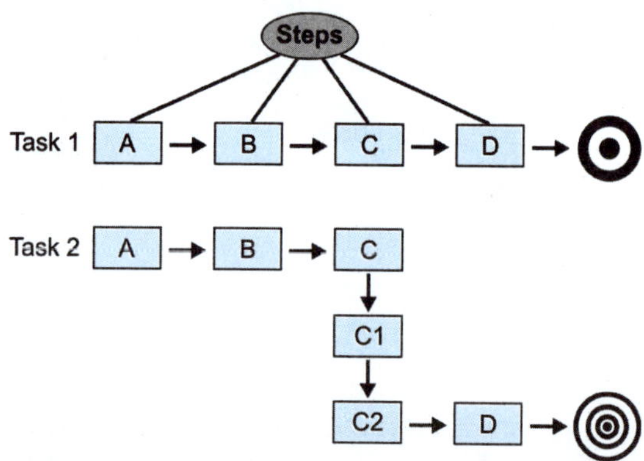

Fig. 20.7: Simple task 1 requires one to follow given steps, A, B, C and D to achieve a target. Task 2 is more difficult and requires executive functions, e.g. planning and decision-making, self-assessment of the results midway, taking corrective measures to achieve the self-directed target

3. Situations where responses are not well-rehearsed or contain novel sequences of actions.
4. Dangerous or technically difficult situations.
5. Situations that require overcoming strong habitual response or resisting temptation.

The executive functions are often evoked when it is necessary to override responses that might otherwise be automatically elicited by stimuli in the external environment. For example, when being presented with a potentially rewarding stimulus, such as a piece of pie, a person might have the automatic response to take a bite. However, where such a response conflicts with internal plans (having decided not to eat pie due to being on a diet), the executive functions might engage and inhibit the response.[5,7]

Development of the Executive Function of PFC Cortex

In addition to being one of the last regions of the brain to evolve, the frontal cortex is one of the last regions to reach maturity during human development. The slow course of the development of the frontal cortex has been charted in several ways, including using the measures noted above for the brain in general.

This research shows that the frontal cortex develops rapidly in early childhood, with important changes occurring at particular ages (at the end of the first year of life, between 3 and 6 years, and around puberty), and then continues to develop into adulthood. Grey matter, for example, does not reach adult levels in the dorsolateral prefrontal cortex until at least the end of adolescence, and myelination of this region continues into the 20s or possibly 30s. At age 20–29, executive functioning skills are at their peak, which allows people of this age to participate in some of the most challenging mental tasks.[8]

References

1. Tandon PN. Not so "silent": The human prefrontal cortex. Neurol India 2012; 61: 578–80.
2. 2. Marya RK. Walter Freeman in History of medicine. 2009. Jaypee, New Delhi.
3. Penfield W. The mystery of the mind: A critical study of consciousness and the human brain. Princeton University Press; 1975.
4. *Brodmann K (1909). "Vergleichende Lokalisation-slehre der Grosshirnrinde" (in German). Leipzig: Johann Ambrosius Barth.*
5. Fuster JM. Frontal lobe and cognitive development . J Neurocytol. 2002; 31: 373–85.
6. Gamo J, Arnsten AFT. Molecular modulation of prefrontal cortex: Rational development of treatments for psychiatric disorders. Behavioral Neuroscience. 2011; 125: 282–96
7. McCalla A. Executive Functioning: Where is it Controlled and How Does it … *https://www.rainbowrehab.com/executive-functioning/*
8. De Luca C, Leventer R. "Developmental trajectories of executive functions across the lifespan". In Anderson, Peter, et al. *Executive functions and the frontal lobes: a lifespan perspective.* Washington, DC: Taylor & Francis. 2008; pp. 3–21.

| Growth Hormone

INTRODUCTION

Although giants and dwarfs have been known since ancient times, it is only approximately a century and a half ago that an association between these disorders and the pituitary was suspected. In 1887, it was reported that in almost all published cases of acromegaly, sella turcica had been found enlarged.[1] In 1912, Harvey Cushing, an American neurosurgeon, published a book, *The Pituitary Body and its Disorders.*[2] Cushing recognized that acromegaly was produced by an excess of a growth factor and considered hypertrophy of the pituitary gland a part of the syndrome. He is credited for introducing the terms hypo- and hyperpituitarism and for suggesting that the pituitary controlled growth. In 1920s, it was demonstrated that injection of pituitary extract to animals caused excessive growth and removal of the pituitary caused growth retardation.[1]

Growth hormone (GH) first was isolated from the human pituitary gland in 1956, by both Li and Papkoff, in California, and Raben, in Massachusetts, but its biochemical structure was not elucidated until 1972. Between 1963 and 1985, about 7700 children in the United States and 27,000 children worldwide were given GH extracted from human pituitary glands to treat severe growth hormone deficiency.[3]

PULSATILE SECRETION OF GROWTH HORMONE

Growth hormone is secreted by somatotroph cells of the anterior pituitary gland. Growth hormone has a very short half-life of 14 minutes. GH is secreted in a pulsatile fashion, mostly at night (during sleep) (Fig. 21.1). In between pulses, the serum GH level is minimal to undetectable. Pulses occur up to 10 times a day, each lasting about an hour and a half and separated by two hours.

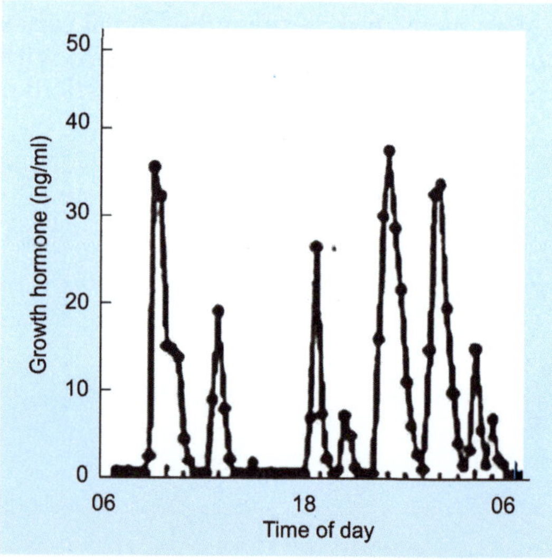

Fig. 21.1: Pulsatile plasma growth hormone levels. Most of the GH secretion occur at night

GH release can be accurately assessed using 24-hour sampling.

The peak daily GH secretion also varies with age and stage of growth and development. During puberty, the peak in daily GH secretory rate is around 150 µg/kg. The augmentation of GH secretion that occurs during puberty has been attributed to changes in sex steroid levels that enhance the frequency and amplitude of GH pulses. This decreases to around 25 µg/kg by age 55 years, paralleling the age-related decline in body mass index.[4, 5]

Regulation of GH Secretion

The regulation of GH secretion is via hypothalamic and peripheral factors which act on somatotrophs. GH secretion is stimulated by growth hormone-releasing hormone (GHRH) and inhibited by somatostatin (GHIH), both released from the hypothalamus. There are specific cell-surface receptors on the somatotroph cells to which these hormones bind. Table 21.1 summarizes factors stimulating or inhibiting growth hormone release.

Table 21.1: Stimulators and inhibitors of growth hormone release	
GH release: Stimulators	*GH release: Inhibitors*
GHRH	Somatostatin
Ghrelin	IGF-1
Sex steroids	Glucocorticoids
Slow wave sleep	REM sleep
Exercise	Hyperglycemia
Starvation	FFA
Protein-rich meal	
Psychological stress	

a. Growth Hormone Releasing Hormone (GHRH)

GHRH is a 44-amino acid polypeptide produced in the arcuate nucleus of the hypothalamus. The neuronal terminals secrete GHRH to reach the anterior pituitary somatotrophs via the portal venous system, which leads to GH transcription and secretion. Moreover, animal studies have clearly demonstrated that GHRH plays a vital role in the proliferation of somatotrophs in the anterior pituitary, whereas the absence of GHRH has been shown to lead to anterior pituitary hypoplasia. This concept of a hypothalamic GHRH was strengthened by observations that the addition of rat hypothalamus extract increased GH secretion from cultured rat pituitaries, but rat cerebral cortex extract had no such effect.[6] The secretion of GHRH is stimulated by several factors including slow wave sleep, hypoglycemia, ghrelin, sex steroids, protein-rich meal, starvation, exercise and psychological stress. It is inhibited by somatostatin, insulin-like growth factor I (IGF-1), REM sleep, hyperglycemia and increased FFA concentration (Fig. 21.2).

b. Ghrelin

Ghrelin is a 28-amino acid peptide that is the natural ligand for the growth hormone secretagogue receptor. In addition, ghrelin targets the arcuate nucleus, from where growth hormone releasing hormone (GHRH) neurons trigger GH secretion (Fig. 21.2). In fact, ghrelin and GHRH have a synergistic effect in increasing circulating growth hormone levels.[7]

c. Sex Steroids

There is a close relationship between rising serum androgen concentrations and increased GH peak amplitude in healthy pubertal boys.[5] In pre- and peripubertal boys, testosterone in physiological and pharmacological doses increases spontaneous and stimulated GH secretion. The GH concentration also rises throughout puberty in females, in whom the increase is proportional to the increase in serum estradiol concentration.

d. Nutritional Factors

Nutritional factors also influence GH secretion. Secretion is increased in malnourished or fasting individuals and following high protein meals and intravenously administered amino acids. Insulin-induced hypoglycemia is also a powerful stimulus to GH release. Conversely, secretion is inhibited by hyperglycemia and leptin. Stress, hypoglycemia and ingestion of protein (high levels of circulating amino acids) stimulates GH secretion, while high levels of glucose and FFA inhibits secretion (Fig. 21. 2).

e. Somatostatin (SST)

Somatostatin is a cyclic peptide, which mostly exerts inhibitory effects on endocrine and exocrine secretions. Many cells in the body including specialized cells in the hypothalamic anterior periventricular nucleus and arcuate nucleus produce SST. These neurons secrete SST into the hypothalamic-hypophyseal portal venous system, via the median eminence, to exert its effect on the anterior pituitary. The secretion of SST by the hypothalamic neurons is induced by high blood glucose, FFA and serum IGF-1 level and REM sleep.

f. Insulin-like Growth Factor-1

Insulin-like growth factor 1 (discussed below) mediates most of GHs peripheral actions, and acts to inhibit GH secretion.

g. Glucocorticoids

GH secretion is inhibited by glucocorticoid excess. In children exposed to excess cortisol secretion, GH secretion has been shown to decrease. Glucocorticoids are well known to inhibit growth and GH secretion in humans and animals, yet *in vitro* these steroids stimulate GH synthesis and secretion. These opposite actions appear to be mediated at different sites. The inhibition involves modulation of hypothalamic somatostatin and the stimulation involves direct actions on the pituitary. Current evidence suggests that the predominant action in vivo is through the inhibitory influences of somatostatin.[8]

Insulin-like growth factor 1 (IGF-1): Growth hormone acts on the liver to stimulate the production and secretion of IGF-1. The liver is the predominant source of circulating

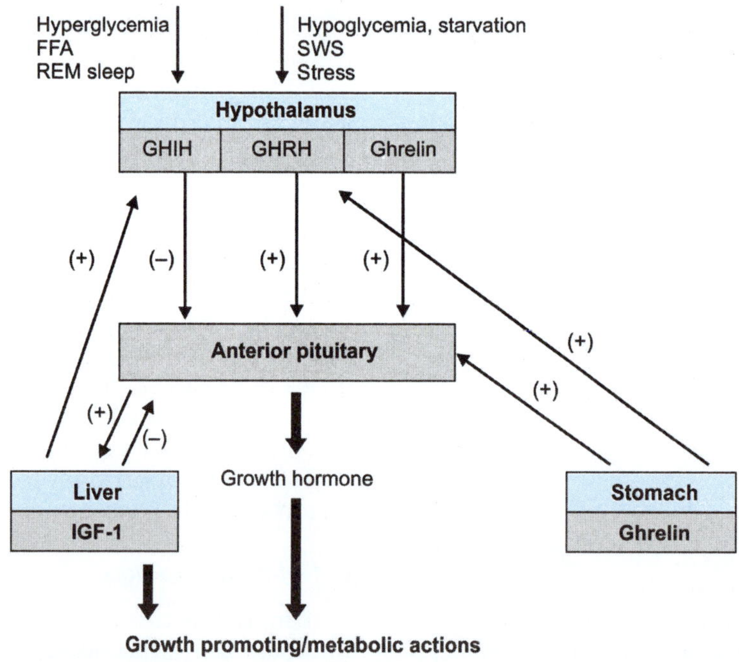

Fig. 21.2: Regulation of secretion of growth hormone and IGF-1

IGF-1; however, *liver-specific IGF-1 deficient* (LID) mice display only a 75% reduction in circulating IGF-1 levels suggesting that extrahepatic tissues contributed to the remaining 25% of the circulating IGF-1 pool. Apart from the endocrine/circulating IGF-1, IGF-1 is also produced locally by tissues and exerts its effects in an autocrine/paracrine manner. IGF-1 plays a critical role in inhibiting GH secretion by feedback mechanisms by stimulating somatostatin and inhibiting GHRH release from the hypothalamus.[9]

Interestingly, insulin and IGF-1 share many structural and functional similarities implying that they have originated from the same ancestral molecule. Both molecules could have been part of the cycle of food intake and consequent tissue growth.

Multiple animal and human studies have analyzed the physiological effects of IGF-1 over the last two decades. Initial studies on hypophysectomized animals showed IGF-1 to promote growth in all tissues. These studies also demonstrated the anabolic effects of IGF-1 by way of increasing GFR, improving wound healing and reversal of catabolic effects of nutritional deprivation.[10]

ACTIONS OF GH/IGF-1

A. Growth Promotion

IGF-1 plays a key role in growth, where it acts not only as a determinant of postnatal growth but also as an intrauterine growth promoter. Total inactivation of the IGF-1 gene in mice produced a perinatal mortality of 80% with the surviving animal showing significant growth retardation compared to controls. More detailed studies on transgenic mice have clearly demonstrated this fact with selective deletion of IGF-1 gene expression only in the liver, showing low serum IGF-1 concentrations with only 6–8% postnatal growth retardation. In contrast, animals with total IGF-1 deletion or only peripherally produced IGF-1 deletion showed marked growth retardation. Therefore, considering growth, both paracrine/autocrine IGF-1 and liver-derived IGF-1 are important in promotion of growth.[11]

In postnatal period, one of the main actions of GH via IGF-1 in children is increase in linear growth of long bones. The site of action is a thin layer of cartilage between epiphysis and metaphysis, called epiphyseal plate or growth plate (Fig. 21.3). Growth plate is a thin layer of cartilage between epiphysis and metaphysis, where growth of long bones occurs by chondrogenesis.[12]

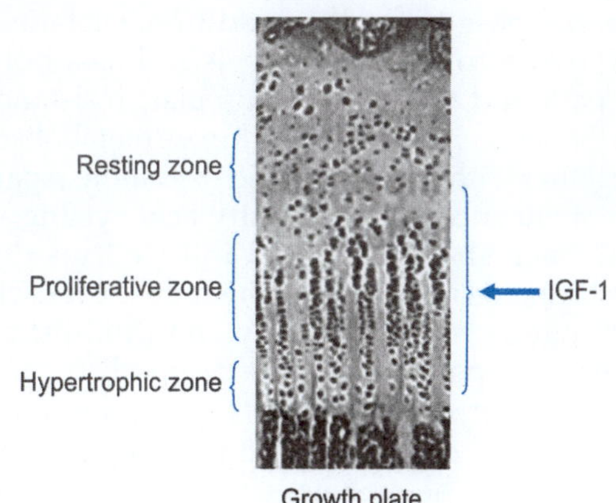

Fig. 21.3: Mechanism of action of GH/IGF-1 on linear growth

B. Metabolic Effects

a. Glucose Metabolism: Anti-insulin Action

The involvement of the pituitary gland in the regulation of glucose metabolism was originally observed by Houssay.[13] Hypophysectomized animals showed profound fasting hypoglycemia, which could be corrected by administration of anterior pituitary extracts. He also reported that diabetes produced by pancreatectomy was alleviated by hypophysectomy.

It is now well established that GH can stimulate hepatic glucose production by two mechanisms: Gluconeogenesis and glycogenolysis. However, it is still controversial whether GH preferentially stimulates gluconeogenesis or glycogenolysis and there is data to support either theory.

Almost forty years ago, it was shown that infusion of high dose GH into the brachial artery of healthy adults reduced forearm glucose uptake in both muscle and adipose tissue. This was paralleled by a drop in RQ and an increase in muscle uptake of FFA, both of which suggested oxidation of FFA by the muscle. This pattern was opposite that of insulin, and co-administration of insulin and GH resulted in only minimal changes in net fluxes of glucose and FFA across the forearm bed. These studies clearly indicated direct insulin antagonistic effects of GH on muscle and adipose tissue.[14] In addition, sustained exposure to high GH levels induces both hepatic and peripheral (muscular) resistance to the actions of insulin on glucose metabolism together with increased lipid oxidation. Apart from enhanced glucose/fatty acid cycling, it has been shown that GH induced insulin resistance is accompanied by reduced muscle glycogen synthase activity and diminished glucose-dependent glucose disposal.[15]

Growth hormone (GH) has been established as one of the counterregulatory hormones. The diabetogenic effect of GH is also supported by the high prevalence of diabetes in acromegaly patients.

b. Lipid Metabolism

Growth hormone plays a major role in lipid metabolism. The most striking effect of a single exogenous GH dose is a marked increase in circulating levels of FFA and ketone bodies, reflecting stimulation of lipolysis and ketogenesis. Plasma FFA values are usually more than double at 2–3 hours and the effect lasts up to 8 hours (Fig. 21.4). GH exerts a lipolytic effect predominantly in the visceral adipose tissue, and to a lesser extent in the subcutaneous adipose tissue. The depot-specific effect of GH could be explained by the fact that GH increases lipolysis by increasing adipose tissue hormone-sensitive lipase activity.[16, 17]

c. Protein Metabolism

The clinical picture of acromegaly and gigantism includes increased lean body mass of which skeletal muscle mass accounts for approximately 50%. Moreover, retention of

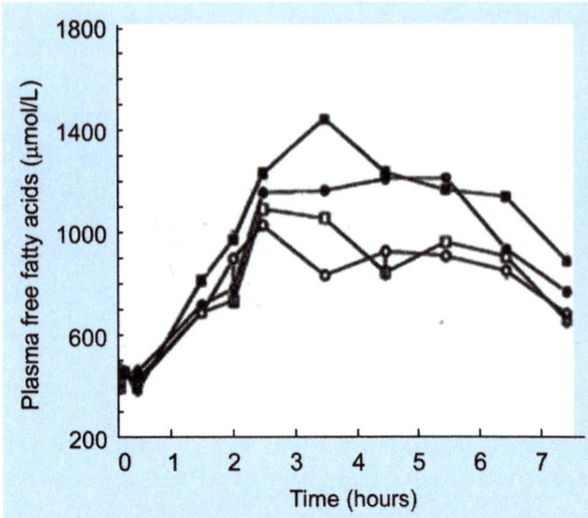

Fig. 21.4: Effect of administration of a single bolus of growth hormone on plasma FFA levels in four normal subjects, after Vijayakumar et al[17]

nitrogen was one of the earliest observed and most reproducible effects of GH administration in humans. Studies with GH administration in GH deficient children, using a variety of classic anthropometric techniques, strongly suggested that skeletal muscle mass increased significantly during treatment.[18] Based on these recent studies, it seems that the nitrogen retaining properties of GH predominantly involve stimulation of protein synthesis.

GH may exert metabolic effects either directly or indirectly through increased production of IGF-1. GH administration increases circulating IGF-1 levels via stimulation of hepatic synthesis and secretion of IGF-1; may also enhance local IGF-1 synthesis, which exerts paracrine or autocrine effects. Administration of GH or IGF-1 in adult humans has been demonstrated to enhance protein anabolism. Combined administration of GH and IGF-1 was observed to be more anabolic than either IGF-1 or GH alone. It seems, protein accretion results mainly from direct effects of GH on tissues; additional indirect effects via IGF-1 production are also likely.[19] This increase in overall protein synthesis is accompanied by an increase in muscle RNA content suggesting an increase in peptide chain initiation. These results confirmed and extended earlier observations that GH increased muscle ribosome content, RNA polymerase activity, and protein synthesis.[20]

Bush et al[21] studied the rate of protein synthesis in fasted and fed state. In fed state, pigs had greater protein synthesis as compared to fasted state. The effect of feeding was further enhanced in pigs who were also administered exogenous GH (Fig. 21.5). They concluded that growth hormone improves the efficiency with which dietary proteins are used for protein synthesis.

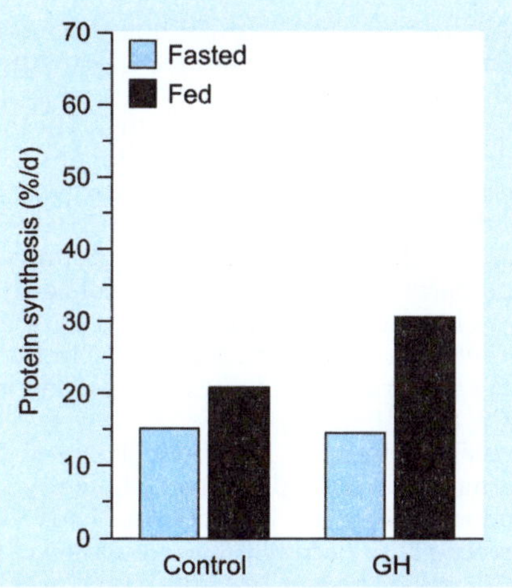

Fig. 21.5: Rate of muscle protein synthesis in fasted and fed state in control and growth hormone administered pigs, after Bush et al[21]

Effect of Age on GH/IGF-1 Level: Somatopause?

Growth hormone (GH) secretion declines progressively with age (Fig. 21.6), and many age-related changes such as a decrease in lean body mass and an increase in body fat, especially in the visceral/abdominal compartment. These observations led Rudman et al[22] to the hypothesis that many of the senescent changes in body composition and organ function were related to or caused by hyposomatotropinemia. The term "somatopause" has been used for Rudman's hypothesis. Many studies done in the late 90s have uniformly documented that adults with severe GH deficiency are characterized by

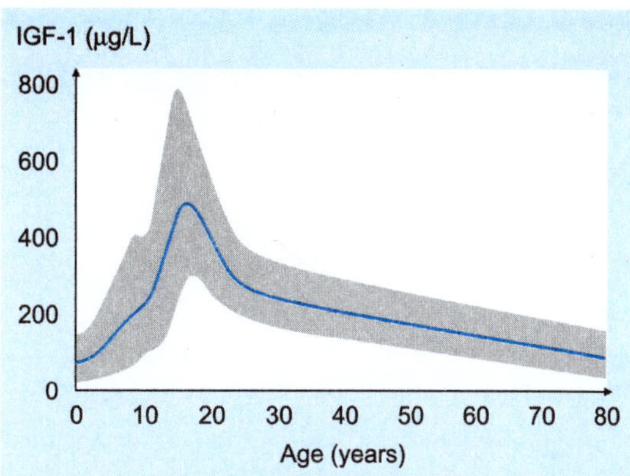

Fig. 21.6: Plasma IGF-1 levels in relation to age

increased fat mass and reduced lean body mass. It is also known that GH substitution in GH deficient adults normalizes body composition.[22]

Increased adiposity is another feature of aging. Spontaneous GH secretion is inversely related to increase in intra-abdominal fat (Fig. 21.7). The response of GH secretion to GHRH is also reduced with increasing BMI, independent of sex and age.

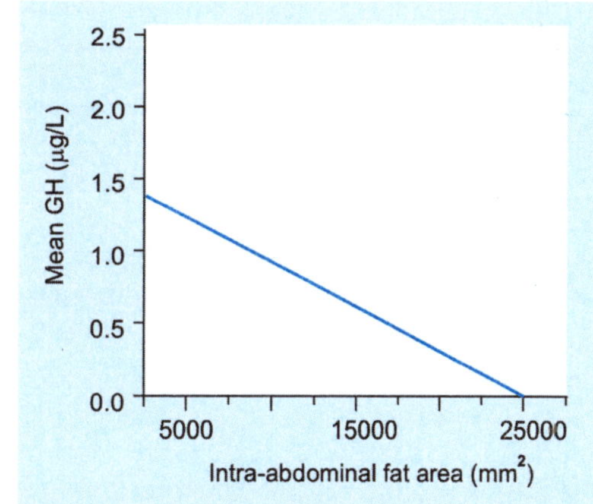

Fig. 21.7: Relation between growth hormone secretion and intra-abdominal fat

The mechanisms underlying this relationship to adiposity are unclear. Whether reduced pituitary GH secretion drives adiposity, in particular visceral fat, or whether the low GH secretion is a result of adiposity remains to be established.[6] Caloric restriction and weight loss restore the GH secretion to normal levels.

References

1. Lindholm, J. Pituitary (2006) 9: 5. https://doi.org/10.1007/s11102-006-7557-4
2. Cushing H. The Pituitary Body and Its Disorders, Clinical States Produced by Disorders of the Hypophysis Cerebri. Philadelphia: J.B. Lippincott, 1912.
3. Ayyar VS. History of growth hormone therapy. Indian J Endocrinol Metab. 2011: 15: S162–65.
4. Van Cauter E, Leproult R, Plat L. Age-related changes in slow wave sleep and REM sleep and relationship with growth hormone and cortisol levels in healthy men. JAMA, 2000; 284: 861–68.
5. Meinhardt UJ, Ho KK. Modulation of growth hormone action by sex steroids. Clin Endocrinol (Oxf). 2006; 65: 413–22.
6. Murray PG, Higham CE, Clayton PE. The hypothalamo growth hormone axis: the past 60 years. J Endocrinol. 2015; 226: 123–40.
7. Osterstock G, Escobar P, Mitutsova V, et al. (2010). Ghrelin Stimulation of Growth Hormone-Releasing Hormone Neurons is Direct in the Arcuate Nucleus. PLoS ONE 5(2): e9159. https://doi.org/10.1371/journal.pone.0009159
8. DOI: https://doi.org/10.1016/1043-2760 (92) 90142-N
9. Ohlsson C, Mohan S, Sjögren K, et al. The role of liver-derived insulin-like growth factor-I. Endocr. Rev. 2009; 30: 494–535.
10. Behringer RR, Lewin TM, Quaife CJ, et al. Expression of insulin-like growth factor 1 stimulates normal somatic growth in growth hormone-deficient transgenic mice. Endocrinology. 1990; 127: 1033–45.
11. Gunawardane K, Hensen TK, Muller N, et al. Normal physiology of growth hormone in adults. De Groot LJ, Chrousos G, Dungan K, et al., editors. South Dartmouth (MA): MDText.com, Inc.; 2000.
12. http://www.ncbi.nlm.nih.gov/books/NKB343487
13. Houssay BA. The hypophysis and metabolism. N Eng J of Medicine. 1936; 214: 961–85.
14. Rabinowitz D, Klassen GA, Zierler KL. Effects of human growth hormone on muscle and adipose tissue metabolism in the forearm of men. J Clin Invest. 1965; 44: 51–61.
15. Ørskov L, Schmitz O, Jørgensen JOL, et al. Influence of growth hormone on glucose induced glucose uptake in healthy men as assessed by the hyperglycaemic clamp technique. J Clin Endocrinol Metab. 1989; 68: 276–82.
16. Vijaykumar A, Novosyadlyy R, Wu Y, et al. Biological effects of growth hormone on carbohydrate and lipid metabolism. Growth Horm IGF Res. 2010; 20: 1–7.
17. Vijayakumar A, Yakar S, LeRoith D. The intricate role of growth hormone in metabolism. Front Endocrinol 2011, https://doi.org/10.3389/fendo.2011.00032
18. Tanner JM, Hughes PCR, Whitehouse RH. Comparative rapidity of response of height, limb muscle and limb fat to treatment with human growth hormone in patients with and without growth hormone deficiency. Acta Endocrinol. 1977; 84: 681–696.
19. Berneist K, Keller U. Metabolic actions of growth hormone: direct and indirect. Baillière's Clinical Endocrinology and Metabolism. 1996 10: 337–52.
20. Fryburg D A, Berret EJ. The regulation of amino acid and protein metabolism by growth hormone. J Nutr. 2006; 136: S212-17.
21. Bush JA, Kimball SR, O' Connor PMJ, et al. Translational control of protein synthesis in muscle and liver of growth hormone treated pigs. Endocrinology. 2003; 144: 1273–83
22. Rudman D, Feller AG, Nagraj HS, et al. Effects of human growth hormone in men over 60 years old. N Engl J Med. 1990; 323:1–6.

Chapter 22

Islets of Langerhans

INTRODUCTION

The islets of Langerhans were discovered in 1869 by Paul Langerhans when he was a medical student in Berlin. A student of the eminent pathologist Rudolf Virchow, Langerhans described the microscopic anatomy of the rabbit pancreas in his MD thesis and reported the presence of "… small cells of almost perfect homogeneous content and of a polygonal form, with round nuclei, mostly lying together in pairs or small groups" (English translation). The function of these cells was, of course, unknown to Langerhans. Except for describing their morphology, he did not give them a name. The term "islets of Langerhans" was introduced in 1893 by Edouard Laguesse, who observed them in the human pancreas and (with remarkable foresight) suggested that they may produce internal secretions that regulate "glycemia".[1]

DISCOVERY OF VARIOUS TYPES OF CELLS IN ISLETS OF LANGERHANS

The first bonafide histochemical differentiation of islet cell types can be attributed to Lane[2], who identified two types of granular cells in islets based on their different solubility in alcohol and a special stain. When the tissue was fixed in alcohol and stained, some cells whose cytoplasm was deeply stained were called "A cells" and those cells with faintly stained cytoplasm were called "B cells" by Lane (Fig. 22.1). This century old observation fits with the now well-known fact that ethanol extracts insulin from the pancreas (the cause of lighter staining property of B cells). A scientist described "C cells" in the islets, but his claim could not be substantiated. In 1931, Bloom using another special stain not only observed A and B cells but also a third type of

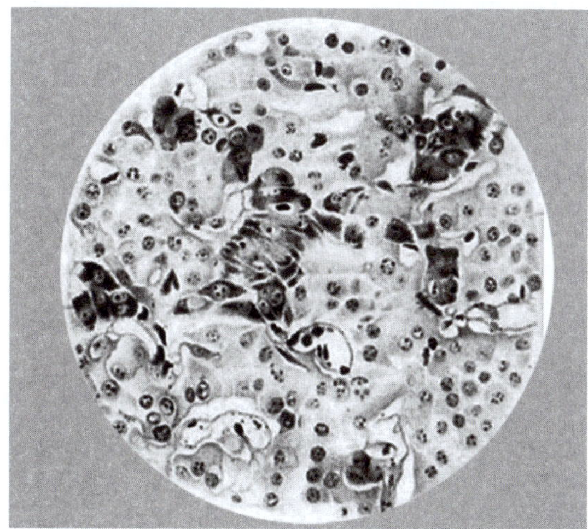

Fig. 22.1: Historical picture of islet of Langerhans published by Lane.[2] Alpha cells have dark stained cytoplasm, whereas beta cells show light stained cytoplasm

cells which he called "D cells." Thus by 1931, A, B, and D types of cells of the islets were firmly established. About 50 years later, development of immunochemistry techniques led to the discovery of F (pancreatic polypeptide) cells in the islets of Langerhans. A, B and D types of cells of the islets cells are also known by Greek symbols, α, β, δ cells, respectively.

A recent study has shown that β, α, and δ cells are scattered throughout the human islets. Thus, human islets do not show the anatomical subdivisions like rodent islets where the β cells are concentrated in the core of the islet, and α and δ cells are located in the mantle (Fig. 22.2).[3] Functional significance of this difference, if any, is not known.

The average adult human pancreas is estimated to contain one to two million islets. In the human pancreas, the concentration of islets is up to two times higher in the tail compared to the head and neck. However, the cellular composition and architectural organization of cell types within the islets is preserved throughout the pancreas. Insulin-producing β cells are the most common cell types, making up 50–70% of islet mass. As the second most abundant islet cell type, α cells make up about 35% of islet mass in humans. Somatostatin-producing δ cells comprise less than 10% of islet mass, and are evenly distributed throughout the pancreas. Pancreatic polypeptide (PP) producing cells, also known as F cells, comprise less than 5% of islet mass. Notably, the islets are highly vascularized—receiving 15% of pancreatic arterial blood flow despite composing only 2% of the pancreatic mass.

Discovery of Insulin and Glucagon

In 1889, Minkowskiand Mering observed that removal of a dog's pancreas immediately resulted in diabetes, but ligation of the pancreatic duct had no effect. They concluded that the pancreas is of crucial importance for maintaining euglycemia and that the pancreas contains one or more factors that regulate blood glucose through secretion to the general circulation. In the years that followed, several independent scientists, such as Kleiner and Paulescu successfully used pancreatic extracts to lower blood glucose in diabetic dogs. These studies collectively paved the path to the isolation of insulin in 1921 by the well-acclaimed team working in Toronto led by Fredrick Banting.[4] They had isolated material from pancreas extracts that dramatically prolonged the lives of dogs made diabetic by removal of the pancreas. The very next year in 1922, Banting and Best treated their first

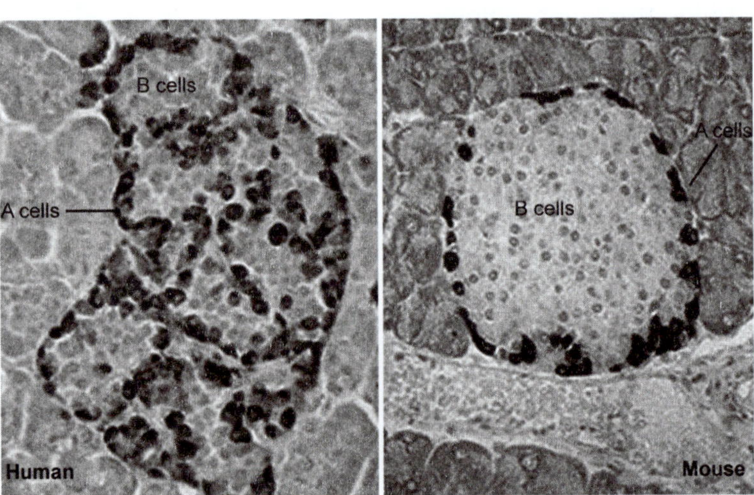

Fig. 22. 2: Pattern of distribution of A and B cells in human and mouse islet of Langerhans. In human islets A cells surround clusters of B cells. In mouse islet, A cells are located at the periphery of the B cells mass

human patient, a diabetic young boy, whose life was saved by the treatment. This is probably a rare example when a research finding was put to clinical use almost immediately.

In 1921, when F. Banting and C. Best tested their first pancreatic extracts in depancreatized dogs, they observed that insulin-induced hypoglycemia was, surprisingly, preceded by a transient, rather mild hyperglycemia. In 1923, Murlin et al suggested that the early hyperglycemic effect of the pancreatic extracts was due to a contaminant with glucogenic properties and named it "glucagon," or the mobilizer of glucose. In a classical paper published in 1948, Sutherland and de Duve established the α cells of the pancreas as being the source of glucagon.[5]

A. REGULATION OF INSULIN SECRETION

The principal level of control on glycemia by the islet of Langerhans depends largely on the coordinated secretion of insulin and glucagon by β and α cells respectively. Both cell types respond oppositely to changes in blood glucose concentration: While β cells release insulin when glucose levels increase whereas hypoglycemic conditions induce α cell secretion. Insulin release from pancreatic β cells is tightly regulated, and allows the sensitive response of insulin levels to plasma glucose and amino acids. In addition, some GI hormones and parasympathetic innervation influence the insulin release.

a. Glucose Stimulated Insulin Secretion

Glucose-stimulated β cell insulin release is the primary mechanism of regulation of insulin secretion. GLUT2 has been traditionally assumed to act as the major mediator of glucose uptake into β cells. It was based on extrapolation from rodent studies and subsequent confirmation of GLUT2 transporters on human β cells. However, more recent studies in human islets suggest that the other insulin-independent glucose transporters GLUT1 and GLUT3 are also involved.[5, 6]

In humans, the effect of hyperglycemia has been studied by use of the hyperglycemic clamp technique. In this technique, individuals are made rapidly hyperglycemic by injection of intravenous glucose, and hyperglycemia is maintained at a predefined target plasma glucose level by variable rate of glucose infusion. Hyperglycemic clamp studies demonstrate a linear relationship between plasma glucose level and rate of insulin secretion in the non-diabetic individual[6] (Fig. 22.3).

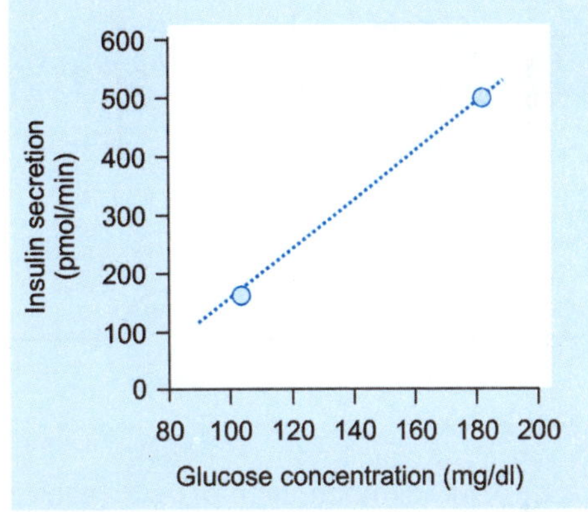

Fig. 22.3: Linear relationship between plasma glucose level and rate of insulin secretion

Mechanism of insulin release: In β cells, the main stimulus for insulin release is elevated blood glucose levels. The circulating blood glucose is taken up by the facilitative glucose transporter 2 (GLUT2), which is located on the surface of the β cells. Once inside the cell, glucose undergoes glycolysis, thereby generating adenosine triphosphate (ATP), resulting in an increased ATP/ADP ratio. This altered ratio then leads to the closure of ATP-sensitive K^+ channels (K_{ATP} channels). Under non-stimulated conditions, these channels are open to ensure the maintenance of the resting potential by transporting positively charged K^+ ions down their concentration gradient out of the cell. Upon closure, the subsequent decrease in the magnitude of the outwardly directed K^+ current elicits the depolarization of the membrane, followed by the opening of voltage-dependent Ca^+ channels. The increase in intracellular calcium concentrations eventually triggers the fusion of insulin-containing granules with the membrane and the subsequent release of their content by exocytosis[6,7] (Fig. 22.4).

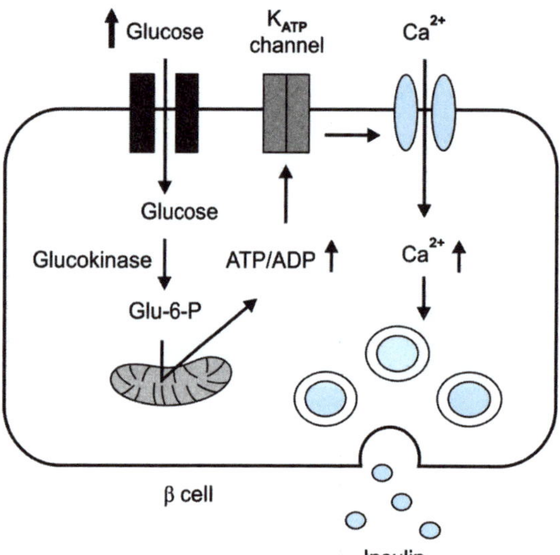

Fig. 22.4: Cellular mechanism of insulin secretion

b. Proteins and Amino Acids

Besides glucose, pancreatic β cells adjust insulin secretion based on many other nutrients including amino acids and fatty acids. Oral protein intake, and subsequent rise in serum amino acids, stimulates insulin release by direct β cell stimulation. The insulinotropic effect varies among amino acids, and there appears to be a synergistic effect of mixed amino acids versus individual administration.[19]

Some amino acids stimulate insulin secretion by acting as substrates in the Krebs cycle, metabolizing glucose-6-phosphate to create ATP. The ATP binds to and closes the potassium channel, leading to cell depolarization and insulin secretion as discussed above.

c. Incretin Hormones

Though glucose concentrations can account for a majority of changes in plasma insulin concentrations, studies evaluating *in vivo* insulin concentrations following meals have identified other factors. Indeed, insulin secretion following oral glucose tolerance test is directly related to blood glucose levels, but is considerably higher than predicted following intravenous glucose infusion (Fig. 22.5). These findings suggest a role for potentiating effects on insulin release by hormones that specifically respond to oral glucose. Intestinal hormones called incretins are credited with facilitating this response. The most active incretins are glucagon-like peptide-1 (GLP-1) and glucose-dependent insulinotropic polypeptide (GIP), but gastrin, secretin, and cholecystokinin may also have minor roles. In response to glucose and other nutrients, intestinal L cells secrete GLP-1 and K cells secrete GIP. These hormones then bind their specific receptors on the pancreatic β cell membrane.[6]

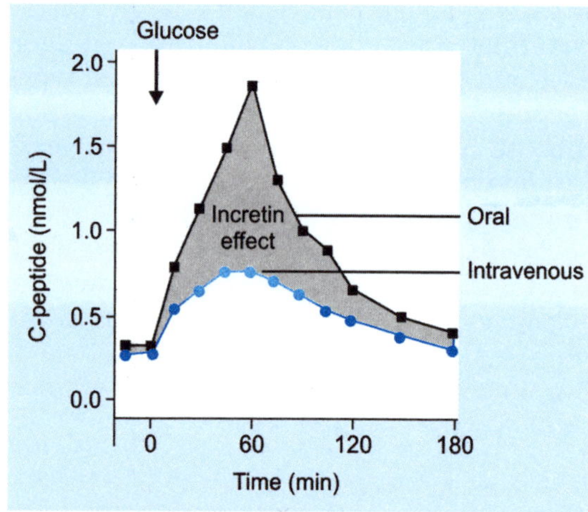

Fig. 22.5: Effect of oral and intravenous glucose administration on C-peptide levels. C-peptide level reflects insulin level

d. Autonomic Innervation

The pancreatic islets are richly innervated by autonomic nerves. The islet *parasympathetic* nerves emanate from intrapancreatic ganglia, which are controlled by preganglionic vagal nerves. The islet *sympathetic* nerves are postganglionic with the nerve cell bodies located in ganglia outside the pancreas. The *sensory* nerves originate from dorsal root ganglia near the spinal cord. Inside the islets, nerve terminals run close to the endocrine cells. The classic neurotransmitters are acetylcholine and norepinephrine. *Activation of the parasympathetic nerves or administration of their neurotransmitter stimulates insulin and glucagon secretion*, whereas *activation of the sympathetic nerves or administration of their neurotransmitter inhibits insulin but stimulates glucagon secretion*. The autonomic nerves contribute to the cephalic phase of insulin secretion, to glucagon secretion during hypoglycemia, and to the inhibition of insulin secretion, which is seen during stress.[8] Various factors regulating insulin secretion are given in Table 22.1.

Table 22.1: Factors regulating insulin secretion

Stimulating factors	*Inhibiting factors*
Glucose	Hypoglycemia
Amino acids	Sympathetic stimulation: Epinephrine
Free fatty acids	
Incretins (GIP, GLP-1)	
Parasympathetic stimulation	

Phases of Insulin Secretion

i. Cephalic Phase

After food intake, a small amount of insulin is released even before there are significant increments in plasma glucose concentrations. This is known as preabsorptive, or cephalic phase of insulin secretion.[9] Even though preabsorptive insulin release is a relatively minor component of total postprandial β cell secretion, it appears to be physiologically important, because studies in both rats and humans have shown that loss of the pre-absorptive insulin response impairs glucose tolerance.[10] Preabsorptive insulin secretion can be prevented by vagotomy or muscarinic blockade with atropine (a muscarinic cho-linergic receptor blocker) or trimethaphan (a nicotinic cholinergic receptor blocker)[9]

(Fig. 22.6). This observation indicates that preabsorptive insulin secretion occurs primarily as a result of activation of the parasympathetic cholinergic nerves innervating the islets (cephalic phase (Cf) of gastric secretion).

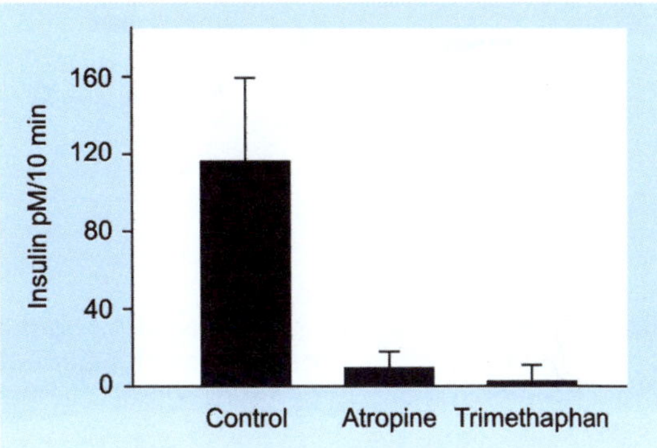

Fig. 22.6: Effect of parasympathetic blocking agents on the rate of insulin secretion during the first 10 minutes after a meal (preabsorptive phase). There is almost complete blockade of insulin secretion, after Alesso et al[11]

ii. Postabsorptive Phase

The traditional view of meal-induced insulin secretion has been that once nutrient absorption has commenced, stimulation of the β cell is mediated predominantly by circulating glucose and the incretin hormones, and that the PNS does not contribute significantly to insulin secretion during this period. However, blockade of cholinergic receptors by atropine or trimethaphan results in suppression in postabsorptive insulin secretion (Fig. 22.7). It is obvious that, besides nutrients, parasympathetic-mediated insulin secretion also contributes to postabsorptive phase of insulin secretion.[11]

B. REGULATION OF GLUCAGON SECRETION

The secretion of glucagon by pancreatic α cells plays a critical role in the regulation of glycemia. This hormone counteracts hypoglycemia and opposes insulin actions by stimulating hepatic glucose synthesis and mobilization, thereby increasing blood glucose concentrations.

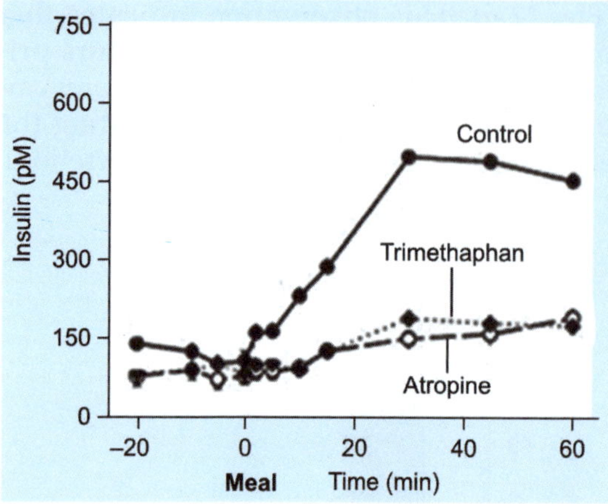

Fig. 22.7: Effect of parasympathetic blocking agents on the plasma insulin concentration after a meal. There is partial blockade of insulin secretion, after Alesso et al[11]

a. Glucose

Both ingestion of glucose and infusion of glucose, which result in hyperglycemia, normally cause a decrease in plasma glucagon concentrations. Conversely, decrements from both normoglycemic and hyperglycemic plasma glucose concentrations stimulate glucagon secretion (Fig. 22.8).

Increments in plasma glucagon concentration are significantly correlated with the magnitude and the rate of the decrease in plasma glucose concentration.[12, 13]

Cellular mechanism of glucagon release: Glucose is taken up by the α cells through the glucose transporter 1 (GLUT1). The cellular machinery controlling the secretion of glucagon in α cells is remarkably similar to that regulating the secretion of insulin in β cells, yet their secretion is reciprocally regulated by blood glucose concentration. In contrast to the secretion of insulin, the release of glucagon is stimulated under conditions of hypoglycemia and subsequently decreases when blood glucose increases (Fig. 22.9).

Similar to the β cells, the α cells comprise a series of ion channels that modulate the membrane potential in a glucose-dependent manner. In contrast to the β cells, however, the α cells require a lower intracellular ATP concentration to inhibit the K_{ATP} channels, and thus to open the voltage-dependent Ca^{2+} channels. As a result, under conditions of high glucose concentration, the K_{ATP} channels depolarize the membrane potential to a point where Na^+ and Ca^{2+} channels are inactive. The resulting lack of Ca^{2+} and Na^+ influx prevents glucagon exocytosis and inhibits the release of glucagon to the general circulation. When blood glucose concentration is low, K_{ATP} channels of the β cells are open, and this results in a membrane potential that leads to closing of the voltage-dependent calcium channels, and subsequently prevents Ca^{2+}

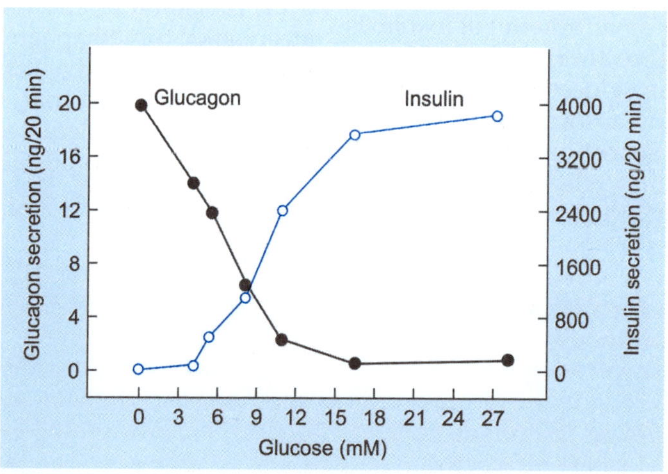

Fig. 22.8: Reciprocal relation between insulin and glucagon secretion

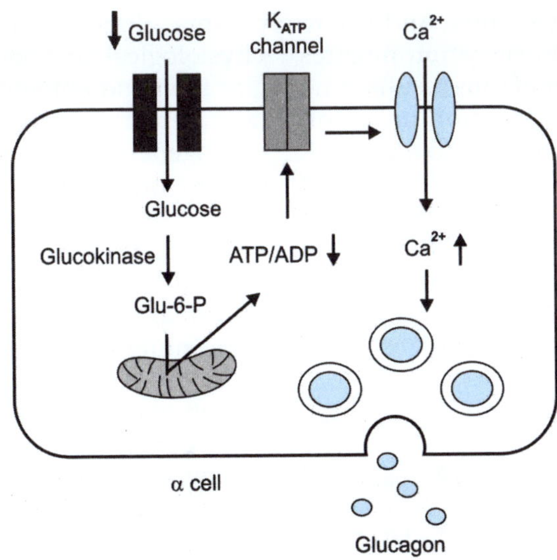

Fig. 22.9: Cellular mechanism of glucagon secretion

influx and insulin exocytosis. In contrast, the K_{ATP} channels of the α cells are closed under conditions of low glucose (and low ATP levels) and the K_{ATP} channels therefore impose a membrane potential that leads to opening of Ca^{2+} channels. The subsequent Ca^{2+} influx triggers the release of glucagon through exocytosis of glucagon granules[4] (Fig. 22.9).

Intrinsic *vs* extrinsic mechanism of α cell response: Glucose is a major determinant of glucagon (and insulin) secretion. Whereas the mechanisms by which glucose regulates insulin secretion from the β cells have been well established, mechanism of glucagon secretion from α cells in response to glucose remains controversial. The moot point is whether glucagon secretion is regulated by glucose acting via an *intrinsic mechanism* (within the α cells) or by factors released from other cells within the islets, i.e. insulin, somatostatin, etc. (*paracrine mechanisms*).

Since glucose stimulates insulin release, and insulin inhibits glucagon secretion, it is looks plausible that inhibition of glucagon secretion from α cells is secondary to stimulation of the β cells. A similar argument could be made for somatostatin, the release of which is also stimulated by glucose. On the other hand, the argument that paracrine factors influence glucagon secretion are supported by reports that *isolated* α cells (that no longer have paracrine input) are unable to respond appropriately (i.e. decreasing their activity and glucagon secretion) to increased glucose concentrations; in fact, the reports demonstrate that glucose *stimulates* glucagon secretion when α cells are removed from their normal paracrine environment.

Insulin released from the β cells was first to be suggested and is still considered as a putative mediator of glucose-inhibited glucagon secretion. At present, the consensus is that *under hypoglycemic conditions α cell-intrinsic glucose sensing mechanism of glucagon release is probably dominating. In hyperglycemia paracrine factors become more important* for glucose regulation of glucagon release.[14]

Based on these data, the physiological construct is as follows[15]: (1) β cell insulin, perhaps among other β cell secretory products, tonically restrains α cell glucagon secretion during postabsorptive euglycemia; (2) a decrease in α cell insulin secretion, in concert with a low plasma glucose concentration, signals an increase in α cell glucagon secretion during hypoglycemia; and (3) an increase in β cell insulin secretion negates direct α cell stimulation and thus results in no change in, or even suppression of, α cell glucagon secretion after a mixed meal. Clearly, insulin is β cell secretory product that, in concert with glucose and among other signals, reciprocally regulates α cell glucagon secretion in humans.[16] The intimate relationship between β cell insulin secretion and α cell glucagon secretion is illustrated in Fig. 22.10.

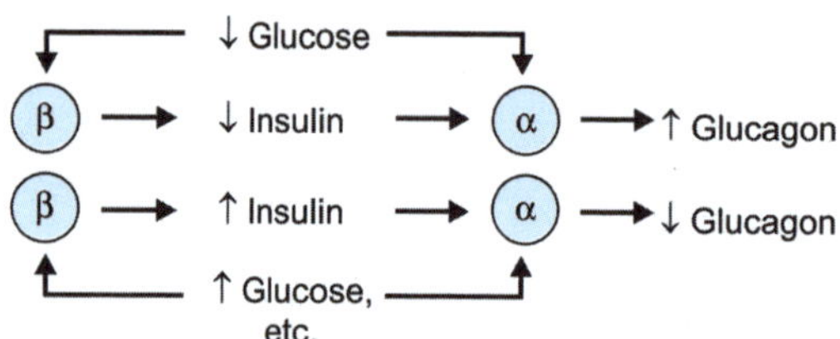

Fig. 22.10: Alpha cell–beta cell interaction in response to glucose, after Cryer[16]

b. Amino Acids

Amino acids stimulate release of both hormones and in general are more effective stimulators of glucagon than of insulin secretion.[17] Individual amino acids differ in their ability to initiate glucagon secretion, with arginine being the most potent. Under physiological conditions, amino acids play an important role in the control of glucagon release. For example, insulin response to a protein meal would cause hypoglycemia if it was not for concomitant increase in amino acid-mediated increase in glucagon secretion. These findings suggest that circulating amino acids play an important role for the regulation of glucagon release *in vivo*.[5]

c. Autonomic Regulation

Increased glucagon release represents the most important glucose counterregulatory factor that is critical to prevent or rapidly correct hypoglycemia. As arterial plasma glucose levels decline in the physiological range, insulin secretion decreases mainly due to glucose sensing in the β cell. This will enhance hepatic glucose production. A fall in

plasma glucose concentration below the physiological range stimulates the secretion of glucagon and epinephrine. Glucagon stimulates hepatic glycogenolysis as well as gluconeogenesis. Epinephrine enhances hepatic glucose output and decreases glucose clearance by tissues such as muscle and fat.

Two main mechanisms have been proposed to mediate the α cell response to hypoglycemia. The first mechanism involves relief from inhibitory paracrine/endocrine influences by neighboring β and δ cells as discussed above. The second mediator of the glucagon release to hypoglycemia is the autonomic input to the α cell. There are three major autonomic influences on the α cell: sympathetic nerves, parasympathetic nerves, and circulating epinephrine. The two classical autonomic inputs, the islet sympathetic and parasympathetic nerves, are also activated by hypoglycemia, and both stimulate glucagon secretion from the α cell. Thus, *all three autonomic inputs are activated by hypoglycemia and are capable of stimulating glucagon secretion.* However, the relative contribution of the different autonomic inputs to enhance glucagon secretion during hypoglycemia is unknown.[5]

According to Taborsky and Mundinger[18] there is progressive recruitment of autonomic inputs as the glucose levels fall from a fasting level (Fig. 22.11) all the way to a near lethal level of 15 mg/dl. The implication is that their relative contributions to the glucagon response depend on the severity of the hypoglycemia. The parasympathetic nervous system is the first autonomic input to the islet activated because its threshold of activation occurs at a glucose level between 85 and 75 mg/dl[20] which is very mild hypoglycemia by any standard. More marked hypoglycemia is required to activate pancreatic sympathetic nerves since such activation is first detected around 35 mg/dl.[20] The hypoglycemic threshold for activation of the sympathoadrenal system falls between those of pancreatic parasympathetic and sympathetic nerves. Thus, there is a modest epinephrine response to insulin-induced hypoglycemia that begins

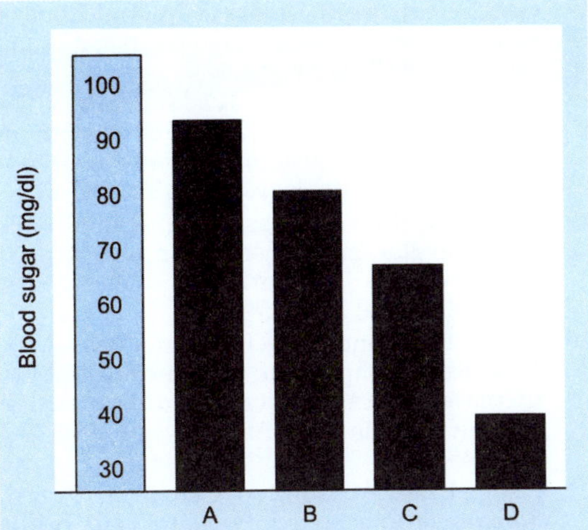

Fig. 22.11: A schematic diagram illustrating the threshold for four factors that cumulatively mediate the glucagon response to hypoglycemia. A: β cell inhibition; B: Parasympathetic neural discharge; C: Epinephrine release; D: Sympathetic neural discharge

between 75 and 65 mg/dl, and increases in magnitude with the severity of hypoglycemia.

In conclusion, because the four main mediators of the glucagon response to hypoglycemia differ in their thresholds and magnitudes of activation throughout various degrees of hypoglycemia, their relative contributions to the resultant glucagon response depends on the level of hypoglycemia. Disinhibition of glucagon release (from insulin) is likely the sole mediator of the small glucagon response seen as the upper end of the mild hypoglycemic range, whereas it is likely that both disinhibition and parasympathetic activation contribute at the lower end of this range. Within the moderate hypoglycemic range, it is likely that parasympathetic nerves and the adrenal medulla overtake disinhibition as the major mediators of the glucagon response. Finally, in the marked hypoglycemic range, it is likely that all three autonomic inputs mediate the large glucagon response.[20]

C. ROLE OF DELTA CELLS IN ISLET CELL SECRETION

The role of δ cells of the islets of Langerhans is still not completely known. There have been reports that somatostatin hormone, secreted by δ cells, suppresses the hormone secretion of both α and β cells. What is then the ultimate result of interactions between α and β cells and the inhibitory action on both by δ cells? The *insulin and glucagon secretory response to nutrient stimuli was enhanced in vivo in SST knock-out mice.* In addition, the glucose-induced suppression of glucagon secretion was absent. However, no defect in secretion of insulin or glucagon was detected under basal conditions.[21] This suggests that SST negatively regulates α and β cell function only under conditions of nutrient stimulus.

References

1. Baskin DG. A historical perspective on the identification of cell types in pancreatic islets of Langerhans by staining and histochemical techniques. J Histochem Cytochem. 2015; 63: 543–58.
2. Lane MA. The cytological characters of the areas of Langerhans. Am J Anat. 1907; 7: 409–22.
3. Longnecker D. Anatomy and Histology of the Pancreas. *Pancreapedia: Exocrine Pancreas Knowledge Base*, 2014. DOI: 10.3998/panc.2014.3
4. Muller TD, Finsn B, Clemmensen RD, et al. The new biology and pharmacology of glucagon. Physiol Rev. 2017; 97: 721–66.
5. Gromada J, Franklin I, Wollenheim CB. α cells of the endocrine pancreas: 35 years of research but the enigma remains.Endocrine Reviews, 2007; 28: 84–116
6. Mann E, Bellin, Melena D. Secretion of Insulin in Response to Diet and Hormones.*Pancreapedia: Exocrine Pancreas Knowledge Base*, 2016. DOI: 10.3998/panc. 2016.3.
7. O'Rahilly SO, Hosker JP, Rudenski AS, et al.The glucose stimulus-response curve of the β-cell in physically trained humans, assessed by hyperglycemic clamps. Metabolism. 1988; 37: 919–23.
8. Ahren B, Wierrup N, Sundler F. Parasympathetic nerves and cephalic phase of insulin secretion. Diabetes 2006 Dec; 55 (Supplement 2): S98–S107.
9. Strubbe JH, Steffens AB. Neural control of insulin secretion. Horm Metab Res. 1993; 25: 507–12.
10. Ahren B. Autonomic regulation of islet hormone secretion: implications for health and disease. Diabetologia. 2000; 43: 393–410.
11. Alesso DA, Keiffer TJ, Taborsky DJ, et al. Activation of the parasympathetic nervous system is

necessary for normal meal-induced insulin secretion in rhesus macaques. J Clin Endo Metab 2001; 86: 1253–59.

12. Ohneda A, Aguilar-Parada E, Eisentraut A, et al. Control of pancreatic glucagon secretion by glucose. Diabetes. 1969; 18: 1–10.

13. Ohneda A, Sato M, Matsuda K, et al. Plasma glucagon response to blood glucose fall, gastro-intestinal hormones and arginine in man. Tohoku J Exp Med. 1972; 107: 241–51.

14. Gelfe E.Glucose control of glucagon secretion—'There's a brand-new gimmick every year'. Ups J Med Sci. 2016; 121: 120–32.

15. Cooperberg BA, Cryer PE. Insulin reciprocally regulates glucagon secretion in humans. Diabetes. 2010; 59: 2936–40.

16. Cryer PE. Minireview: Glucagon in the pathogenesis of hypoglycemia and hyperglycemia in diabetes. Endocrinology. 2012; 153: 1039–48.

17. Gerich J, Charles M, Grodsky GM. Regulation of pancreatic insulin and glucagon secretion. Annu Rev Physiol. 1976; 38: 353–88.

18. Taborsky GD, Mudinger TO. Minireview: The role of the autonomic nervous system in mediating the glucagon response to hypo-glycemia. Endocrinology. 2012; 153: 1055–62.

19. Patel DG. Role of parasympathetic nervous system in glucagon response to insulin-induced hypoglycemia in normal and diabetic rats. Metabolis. 1984; 33: 1123–27.

20. Havel PJ, Mundinger TO, Taborsky GJ, Jr. Pancreatic sympathetic nerves contribute to increased glucagon secretion during severe hypoglycemia in dogs. Am J Physiol. 1996; 270: E20–26.

21. Hauge-Evans AC, King AJ, Carmignac D, et al. Somatostatin secreted by islet δ cells fulfils multiple roles as a paracrine regulator of islet function. Diabetes. 2009; 58: 403–11.

Chapter 23

Vitamin D Hormone (Dihydroxycholecalciferol)

INTRODUCTION

Like most of the vitamins, the discovery of vitamin D was a consequence of the knowledge of its deficiency disorders. The first clinical description of rickets appeared in 1650 after the widespread appearance of rickets in northern Europe due to industrialization. The incidence of rickets reached serious proportions with the development of urbanized industrial population. Smoky skies coupled with relatively indoor life necessitated by this environment drastically reduced the solar exposure of the people, thereby curtailing the chief source of vitamin D. Cod-liver oil was recognized as a therapeutic measure for curing rickets in 1811. By the beginning of 20th century, the relation between dietary deficiency and many diseases such as beriberi and scurvy was demonstrated and the term vitamin was introduced by Funk in 1912. In a series of publications, Mellanby demonstrated that rickets is a deficiency disease and it could be cured by cod-liver oil or butter fat but attributed the cure to "fat-soluble A." Huldscginsky in 1920, provided the experimental proof of the curative effects of UV radiation on rickets. In 1925, McCollum had the honor of naming the fourth discovered vitamin as vitamin D. In 1924, Steenbock showed that phytosterol and ergosterol became rich in vitamin D after UV radiation.

Subsequently, vitamin D was crystallized from irradiated ergosterol and the compound was named calciferol. In 1935, Windaus et al determined the chemical structure of calciferol, and 7-dehydrocholesterol was shown to be a provitamin D. After this, except for official adoption of the name vitamin D_2, for ergocalciferol and vitamin D_3, for cholecalciferol, the research activity on vitamin D almost came to standstill. Only in late sixties, the role of the liver and the kidney in vitamin D metabolism was elucidated.[1]

VITAMIN D METABOLISM

i. Cutaneous Production of Vitamin D

Diet is a very poor source of vitamin D. Vitamin D does not occur in vegetable kingdom. Cutaneous synthesis of vitamin is an important source of vitamin D. Most of the vitamin D synthesis occurs in the actively growing layers of the epidermis (strata spongiosum and basale) by exposure to sunlight. Radiation energies between 290 and 320 nm are most effective. 7-dehydrocholesterol present in the epidermis acts as a provitamin D. Ultraviolet radiation produces a cleavage of B ring thereby forming previtamin D (9, 10-secosteroid). Previtamin D undergoes a temperature-dependent isomerization to form vitamin D_3 (also called cholecalciferol), taking 2–3 days for completion of the process (Fig. 23.1). The unique thermally regulated synthesis of vitamin D_3 ensures a gradual release of the vitamin from the epidermis into circulation. This concept is confirmed by the observation that subjects exposed to whole body UV radiation have a significant increase in the circulating concentrations of vitamin D_3, about 6–9 hours after the exposure that reaches a peak 24–48 hours after the exposure, before gradually returning to baseline by 7 days.[2] Once vitamin D is formed, vitamin D-binding protein in the dermal capillary circulation helps to translocate the vitamin from blood-less epidermal tissue into circulation.

Melanin pigment present in the epidermis interferes with the synthesis of vitamin D by absorbing UV radiation. The concept that skin pigmentation is evolved for the control of vitamin D synthesis in the skin is supported by excessive cutaneous melanin seen in populations exposed to greater UV radiation. Moreover, it has been observed that when surgically excised skin from Blacks and Caucasians was exposed to solar radiation, greater amount of vitamin D was produced in the latter. However, now it is clear that melanin is only one of the many factors that regulate photosynthesis of vitamin D in the skin. Cutaneous production of vitamin D seems to be under an autoregulatory control. Excessive exposure of even Caucasian skin to sunlight does not cause vitamin D intoxication. Continuous exposure to UV radiation depletes cutaneous provitamin D, but does not increase production of previtamin D. Holick et al[3] have reported the effect of exposure of skin for different durations to sunlight. During the first 10–15 minutes of exposure, approximately 15% of provitamin D changed to previtamin D. After one hour of exposure, 40% provitamin D was depleted but only 15% increase in previtamin D was observed. The remaining 25% of photolyzed provitamin D was accounted for by the presence of inactive isomers, tachysterol and lumisterol. Further solar exposure depleted the stores of provitamin D in the epidermis,

but the concentration of previtamin D_3 or vitamin D_3 did not increase.[3]

The amount of solar exposure required for providing vitamin D adequate for the body's requirements varies in different individuals and under different conditions. The photosynthesis of vitamin D_3 depends upon (i) the surface area of the skin exposed to sunlight, (ii) the time of the day of exposure (UV radiation is most intense between 11 AM and 2 PM, (iii) the amount of melanin pigment presents in the epidermis, (iv) latitude (UV radiation is most intense at the equator), (v) season in winter less UV radiation reaches the surface of the earth, (vi) environmental pollution such as smoke, fog and dust prevents UV radiation from reaching the earth. However, prolonged exposure to sunlight does not necessarily mean greater production of vitamin D since as mentioned earlier. Solar radiation can isomerize previtamin D_3 to inactive, isomers, tachysterol and lumisterol as well as produce photodegradation of vitamin D_3.[3]

Vitamin D-binding protein has no affinity for tachysterol or lumisterol and hence translocation of these isomers into circulation does not occur. These products are sloughed off during natural turnover of skin. Patients with uremia seem to be unable to produce vitamin D in the skin. It is believed that one or more substances present in the skin of a patient with chronic renal failure act like melanin and absorb UV radiation.

ii. Hepatic Metabolism of Vitamin D

Vitamin D_3, synthesized in the skin, enters the circulation bound to vitamin D-binding protein. Dietary vitamin D_2 or D_3 enters the circulation through lymphatic system. Subsequently, both vitamins D_2 and D_3 are metabolized similarly.

In the liver, vitamin D is metabolized by vitamin D-25-hydroxylase to form 25-hydroxyvitamin D {25(OH)D} (now also known as calcidiol) (Fig. 23.1). The enzyme is located in the mitochondrial and microsomal fractions of the hepatocytes; the liver seems to be the

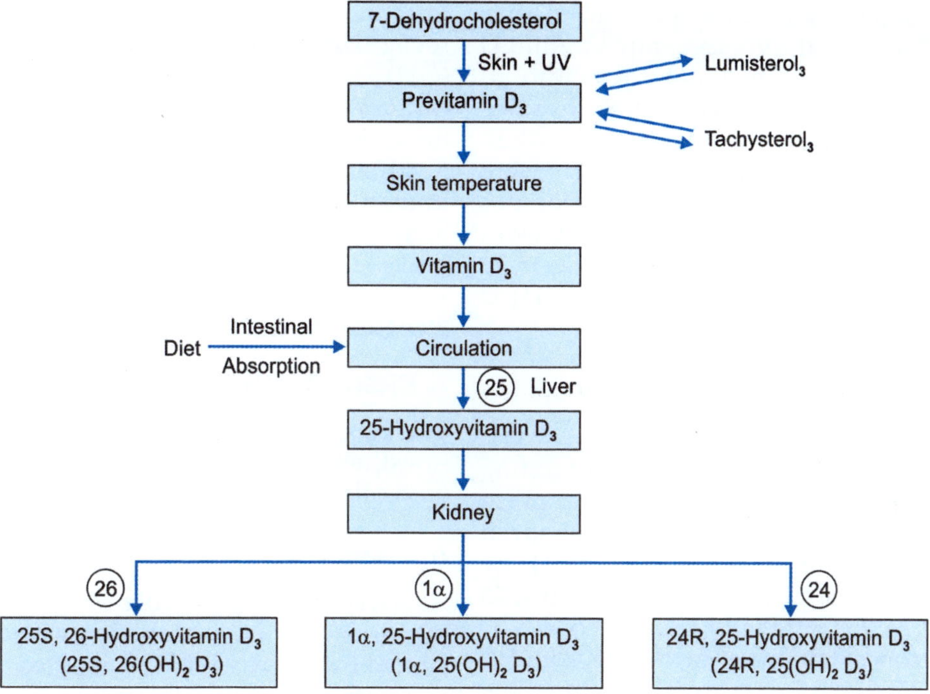

Fig. 23.1: Vitamin D metabolism

only site of 25(OH)D synthesis in humans. The reserve capacity of vitamin D-25-hydroxylase in the liver is substantial. Severe parenchymal damage is required to lower the level of plasma 25(OH)D. The enzyme vitamin D-25-hydroxylase does not seem to be tightly regulated since circulating levels of 25(OH)D vary with the amount of dietary intake of vitamin D or with the degree of solar exposure. Decreased plasma 25(OH)D levels are observed in patients with nephrotic syndrome having proteinuria greater than 4 g/day, due to renal loss of vitamin D tagged to vitamin D-binding protein.[4]

iii. Renal Metabolism of Vitamin D

It was only in 1970 that Fraser and Kodicek.[5] demonstrated the intimate relation between the kidney and vitamin D metabolism. These workers demonstrated that homogenates of chick kidney could metabolize 25(OH)D to a biological active metabolite (Fig. 23.1). These workers identified the active metabolite as 1, 25-dihydroxycholecalciferol. 1, 25(OH)$_2$D$_3$

is now also known as calcitriol. The important role of kidney in the activation of 25(OH) was demonstrated by another experiment Twenty-four hours after injection of ^{3}H-25 (OH)D, ^{3}H-1, 25(OH)$_2$ D could be detected in the blood and tissues of vitamin D deficient rats but not in vitamin D deficient rats that had undergone bilateral nephrectomy before receiving radioactive 25(OH)D. It was also shown that physiological concentrations of 25(OH)D could not stimulate intestinal calcium transport in anephric rat.[5] The renal 25(OH)-1α-hydroxylase is located in the proximal convoluted tubules.[6] It is now accepted that 1, 25-dihydroxy metabolites of vitamin D$_2$ or D$_3$ are the biologically active forms of vitamin D$_2$ and D$_3$, respectively. These metabolites are 10 times more active than vitamin D$_2$ or D$_3$ in healing rickets or stimulating intestinal calcium absorption.

The activity of renal 25-OH-1α hydroxylase appears to be tightly regulated since plasma 1, 25(OH)$_2$ D$_3$ concentration remains constant over a wide range of substrate 25(OH)D$_3$. Parathormone (PTH) seems to play a crucial role in the synthesis of calcitriol since it was found that hypocalcemic vitamin D deficient rats could metabolize calcidiol to calcitriol more effectively than normo-calcemic vitamin D replete rats. But, when vitamin D-deficient hypocalcemic rats were thyroparathyroidectomized, the difference was lost. In pregnancy, and lactation, growth hormone, estrogens and prolactin seem to enhance renal production of 1, 25(OH)$_2$D directly or indirectly.[7, 8]

iv. Alternate Renal Metabolic Pathways for 25(OH)D

When vitamin D nutrition and circulating plasma concentrations of calcium and phosphorus are normal, 25(OH)D is metabolized into a variety of products (Fig. 23.1), by hydroxylation at C 1, 24, and 26 to form 1, 25 (OH)$_2$D, 24, 25(OH)$_2$D and 25, 26 (OH)$_2$ D. The plasma concentrations of each of 24, 25(OH)$_2$ D and 25, 26(OH)$_2$D are 50–100 times the concentration of 1, 25(OH)D. The metabolites other than 1, 25(OH)$_2$D have no biological activity. Production of 25(OH)D is uncontrolled. Its plasma concentrations vary directly with the dietary intake/cutaneous production of vitamin D. When plasma concentration of 1,25(OH)$_2$D is adequate, remaining 25(OH)D is converted to 24, 25(OH)$_2$D or 25, 26 (OH)$_2$D. Renal 25(OH)-1α-hydroxylase converts two inert metabolites mentioned above to trihydroxy metabolites 1, 24, 25-trihydroxycholecalciferol (1, 24, 25(OH)$_3$ D) and 1, 25, 26-trihydroxy cholecalciferol (1, 25, 26(OH)$_3$D). The trihydroxy metabolites again have no biological activity[8, 9] (Fig. 23.1).

Vitamin D Receptor

Vitamin D receptor (VDR) was discovered in 1968. In 1980s, VDR was found to be widely distributed in different tissues of the body such as gonads, stomach, epidermis, pituitary gland, pancreas, breast, parathyroid gland, thymus, cardiac muscle, skeletal muscle, placenta, etc.[10] Initially no physiologic significance was attached to such reports; only intestine, bone and kidney continued to be recognized as the target tissues of 1, 25 (OH)$_2$D. Subsequently, reports indicated calcium-binding protein synthesis in many of these tissues, including brain. *In vitro*, 1, 25 (OH)$_2$D was found to inhibit proliferation of human fibroblast cells and keratinocytes, increase TSH synthesis, inhibit PTH synthesis.[11, 12]

BIOLOGICAL ACTIONS OF VITAMIN D

A. Classical Actions

The classical actions of vitamin D on the intestine, bone and kidneys are concerned with calcium homeostasis.

a. Intestine

As early as 1937, the role of vitamin D in intestinal calcium absorption was demonstrated by Nicolaysen.[13] He injected varying amounts of calcium (chloride) into isolated jejunum loops and estimated amount of calcium absorbed in five hours (Fig. 23.2). The results clearly demonstrated deficient calcium absorption in vitamin D deplete rats.

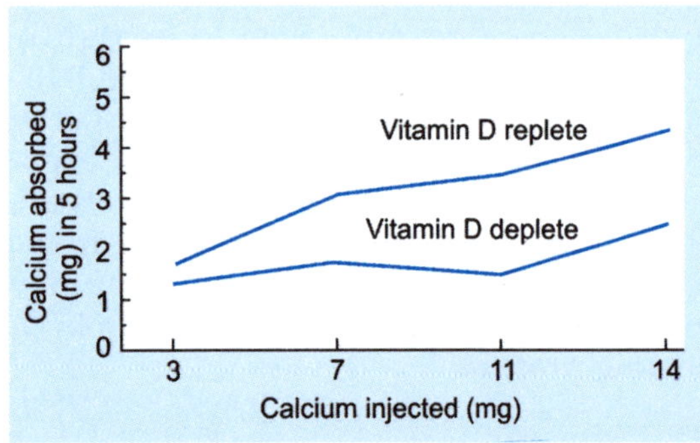

Fig. 23. 2: Absorption of calcium injected in an isolated jejunal loop, after Nicolaysen[13]

In the enterocytes of the small intestine, the genomic action of 1, 25(OH)$_2$D results in greater production of calcium-binding protein, *calbindin*. The net result is greater absorption of dietary calcium and phosphates. The mechanisms by which 1, 25(OH)$_2$D regulates transcellular calcium transport are best understood in the intestine (Fig. 23.3). Here 1, 25(OH)$_2$ D stimulates (1) calcium entry across the brush border membrane into the cell, (2) calbindin-mediated transport of calcium through the cell cytosol, and (3) removal of calcium from the cell at the basolateral membrane by active transport (Ca^{2+}-ATPase, calcium pump). If the luminal Ca^{2+} concentration is high, Ca^{2+} can be absorbed passively through paracellular transport[14] (Fig. 23.3).

b. Bone

Bone and muscle accumulate about 60% of injected dose of vitamin D. Though gross skeletal abnormalities have been observed in vitamin D deficient animals, no direct effect of 1, 25(OH)$_2$D on the process of ossification has been observed. 1, 25(OH)$_2$D does not seem to be essential for ossification of bone. When plasma calcium and phosphate levels were maintained at normal range in vitamin D deficient rats by dietary manipulation, the skeletal histology was found to be normal.[15] However, in cultured rat osteosarcoma cells, 1, 25(OH)$_2$D stimulates the synthesis of osteocalcin, the bone derived protein, in a dose dependent manner.[16] In patients with postmenopausal osteoporosis, 1, 25(OH)$_2$D administration has been shown to increase circulating osteocalcin levels.

Mobilization of calcium from the bone is another well-known action of vitamin D, especially when administered in pharmacologic doses. At physiologic concentrations, 1, 25(OH)$_2$D acts in concert with parathormone to stimulate osteoclastic activity.[17] At pharmacologic concentrations, it was found to stimulate osteoclastic activity by inducing stem cells to differentiate into osteoclast cells. Exposure to human peripheral monocytes that possess receptors for 1, 25(OH)$_2$ D result in their differentiation into multinucleated giant cells capable of mobilizing calcium from bone chips.[18]

c. Kidneys

The most important endocrine effect of 1, 25 (OH)$_2$D$_3$ in the kidney is a tight control of its own homeostasis through simultaneous suppression of 1α-hydroxylase and stimulation of 24-hydroxylase (Fig. 23.4).

In the kidneys, 1, 25(OH)$_2$D increases reabsorption of calcium in the distal tubules through a cytosolic transport protein calbindin-D28K. However, 1, 25(OH)$_2$D$_3$ involvement in the renal handling of calcium and phosphate continues to be controversial due to the simultaneous effects of 1, 25(OH)$_2$D$_3$ on plasma PTH and on intestinal calcium and phosphate absorption, which affect the filter load of both ions.

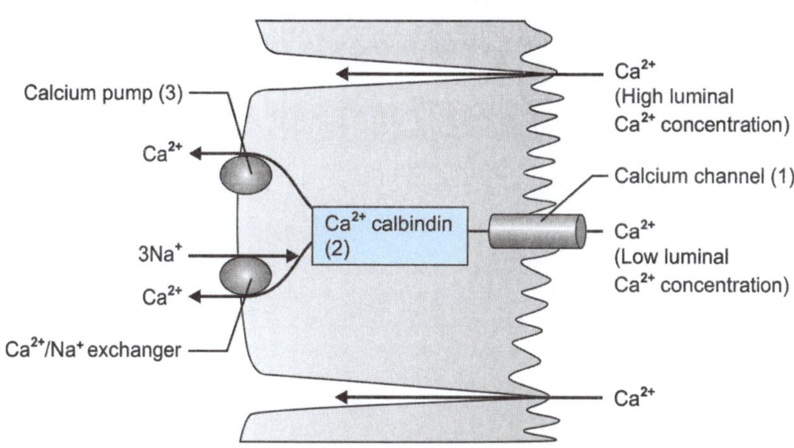

Fig. 23. 3: Sites of action of 1,25(OH)$_2$D in calcium absorption in the enterocytes. Please *see* text

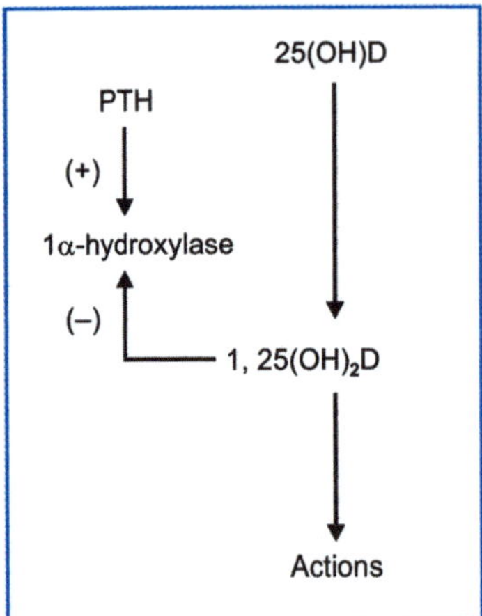

Fig. 23.4: Regulation of synthesis of 25(OH)D-1α-hydroxylase

B. Non-classic Actions of Vitamin D

1. Role of Vitamin D Hormone in Insulin Secretion

The presence of VDR in the cells of islets of Langerhans is now well accepted. Initial results revealed that vitamin D-deficient rats were unable to respond to a glucose challenge by secreting appropriate amounts of insulin, which could be corrected by the administration of 1, 25(OH)₂D₃.[19] Moreover, onset of experimental diabetes can be delayed by administration of 1, 25(OH)₂ D.[20] Thus, a role of vitamin D on islet insulin release is most likely; either direct or indirectly through its effect on plasma calcium concentrations. 1, 25-dihydroxyvitamin D not only improves insulin sensitivity of the target cells (liver, skeletal muscle, and adipose tissue) but also enhances and improves β cell function.[21] The association of vitamin D status and cardio-metabolic disorders (cardiovascular disease, diabetes, and metabolic syndrome) was reviewed recently in a meta-analysis of 28 independently published studies.[22] The findings showed a significant 55% reduction in the risk of diabetes (9 studies), a 33% reduction in the risk of cardiovascular diseases (16 studies), and a 51% reduction in metabolic syndrome (8 studies) associated with a high serum 25-dihydroxyvitamin D concentration.[23]

2. Role of Vitamin D Hormone in the Immune System

The presence of the VDR in activated T lymphocytes was reported by Provvedini, et al.[24] This report suggested a role for 1, 25(OH)₂D₃ in the immune system, but the role is just now beginning to be defined. VDR have also been reported in thymus, a repository of immature lymphocytes. Vitamin D deficiency markedly reduces the ability of mouse to develop delayed hypersensitivity reaction. These results suggest that T-helper cell lymphocyte is vitamin D responsive, but both immunostimulation and immunosuppression can be found in *in vivo* conditions. Currently, there is no evidence that B lymphocyte-mediated immunity is influenced by 1, 25(OH)₂D.

The most dramatic results obtained to date in the immune system are those found in experimental autoimmune encephalomyelitis (EAE) that can be induced in mice. Administration of 1, 25(OH)₂D₃ suppressed the development of the disease in experimental animals[25] (Fig. 23.5). Current results strongly suggest that 1, 25(OH)₂D₃ or its analogues function by stimulation TH-2 T-helper cells to

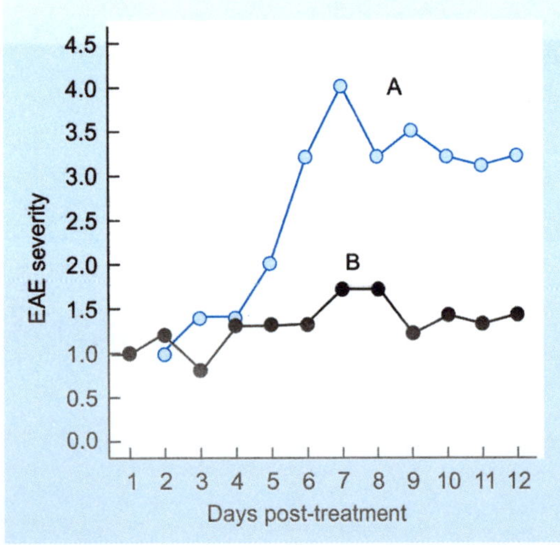

Fig. 23.5: Effect of administration of 1, 25(OH)₂D on the severity of experimental autoimmune encephalitis (EAE) in rats (B) as compared to untreated rats (A), after Cantorna et al[25]

produce transforming growth factor-β1 and IL-4.

Of some interest is the idea that immunomodulation action of vitamin D might be useful in the management of transplant rejection. The possible use of vitamin D in the treatment of autoimmune diseases such as diabetes mellitus, rheumatoid arthritis is being investigated.

3. Role of Vitamin D Hormone in the Parathyroid Gland Function

Perhaps the most well-established non-classic function of 1, 25(OH)$_2$D$_3$ is in the parathyroid gland. Specific localization of 1, 25(OH)$_2$D$_3$ in the parathyroid gland[10] and the presence of VDR strongly suggested that 1, 25(OH)$_2$D$_3$ may have a direct action through its receptor in the parathyroid glands. Vitamin D deficiency is well known to result in secondary hyperparathyroidism via hypocalcemia. In addition, vitamin D may have a direct action on parathyroid gland. PTH secretion by isolated parathyroid glands or cells could be suppressed by the direct administration of 1, 25(OH)$_2$D$_3$.[26]

4. Role of Vitamin D Hormone in Skin

As discussed in detail earlier, the epidermis is the major source of vitamin D for the body. Epidermis of the skin consists of four layers: Stratum basale, stratum spinosum, stratum granulosum, and stratum corneum. The cells in all of the layers except the stratum basale are called keratinocytes (Fig. 23.6). Like many other cells in the body, keratinocytes also express the vitamin D receptor (VDR) and so are capable of responding to the 1, 25 (OH)$_2$D.[27]

1, 25(OH)$_2$D, acting through the VDR, regulates proliferation in the stratum basale of the epidermis and promotes the sequential differentiation of keratinocytes as they form the upper layers of the epidermis. Loss of VDR or loss of the capacity to produce 1, 25(OH)$_2$D disrupts differentiation of the epidermis and results in hyperproliferation of the basal layers.[27] Hosomi et al[28] provided evidence for the first time that 1, 25(OH)$_2$D$_3$ induces keratinocyte differentiation. The differentiation of the keratinocyte is associated with an inhibition of proliferation.

Psoriasis is a chronic immune-mediated inflammatory skin disease. Psoriasis lesions are characterized by hyperproliferation of epidermal keratinocytes. Indeed, significant associations between low vitamin D status and psoriasis have been systematically observed. Due to its role in proliferation and maturation of keratinocytes, vitamin D has become an important local therapeutic option in the treatment of psoriasis.[29]

5. Role of Vitamin D in Reproductive Function

Halloran and DeLuca, in 1980, reported that female reproduction is markedly diminished in vitamin D deficiency. An 80% reduction in fertility was found and could not be corrected

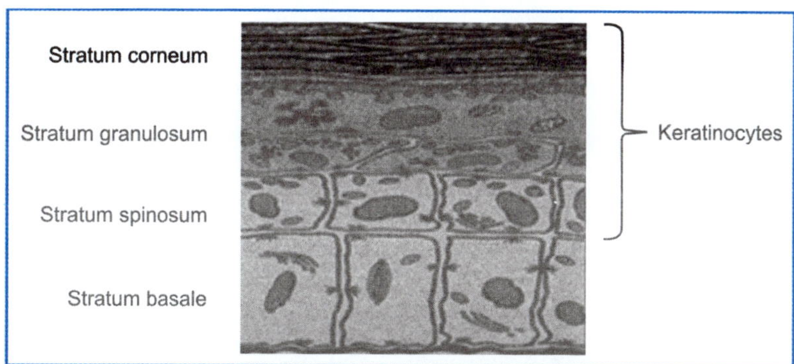

Fig. 23.6: Keratinocytes in the epidermis of the skin

by correcting the hypocalcemia.[30] This defect, therefore, is quite clearly one related to an absence of the vitamin D molecule. The infertility brought about by vitamin D deficiency in the female rat can be easily corrected by the administration of 1, 25(OH)$_2$D$_3$.[30] All such reports suggested that the ovary is a target of vitamin D action. Moreover, the observations that ovarian cells contain VDR and 1, 25 (OH)$_2$D$_3$ accumulates in the ovarian cells[10, 12] lends further support to the view.

In the case of male reproduction, vitamin D deficiency also reduces the effectiveness of the male. A significant reduction found in sperm count in vitamin D rats which could be reversed by vitamin D repletion.[31]

References

1. Marya RK. Vitamin D Revisited: Metabolism and Actions. Intern J of Basic and Applied Physiol. 2015; 4: 1–13.

2. MacLaughlin JA, Anderson RR, Holick MF. Special character of sunlight modulates the photosynthesis of previtamin D$_3$ and its photoisomers in human skin. Science 1982; 216: 1001–03.

3. Holick MF, MacLaughlin JA, Doppelt SH. Factors that influence the cutaneous photosynthesis of previtamin D3. Science. 1981; 211: 509–21.

4. Holick MF, Potts JT, Krane SM. Calcium, phosphorus and vitamin D metabolism.In: Braunwaldd E, ed. Harrison's Principles of Internal Medicine. New York, McGraw-Hill 1986.

5. Fraser DR, Kodicek E. Unique biosynthesis by kidney of a biologically active vitamin D metabolite. Nature. 1970; 228: 764–66.

6. Boyle IT, Miravet L, Gray RW, et al. The response of intestinal calcium transport to 25-hydroxy and 1, 25-dihydtoxyvitamin D in nephrectomised rats. Endocrinology. 1972; 90: 605–08.

7. Garabedian M, Holick MF, DeLuca HF. Control of 25-hydroxycholecalciferol metabolism by parathyroid glands. Proc Natl Acad Sci, USA. 1972; 69: 1673–76.

8. Baksi SN, Kenny AD. Vitamin D metabolism in immature Japanese quail: Effect of ovarian hormones. Endocrinology. 1977; 101: 1216–20.

9. Gkonon PJ, London R, Hendler ED. Hypercalcemia and elevated 1, 25-dihydroxyvitamin D levels in a patient with end-stage renal disease and active tuberculosis. N Engl J Med. 1984; 311: 1683–85.

10. Stumpf WE, Sar M, Reid FA, et al. Target cells for 1, 25-dihydroxyvitamin D$_3$ in intestinal tract, stomach, kidney, skin, pituitary, and parathyroid. Science. 1979; 206: 1188–90.

11. Mayer E, Kadowaski S, Williams G, et al. Mode of action of 1, 25-dihydroxyvitamin D$_3$. In: Kumar R, ed. Vitamin D, Basic and Clinical Aspects. Boston, Martinus Nijhoff, 1984.

12. Whitesett JA, Tsang RC, Norman EJ, et al. Synthesis of 1, 25-dihydroxyvitamin D$_3$ by human placenta in vitro. J Clin Endocrinol Metab. 1981; 53: 484–88.

13. Nicolaysen. Studies upon the mode of action of vitamin D. Biochem J. 1937; 31: 323–29.

14. Christakos S, Dhawan P, Porta A, et al. Vitamin D and intestinal calcium absorption. Moll Cell Endocrinol. 2011; 347: 25–29.

15. Bickle DD, Munson S. 1, 25-dihydroxyvitamin D$_3$ increases calmodulin binding to specific proteins in the chick duodenal brush border membrane. J Clin Invest. 1985; 76: 2316–24.

16. DeLuca HF. Recent advances in our understanding of vitamin D endocrine system. J Lab Cli Med. 1976; 87: 7–26.

17. Holtrop ME, Cox KA, Carnes DL, Holick MF. Effects of serum calcium and phosphorus on skeletal mineralization in vitamin D-deficient rats. Am J Physiol. 1986; 251: E234-40.

18. Zerwekh Je, Sakhee K, Pak CYC. Short-term 1, 25-dihydroxyvitamin D$_3$ administration raises serum osteocalcin in patients with postmenopausal syndrome. J Clin Endocrinol Metab. 1985; 60: 615–17.

19. Haussler MR. Vitamin D receptors: Nature and functions. Ann Rev Nutr. 1986; 6: 527–62.

20. Cherotow BS, Sivitz WI, Baranetsky NG, et al. Cellular mechanisms of insulin release. The effect of vitamin D deficiency and repletion. Endocrinology. 1983; 113: 1511–18.

21. Chertow BS, Sivitz WI, BarenetskyMG, et al. Islet insulin release and net calcium retention in vitro in vitamin D-deficient rats. Diabetes 1986; 35: 771–75.

22. Sung C-C, Liao M-T, Lu KC, et al. Role of Vitamin D in Insulin Resistance. Journal of Biomedicine and Biotechnology, (2012), Article ID 634195, 2012. https: //doi.org/10.1155/2012/634195.

23. Parker J, Hashmi O, Dutton D, et al. Levels of vitamin D and cardiometabolic disorders: systematic review and meta-analysis. Maturitas. 2010; 65: 225–36.

24. Provvedini DM, Rulot CM, Sobal RE, et al. 1, 25-Dihydroxyvitamin D$_3$ receptors in human thymic and tonsillar lymphocytes. J Bone Miner Res. 1987; 2: 239–41.

25. Cantorna MT, Hayes CE, DeLuca HF. 1, 25-Dihydroxyvitamin D_3 reversibly blocks the progression of relapsing encephalomyelitis. Proc Natl Sci USA. 1996; 93: 7861–64.

26. Henry HL, Norman AW. Studies on the mechanism of action of calciferol VII. Localization of 1, 25-dihydroxyvitamin D_3 in chick parathyroid glands. Biochem Biophys Res Commun. 1975; 62: 781–88.

27. Bikle DD. Vitamin D and the skin: Physiology and pathophysiology. Rev Endocr MetabDisord. 2012; 13: 3–19.

28. Hosomi JJ, Hosoi E, Suda AT, et al. Regulation of terminal differentiation of cultured mouse epidermal cells by 1, 25-dihydroxyvitamin D_3. Endocrinology. 1978; 113: 1950–57.

29. Barrea L, Cristina M, Somna CD, et al. Vitamin D and its role in psoriasis: An overview of the dermatologist and nutritionist. Rev Endocr Metab Disord. 2017; 18: 195–205.

30. Halloran BP, DeLuca HF. The effect of vitamin D-deficiency on fertility and reproductive capacity in female rat. J Nutr. 1980; 110: 1573–80.

31. Sood S, Reghunandanan R, Reghunandanan V, et al. Effect of vitamin D repletion on testicular function in vitamin D-deficient rats. Ann Nutr Metab. 1995; 39: 95–98.

Chapter 24

Endocrine Functions of Bone

INTRODUCTION

Bone has traditionally been considered as a structure that is necessary for mobility, calcium homeostasis, and maintenance of the hematopoietic niche. Recent advances in bone biology using complex genetic manipulations in mice have highlighted the importance of bone not only as a structural scaffold to support the human body, but also as a regulator of a number of metabolic processes that are independent of mineral metabolism. These advances point to the role of skeleton as an endocrine organ that modulates glucose tolerance and testosterone production by secretion of a bone-specific protein, osteocalcin.[1]

Bone structure is actually not inert. Bone is continually changing through an elegant process called remodeling, where new bone cells replace older or damaged bone cells. This, in its simplest terms, is accomplished through the interaction of two different types of bone cells: Osteoclasts break down and remove the older bone making way for osteoblasts to build new bone. It is said that through this process, the skeleton completely remodels itself every ten years.[2] Besides osteoblasts and osteoclast cells, the bone contains another type of living cells called osteocytes. In mature bone, osteocytes and their processes reside inside space called lacunae and canaliculi, respectively. When osteoblasts become trapped in the matrix that they secrete, they become osteocytes. Osteocytes are networked to each other via long cytoplasmic extensions that occupy tiny canals called canaliculi, which are used for exchange of nutrients and waste through gap junctions (Fig. 24.1).

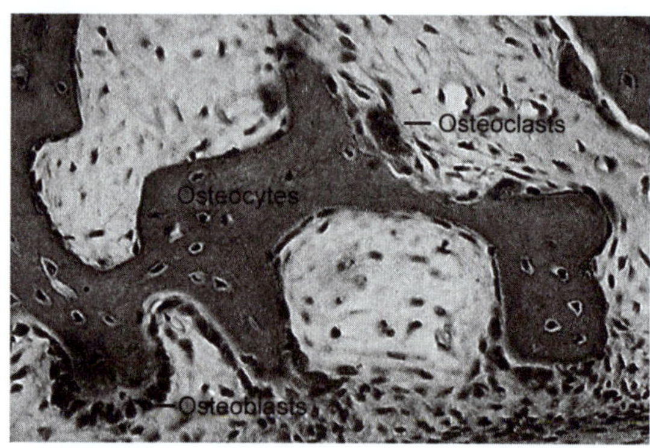

Fig. 24.1: Various types of cells in the bone

BONE IS METABOLICALLY ACTIVE

In contrast to the large number of studies on energy metabolism in muscle and adipose tissue, a little attention has been paid to understanding the bioenergetics of bone metabolism. The sheer size of the skeleton alone implies that its energy requirements should have an impact on total metabolic demands, particularly during growth and remodeling. Indeed, mature osteoblasts that are actively synthesizing and mineralizing matrix exhibit abundant mitochondria, consistent with the increased metabolic demand during this active phase of their lifespan. Likewise, osteocytes, which represent terminally differentiated osteoblasts, survive for years embedded in mineralized bone at densities of greater than 10,000 cells per cubic millimeter. These cells are metabolically active.[3]

Interaction Between Bone and Insulin

a. Role of Insulin in Bone Metabolism

Insulin has been demonstrated to be an osteogenic hormone both *in vitro* and *in vivo*. Osteoblast cells express abundant insulin receptors and respond to insulin treatment by increasing cell proliferation, collagen synthesis, and glucose uptake. Mice knocked out for insulin receptors in their osteoblasts have decreased trabecular bone volume due to reduced bone formation. Moreover, the treatment with insulin has been shown to be effective in determining the reversibility of skeletal alterations of rodent model with type 1 diabetes and also favoring the healing of fractures.[4]

b. Role of Bone in Glucose Metabolism

Blood glucose homeostasis is maintained by a number of hormones, especially insulin. Insulin acts by binding with its receptors. Insulin sensitive receptors (InsR) are well known to be located in hepatocytes, skeletal muscle and adipose tissues (insulin-sensitive tissues). *Surprisingly, InsR have been detected in the bone as well.*[5] The importance of bone in regulation of blood glucose level is demonstrated by the fact that deletion of InsR in skeletal muscle, the bulkiest of all insulin sensitive tissues does not disturb plasma glucose level, or glucose tolerance.[6] The reason probably lies in the fact that another insulin-sensitive organ has an important role in glucose homeostasis. The discovery of InsR on osteoblast cells was the first clue in this direction. *Osteoblasts secrete osteocalcin, a small protein, which has not only osteogenic action but also an endocrine action on β cells of islets of pancreas as well as testis.*[7]

Osteocalcin Biology

Osteoblasts generate two types of proteins called osteocalcin, each having a specific role: (i) Carboxylated osteocalcin, with osteogenic function, and (b) undercarboxylated osteocalcin, with metabolic function.

Carboxylated Osteocalcin

Osteocalcin (OCN) is a small specific protein synthesized by osteoblasts containing three glutamic acid residues (17, 21, and 24 Glu in human) and modified by carboxylation into γ-carboxyglutamic acid residues, generating *carboxylated osteocalcin (cOCN)* (Fig. 24.2). Osteocalcin comprises up to 15% of the non-collagenous protein in bone. Osteocalcin was initially reported to be virtually exclusive to bone and was considered the only bone-specific protein. It is now known to be expressed in other connective and non-connective tissues as well, including calcified cartilage, osteoarthritic articular cartilage, and the nucleus pulposus of the intervertebral disc. cOCN has high affinity for hydroxyapatite, the mineral component of bone extracellular matrix. Osteocalcin plays an important role in later stages of bone formation and remodeling. During bone development, osteocalcin production is very low and does not reach maximal levels until late stages of mineralization. In the osteocalcin knockout mouse, the mineralization of the osteoid fails to occur. cOCN also seems to stimulate osteoblastic differentiation and osteocytic maturation.[8]

Osteocalcin is an important marker of bone turnover in physiological and pathological conditions. Physiologically, serum osteocalcin level is increased in children, particularly during puberty, when the osteocalcin

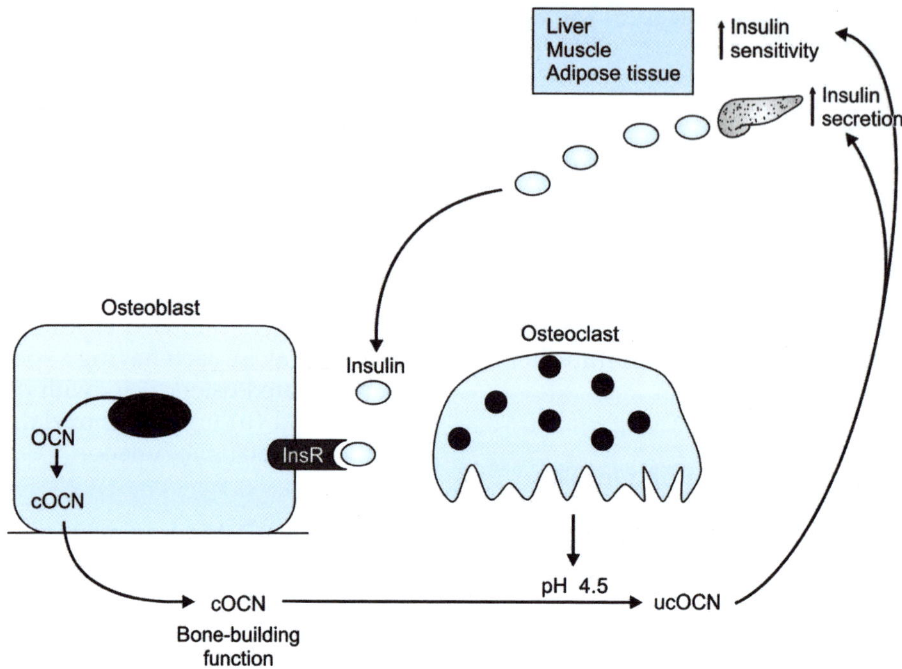

Fig. 24.2: Osteocalcin: Synthesis and functions

concentration is related to the rapidity of physical growth (Fig. 24.3). Serum osteocalcin levels decline gradually with age, reflecting age-related decrease in bone turnover.

Undercarboxylated OCN

There is another form of osteocalcin, known as undercarboxylated osteocalcin (ucOCN), the form of OCN that acts as a hormone. It has little affinity to bone so that most of ucOCN circulates in the blood, while most of cOCN is stored in the bone because of its strong affinity to bone matrix. Studies suggest that the activation of osteocalcin depends on the coordination between bone formation by osteoblasts and bone resorption by osteoclasts. The osteoclast-mediated bone resorption promotes the low pH necessary for osteocalcin decarboxylation and activation (Fig. 24.2). The acid pH is sufficient to decarboxylate the first glutamic acid residue of osteocalcin resulting in a undercarboxylated OCN fragment.[9]

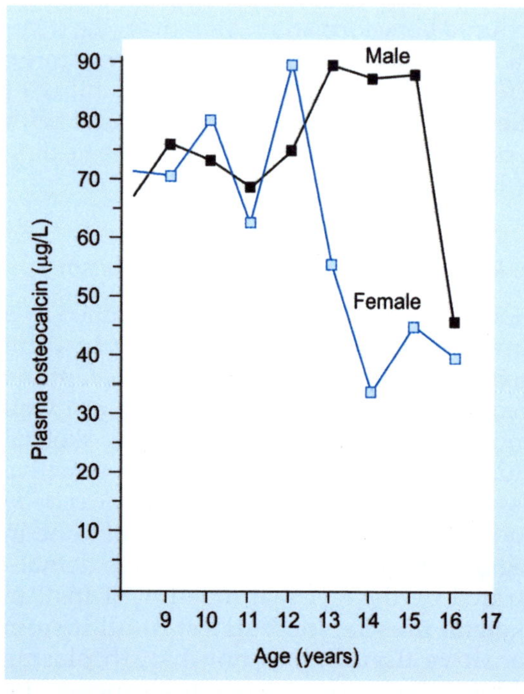

Fig. 24.3: Plasma osteocalcin level in relation to age and sex at puberty

1. ROLE OF OSTEOCALCIN IN REGULATION OF GLUCOSE HOMEOSTASIS

Undercarboxylated osteocalcin increases insulin sensitivity and secretion. Serum levels of unOCN negatively correlate with insulin resistance, obesity, and diabetes. Therefore, unOCN appears to mediate the metabolic effects such as increased β cell proliferation, insulin secretion, insulin sensitivity and adiponectin expression. Recently, it was suggested that insulin signaling in osteoblasts is involved in the synthesis of carboxylated osteocalcin[10–12] (Fig. 24.2).

In OCN knocked out mice have impaired glucose-stimulated insulin secretion and poor glucose tolerance.[11] At the same time the serum adiponectin, a protein hormone that modulates several metabolic processes such as glucose regulation and fatty acid oxidation, was also reduced. All these results suggest that OCN may target β cells and insulin targeting tissue such as muscle, liver, and adipocyte in order to regulate both insulin secretion and sensitivity.[13, 14]

Beta cells proliferation and insulin secretion can be significantly affected at low concentrations of ucOCN. Thus, insulin signaling and OCN form a feed-forward loop, in which insulin affects osteoblasts and in turn increases its own secretion and sensitivity via ucOCN. This feed-forward model explains why the osteoblastic InsR (insulin receptor) knocked out mice have low serum insulin level and high insulin resistance.[15] In human subjects, several clinical studies indicate that serum levels of osteocalcin correlate negatively with insulin resistance, obesity and diabetes. Moreover, serum osteocalcin levels have observed to be positively correlated with improved blood glucose control in subjects with type 2 diabetes. In obese subjects, significant weight loss is associated with a decrease in insulin resistance as well as increased serum osteocalcin levels.[16]

2. ROLE OF OSTEOCALCIN IN REGULATION OF TESTICULAR FUNCTION

It has been known for a long time that the growth and integrity of both the female and the male skeleton are influenced by sex steroid hormones. The biological importance of this regulation is best exemplified by the fact that gonadal failure triggers bone loss in both genders and leads to osteoporosis in post-menopausal women. Feedback rule that applies to most endocrine regulations suggested that given the fact that sex steroid hormones regulate bone mass in both genders, a feedback regulation of testicular function by the bone might exist.[18]

As observed for energy metabolism, the demonstration that osteocalcin plays a biological function on the testes (through the regulation of androgen production) was derived from experimental studies on genetically engineered *Esp⁻/⁻* and *Ocn⁻/⁻* mice, *characterized respectively, by excessive or absent osteocalcin production.* While female mice deficient in osteocalcin were fertile and did not show any gonadal abnormality, male mice with the same defect showed a poor reproductive activity, associated with a decreased volume of the testes, epididymis, and seminal vesicles. These alterations were also associated with a 50% decrease in sperm count. On the contrary, *male Esp⁻/⁻* mice had an increase in testicular volume and a 30% increase in sperm count. All these observations strongly suggest that osteocalcin production by the skeleton is directly involved in endocrine regulation of male reproduction.[17, 18] Moreover, these mice models also demonstrated that *osteocalcin deficiency impairs Leydig cells maturation and reduces testosterone synthesis.* In fact, circulating androgen levels were significantly reduced in *Ocn⁻/⁻* mice and increased in *Esp⁻/⁻* animals. In particular, in cultured Leyding cells deficient in osteocalcin, the treatment with supernatant of wild-type osteoblasts (containing osteocalcin) increased the maturation and the production of testosterone, while this effect did not occur with other cell lines.[18]

A major question raised by these experimental observations is whether the skeleton also regulates testosterone production and male fertility in humans. Even though the available clinical information is limited, there

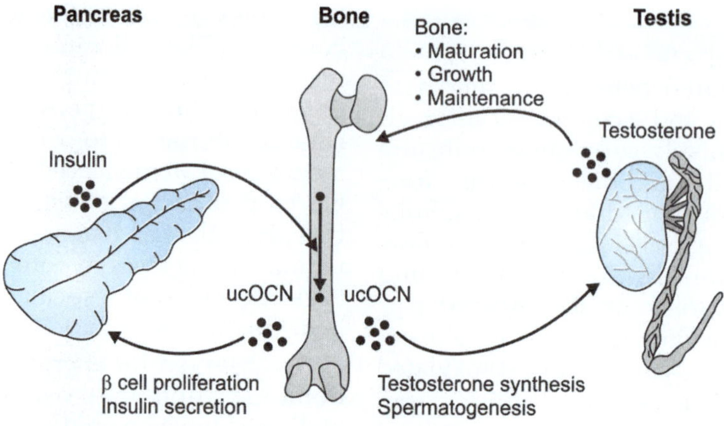

Fig. 24.4: Bone–testis and bone–pancreas interaction

have been some recent studies which seem to support the hypothesis of an endocrine role of osteocalcin on the male gonad in humans. In the first study, carried out on young males during skeletal growth, osteocalcin levels significantly correlated with circulating testosterone which was the major determinant of periosteal circumference. Similar correlations were demonstrated with the assessment of undercarboxylated osteocalcin. Moreover, the correlations between osteocalcin, testosterone, and periosteal circumference were highest in subjects with bone age between 11 and 14 years, during the phase of maximal skeletal growth. Thus, during this phase, the rise in osteocalcin levels (due to the rapid skeletal growth) may further stimulate testicular testosterone production, which, in turn, contributes to an increase in bone size. This could at least in part explain the fact that at the end of skeletal growth males show a larger bone size than females, despite a similar volumetric bone density.[19]

Hyperthyroidism is characterized by increased bone turnover. In such adult males, a close and significant correlation has been observed between plasma osteocalcin and testosterone levels[20] (Fig. 24.5).

In another study, performed in a cohort of type 2 diabetic patients, circulating undercarboxylated osteocalcin positively correlated with free testosterone and negatively with

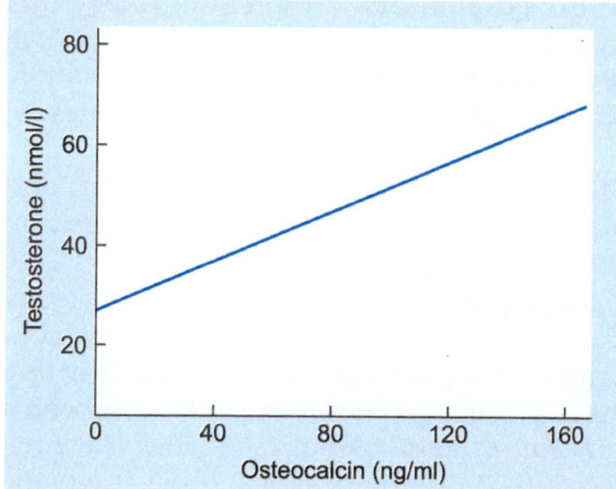

Fig. 24.5: Plasma testosterone and osteocalcin relationship in hyperthyroid patients, after Zhong et at[20]

glycosylated hemoglobin levels.[21] Taken together, results from this study further confirmed the existence of a direct action of osteocalcin on testicular testosterone production in humans and provided a potential explanation for the low testosterone levels frequently described in diabetic subjects.[22]

References

1. Guntur AR, Rosen C. Bone as an Endocrine Organ.EndocrPract. 2012; 18: 758–62.
2. https://www.ncbi.nlm.nih.gov/books/NBK45504/
3. Pritchard JJ. A cytological and histochemical study of bone and cartilage formation in the rat. J Anat. 1952; 86: 259–77.

4. Faienza MF, Luce V, Ventura A, et al. Glucose Metabolism: A Bone-Pancreas Loop. Internal J Endocrinol. 2015, Article ID 758148, 7 pages, 2015. doi: 10.1155/2015/758148.

5. Levy JR, Murray E, Manogalas S, et al. Demonstration of insulin receptors and modulation of alkaline phosphatase activity by insulin in rat osteoblast cells. Endocrinol. 1986; 119: 1786–92.

6. Bruning JC, Michael MD, Winnay JN, et al. A muscle-specific receptor knock out exhibits features of the metabolic syndrome of NIDDM without altering glucose tolerance. Mol Cell 1998; 2: 559–69.

7. Lacombe J, Karsenty G, Ferron M. *In vivo* analysis of the contribution of bone resorption to the control of glucose metabolism in mice. Molecular Metabolism. 2013; 2: 498–504.

8. Boskey A, Robey G. The Regulatory Role of Matrix Proteins in Mineralization of Bone. Ch 11, in Osteoporosis. Fourth Edn, Marcus R, Dampster D, Caule J, et al. Academic press, 2013.

9. Kanazawa I, Yamaguchi T, Yamauchi M, et al., Serum undercarboxylated osteocalcin was inversely associated with plasma glucose level and fat mass in type 2 diabetes mellitus. Osteoporosis International. 2011; 22: 187–94.

10. Lee, NK, Sowa H, Hinoi E et al. Endocrine regulation of energy metabolism by the skeleton. Cell. 2007; 130: 456–69.

11. Ferron M, Wei J, Yoshizawa T, Yoshizawa T, et al. Insulin signaling in osteoblasts integrates bone remodeling and energy metabolism. Cell. 2010; 142: 296–308.

12. Kanazawa I, Sugimoto T. The relationship between bone and glucose/lipid metabolism. Clinical Calcium. 2013; 23: 181–88.

13. Ferron M, Hinoi E, Karsenty G, et al. Osteocalcin differentially regulates β cell and adipocyte gene expression and affects the development of metabolic diseases in wild-type mice. Proceed Nat Acad Scien of the USA. 2008; 105: 5266–70.

14. Booth SL, Centi A, Smith SR, et al. The role of osteocalcin in human glucose metabolism: marker or mediator? Nature Reviews Endocrinology. 2013; 9: 43–55.

15. Kindblom JM, Ohlsson, C Ljunggren O, et al. Plasma osteocalcin is inversely correlated with fat mass and plasma glucose in elderly Swedish men. J Bone Miner Res. 2009; 24: 785–791.

16. Bao YQ, Zhou M, Zhou J, et al. Relationship between serum osteocalcin and glycemic variability in Type 2 diabetes. Clin Expt Pharma Physiol. 2011; 38: 50–54.

17. Kersenty G, Oury F. Regulation of male fertility by the bone-derived hormone osteocalcin. Mol Cell Endocrinol. 2014; 382: 10.1016/j.mce.2013.10.008.

18. Oury F, Sumara G, Sumara O, et al. Endocrine regulation of male fertility by the skeleton. Cell. 2011; 144: 796–809.

19. Kirmani S, Atkinson EJ, Melton LJ, et al. Relationship of testosterone and osteocalcin levels during growth. J Bone and Mineral Res. 2011; 26: 2212–16.

20. Zhong N, Xu B, Cui R, et al. Correlation between serum osteocalcin and testosterone in male hyperthyroidism patients with high bone turnover. Exp Clin Endocrinol Diabetes. 2016; 124: 452–56.

21. Kanazawa I, Tanaka K, Ogawa N, et al. Undercarboxylated osteocalcin is positively associated with free testosterone in male patients with type 2 diabetes mellitus. Osteopor Intern, 2013; 24: 1115–19.

22. Patti A, Gennari L, MerlottiD, et al. Endocrine Actions of Osteocalcin. Int J Endocrinol. 2013 Article ID 846480, 10 pages.http://dx.doi.org/10.1155/2013/846480

Chapter 25

Blood–Testis Barrier

INTRODUCTION

The blood–tissue barrier is a concept originally based on observations reported in early twentieth century. When dyes were administered to laboratory animals, they failed to stain the brain and testis. These findings led to the concept of the blood–brain barrier (BBB) and the blood–testis barrier (BTB). The BBB is constituted almost exclusively by the tight junctions between endothelial cells of the small capillaries in the brain. However, the BTB is constituted by specialized junctions between adjacent Sertoli cells near the basement membrane in epithelium of the seminiferous tubule. The blood–testis barrier is not only comprised of tight junctions but also coexistence and cofunction with ectoplasmic specializations, desmosomes, and gap junctions to create a unique microenvironment for the completion of meiosis and the subsequent development of spermatids into spermatozoa.[1]

STRUCTURE OF TESTIS

The testis consists of two compartments: The highly convoluted seminiferous tubules, and the interstitial space. The interstitium includes Leydig cells, blood vessels, macrophages, lymphocytes, connective tissue, and lymphatic vessels (Fig. 25.1). The Leydig cells

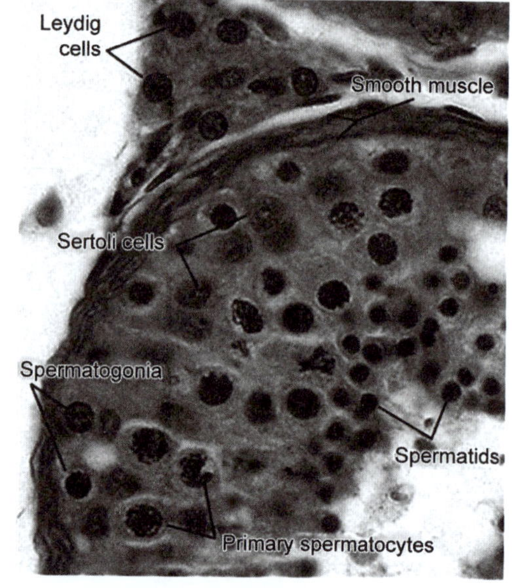

Fig. 25.1: Histological appearance of seminiferous tubule and Leydig cells in the testicular interstitium

within the interstitial space produce testosterone, which is important to maintain spermatogenesis and the male secondary sex characteristics. The seminiferous epithelium is the site of spermatogenesis and includes the germ cells and Sertoli cells. The Sertoli cells extend from the basement membrane to the lumen, surrounding the germ cells, and are responsible for providing structural and functional support to the developing germ cells. Adjacent Sertoli cells are connected by tight junctions, which are located in the basal third of the seminiferous epithelium. The junctions separate the epithelium into *adluminal* and *basal* compartments, thereby separating the germ cells located in the adluminal compartment from the blood constituents. This allows the Sertoli cells to control the microenvironment within the adluminal compartment of the seminiferous tubules. It may be reiterated that the seminiferous tubule basal lamina is not penetrated by blood vessels, lymph vessels, or nerves. Instead, these structures are found only in the interstitium between tubules.

Development and Maturation of Sertoli Cells

The Sertoli cell performs two functionally separate roles, one in the process of testis formation or sexual differentiation in the fetal-neonatal life and the other in spermatogenesis in adult life. These events are separated in time and by function.

The Sertoli cell plays a central role in development of a functional testis, and hence in the expression of a male phenotype in the fetal life. Sertoli cells are the first cells to differentiate recognizably in the indifferent fetal gonad, an event which enables seminiferous cord formation, and differentiation and function of the Leydig cells. The secretions of the Leydig cells (testosterone), then play vital roles in downstream masculinization events. The Sertoli cells also ensure regression of the müllerian ducts via secretion of anti-müllerian hormone. In perinatal period, testosterone plays an important role in descent of the testes into the scrotum.[2] Many years later at puberty, the role of the Sertoli cell switches to the support of spermatogenesis. Without the physical and metabolic support of the Sertoli cells, germ cell differentiation, meiosis and transformation into spermatozoa would not occur in adults.[2] Moreover, the number of Sertoli cells will determine the number of germ cells that can be supported through spermatogenesis and hence will numerically determine the extent of sperm production[3], a factor with obvious bearing on fertility. The switch from 'fetal' to 'adult' Sertoli cell appears to occur during puberty and is termed *functional maturation* of Sertoli cells. Maturation involves loss of proliferative ability, formation of inter-Sertoli cell tight junctions and expression of functions not present in immature Sertoli cells.[4]

Proliferation of Sertoli Cells

The number of Sertoli cells in the adult testis determines both testis size and daily sperm production. This relationship occurs because each Sertoli cell has a fixed capacity for the number of germ cells that it can support.[5] Only immature Sertoli cells proliferate, so the final number of Sertoli cells is determined before adulthood. Sertoli cells proliferate during two periods of life, in fetal or neonatal life and in the peripubertal period in all species (Fig. 25.2).

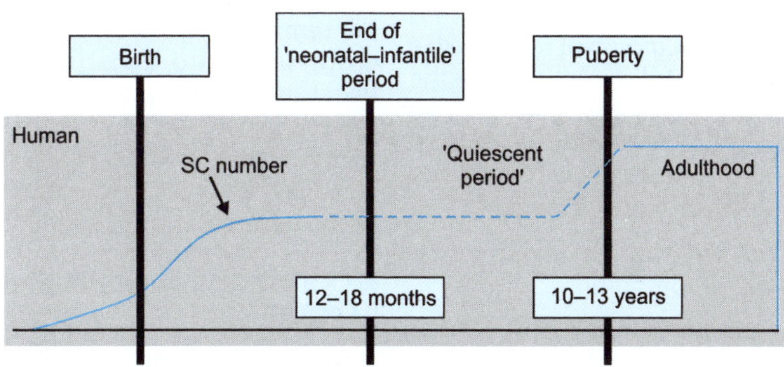

Fig. 25. 2: Two phases of proliferation of Sertoli cells, after Sharpe[3]

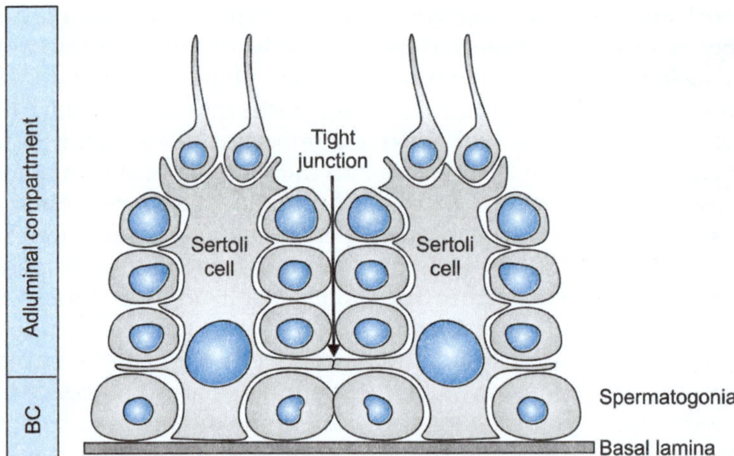

Fig. 25.3: Seminiferous tubule. The Sertoli cell barrier divides the lumen into basal compartment (BC) containing spermatogonia and adluminal compartment containing spermatocytes in various stages of development

At around the onset of puberty, Sertoli cells undergo a radical change in their morphology and function, heralding the switch from an immature, proliferative state to a mature, non-proliferative state. The nucleus enlarges and becomes tripartite. Adjacent Sertoli cells form tight junctions with each other to create a unique adluminal compartment in which meiotic and post-meiotic steps of spermatogenesis can proceed, as well as allowing formation of a fluid-filled lumen. As a result, the germ cells developing in the adluminal compartment become effectively sealed off from direct access to many nutrients and thus become dependent on the secretion of such factors by the Sertoli cells[4] (Fig. 25.3).

Spermatogenesis

Before taking up the discussion of blood testes barrier, it would be useful to recapitulate the stages of spermatogenesis. Spermatogenesis that takes place in the seminiferous epithelium is composed of a series of cellular events and can be divided as follows (Fig. 25.4):

1. Spermatogonial renewal and proliferation via mitosis and differentiation.

2. Cell cycle progression from type B spermatogonia to preleptotene spermatocytes stage of primary spermatocyte, which takes place in the basal compartment, lying outside of the blood–testis barrier (BTB).

3. Cell cycle progression from preleptotene to leptotene, zygotene, and pachytene stages of meiosis I to secondary spermatocyte.

4. Secondary spermatocyte undergoes meiosis II and becomes spermatid.

5. Development of round spermatids to elongated spermatids, and then spermatozoa via spermiogenesis.[6]

Spermatogenesis begins with differentiation of spermatogonial stem cells residing within the spermatogonial stem cell niche, which is located in the basal compartment where several tubules converge and borders the interstitial tissue. The Sertoli cell barrier (blood–testis barrier), which is essential for spermatogenesis, is located near the base of the seminiferous tubule, where it divides the epithelium into 2 distinct compartments, basal and adluminal. Spermatogonia and preleptotene spermatocytes reside in the basal compartment, whereas leptotene and all subsequent stages of spermatogenesis reside in the adluminal compartment.[7]

BLOOD–TESTIS BARRIER (BTB) (SERTOLI CELL BARRIER)

Like other blood–tissue barriers, the BTB is not an impenetrable wall; it is rather restrictive in

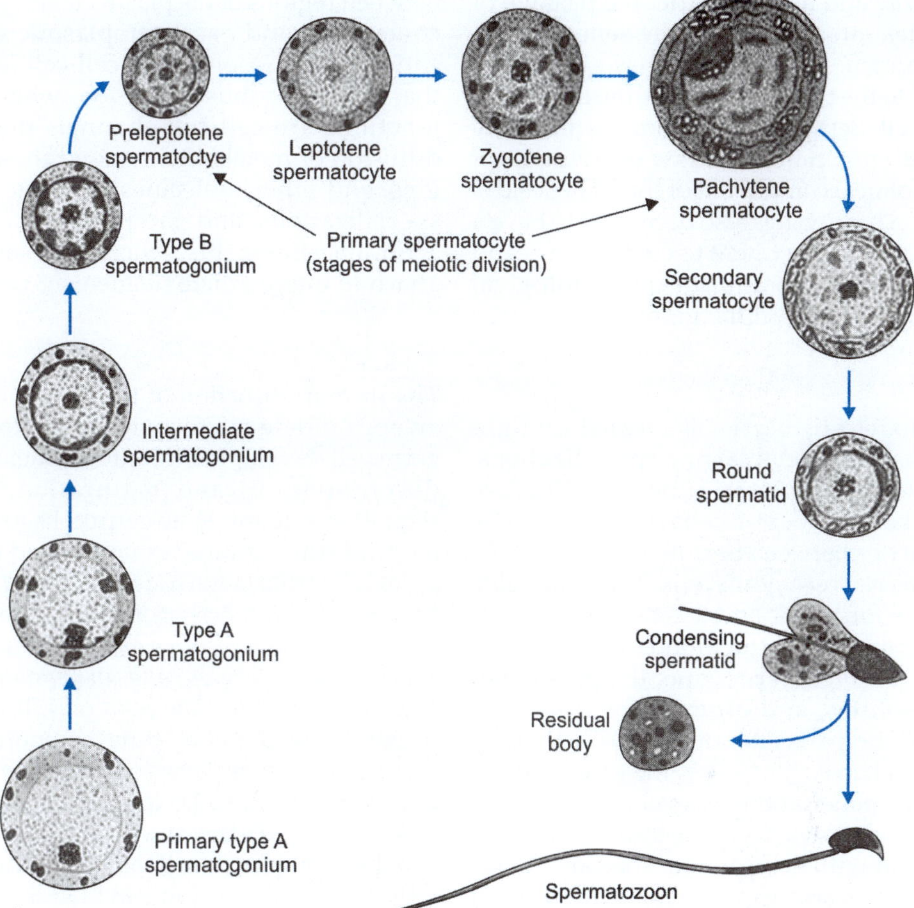

Fig. 25.4: Various stages of spermatogenesis, modified from Bellieve et al[6]

nature. *Blood–tissue barriers restrict or control the entry of molecules present in circulating blood/ interstitium from entering a tissue or compartment. Blood–testis barrier exists between the blood vessels and the adluminal compartment of the seminiferous tubules.* In contrast to the blood–brain barrier that is located at the endothelial tight junctions lining the blood vessels, the blood testis barrier is formed by the tight junctions formed between adjacent Sertoli cells. While this physical barrier (i.e. tight junctions between epithelial cells) is very important for the functions of the BTB, the common misconception is that this physical barrier is the sole component of the BTB. However, the *fully functional BTB* is much more complex, and consists of *anatomical/physical* *(tight junctions, basolateral and apical membranes), physiological (transporters, channels at the basolateral and apical membranes, and the paracellular route) and immunological components.*

In an attempt to more precisely describe the BTB, Russell and Peterson[8], used the term the "Sertoli cell barrier." This term more adequately defined the barrier location at the seminiferous epithelium instead of, as the name BTB implies, at the blood vessels. This description also takes into account that the barrier does not have direction, and is a permeability barrier or "physiological" barrier. They explained that it was both the *occluding junctions, together with the body of the Sertoli cell, that ultimately forms the Sertoli cell*

(SC) barrier and acts to restrict the passage of molecules into and out of the seminiferous tubule lumen. Furthermore, they specifically referred to the tight junctions of the BTB as the Sertoli cell. Sertoli cell junctions, which they believed are a more accurate description of the anatomical component of the BTB. Besides a physical barrier, it is also necessary to have a physiological component to control entry and exit of molecules, and an immunological component to protect the germ cells.

Anatomy of Sertoli Cell Barrier

The blood–testis barrier is created by tight junctions (TJs), ectoplasmic specializations, desmosomes, and gap junctions that are present between Sertoli cells (Fig. 25.5). There is no barrier between Sertoli and germ cells, as well as between germ cells. The TJ has gate and fence functions, and it is the most important component of the blood–testis barrier. The fence function prevents the passage of water, solutes, and other large molecules between the paracellular space (i.e. thereby creating a barrier), whereas the gate function allows the penetration of developing spermatocytes (*see* below). TJs in the testis coexist and cofunction with ectoplasmic specializations. Desmosomes and gap junctions are present between regions of the plasma membrane that contain TJs and basal ectoplasmic specializations. Desmosomes are cell-cell junctions that mediate robust adhesion, whereas gap junctions are cell-cell channels that allow diffusion of metabolites, second messengers, ions, and other molecules. TJs, ectoplasmic specializations, and gap junctions attach to actin microfilaments, whereas desmosomes attach to intermediate filaments[8, 9] (Fig. 25.5).

Physiology of Blood–Testis Barrier

The primary function of the BTB is to create an appropriate microenvironment to control germ cell development (mitosis, meiosis, and differentiation) and maturation, and to sequester autoantigenic germ cells to prevent harmful autoimmune responses, and to shield germ cells from cytotoxic molecules. *Therefore, besides a physical barrier, it is also necessary to have a physiological component to control entry and exit of molecules, and an immunological component to protect the germ cells.* It needs to be emphasized that all three components are important for complete BTB. The components that make up the BTB are (i) the basolateral membrane, (ii) the apical membrane, and (iii) the tight junctional complex between the cells. Each will contribute in their unique way

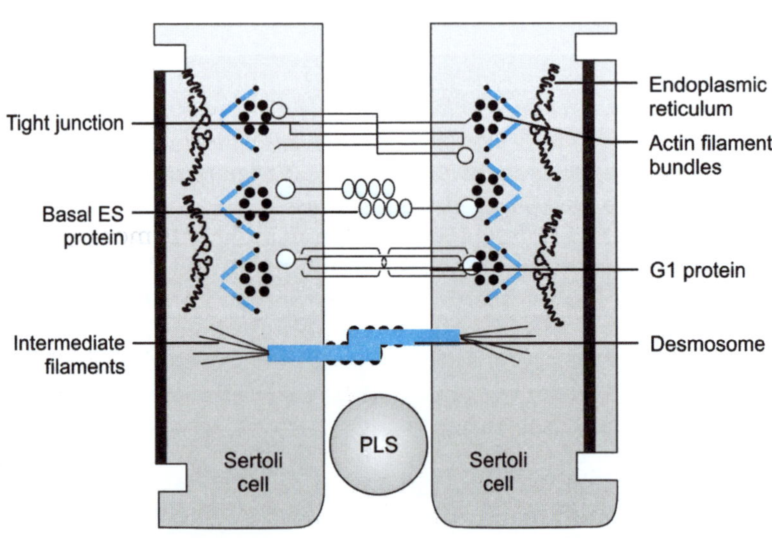

Fig. 25.5: Sertoli cell barrier. Electromagnetic representation. PLS: Preleptotene spermatocyte, modified from Li et al[9]

to restrict the passage of molecules from entering and exiting the lumen and allowing each tissue to form a very unique luminal fluid milieu.[10] The permeability properties of the basolateral and apical membranes are being considered here as the physiological barrier. In BBB (Fig. 25.6A), the tight junctions between the adjacent endothelial cells preclude any paracellular transport. All the transport has to be transcellular. Consequently, only lipid-soluble substances such as O_2 and CO_2, anesthetic agents, caffeine, nicotine and alcohol can easily diffuse across the BBB. All other non-polar solutes and charged ions can cross the BBB either by carrier-mediated diffusion, active transport or by endocytosis.[10] *In Sertoli cell barrier, whereas the anatomical barrier restricts passage of molecules into the lumen, specific transporters located along the basolateral and apical membranes of the Sertoli cells regulate the movement of molecules in or out of the lumen*[10] (Fig. 25.6B). This physiological barrier is required to create the appropriate microenvironment within the seminiferous epithelium and lumen for development (meiosis and differentiation) and

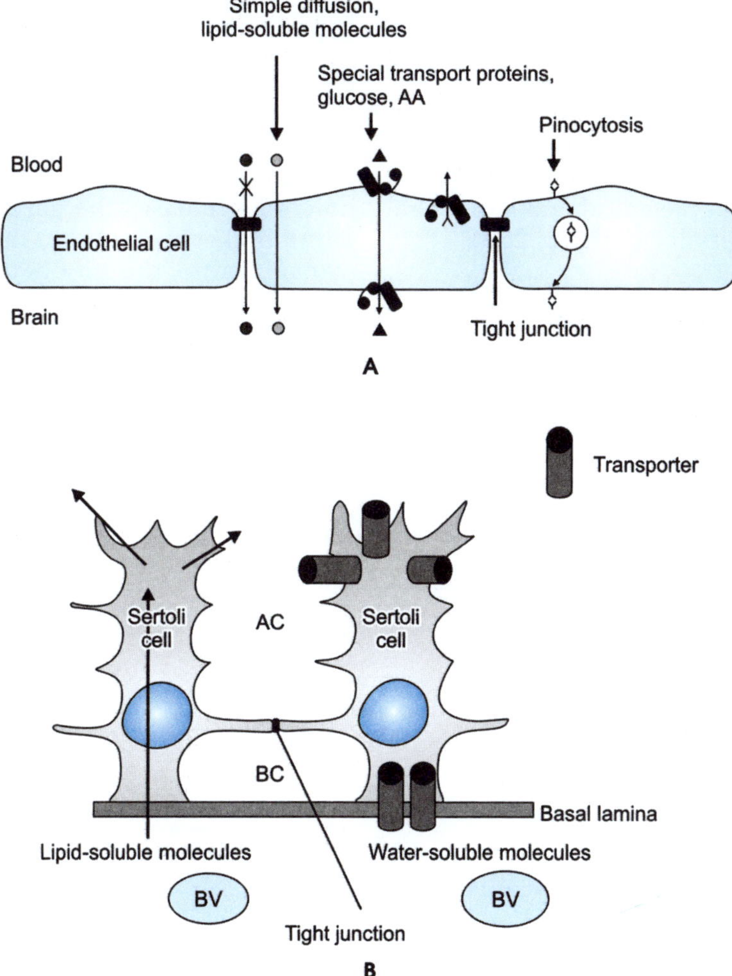

Fig. 25.6: Comparison of blood–brain barrier (A) with anatomical and physiological Sertoli cell barrier (B). In both barriers, paracellular transport is blocked and transporter proteins are required for transcellular transport of water-soluble molecules

maturation of the germ cells.[11] The whole idea of a physiological barrier within the testis originated with the paper by Setchell et al in 1969[11], although it was well known before that time that vascular perfused dyes were excluded from the seminiferous tubular lumen. They performed experiments to examine the permeability of the testis to molecules of differing molecular weight. Radioactive substances were injected into blood and rete testis fluid was collected. The amount of radioactivity in the rete testis fluid collected was compared to that injected into blood. From this study, the authors stated, "It is suggested that this permeability barrier will regulate the access to the seminiferous epithelium of some constituents of blood plasma, isolate the germinal cells immunologically and help to maintain the concentration differences between rete testis and lymph or blood plasma".[11–13] Various types of transporters have been identified at the apical and basolateral borders of the Sertoli cells[14] (Fig. 25.7).

Sertoli Cell Anatomic Barrier is a Dynamic Barrier

During spermatogenesis in the mammalian testis, preleptotene/leptotene spermatocytes differentiate from type B spermatogonia and traverse the Sertoli cell barrier for further development. This timely movement of germ cells involves extensive junction restructuring at the Sertoli cell barrier. In the mammalian testis, the Sertoli cell barrier, unlike the blood–brain barrier, is composed of coexisting tight junctions (TJs) and adherens junctions (AJs). Yet these junctions must open (or disassemble) to accommodate the migration of preleptotene spermatocytes across the BTB during spermatogenesis while maintaining its integrity. The Sertoli cell barrier utilizes a unique "engagement" and "disengagement" mechanism to permit the disruption of AJ that facilitates germ cell movement without compromising the Sertoli cell barrier integrity[15] (Fig. 25.8). Spermatocyte cysts become enclosed within a network of transient compartments fully bounded by old and new tight junctions. Dissolution of the old tight junctions releases the germ cells into the adluminal compartment, thus completing transit across the blood–testis barrier.[16] Studies have shown that these events are regulated by testosterone and cytokines [e.g. the transforming growth factor (TGF)-βs], which promote and disrupt the Sertoli cell barrier assembly, respectively.[17]

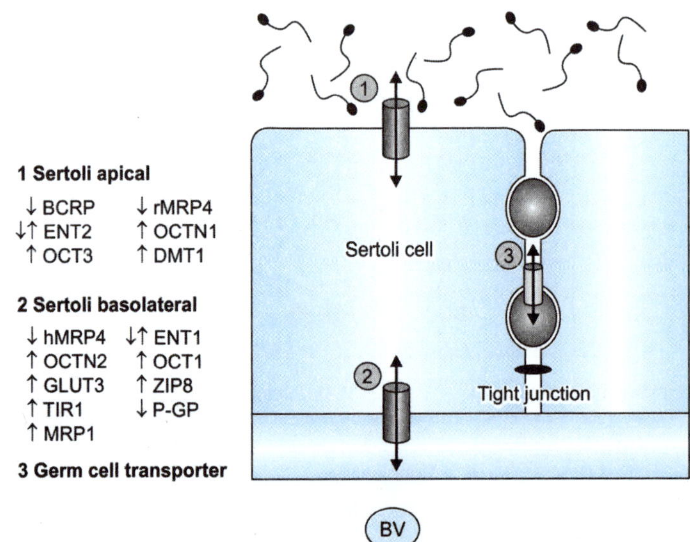

Fig. 25.7: Various transporters identified at the apical and basolateral border of Sertoli cells, modified from Klein et al[14]

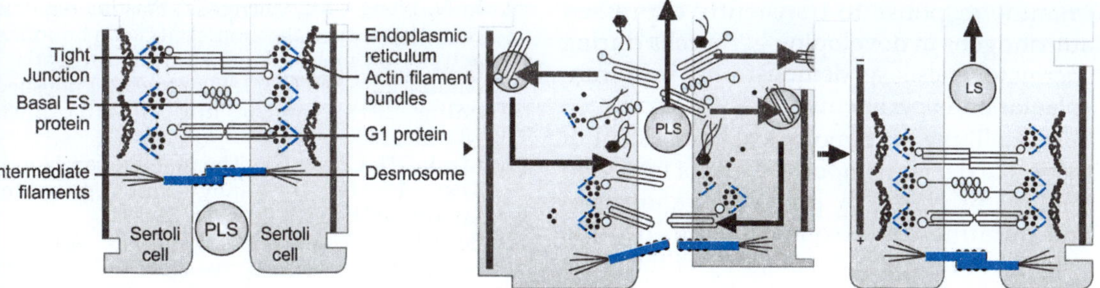

Fig. 25.8: Movement of spermatocyte across Sertoli cell anatomic barrier showing engagement" and "disengagement" mechanism to permit the transient disruption of anatomical barrier. PLS: Preleptotene spermatocyte; LS: Leptotene spermatocyte (modified from Li et al[9])

Immunological Barrier in Testis

Besides the eye and fetoplacental unit, the mammalian testis belongs to a small group of tissues that are "immunologically privileged". The mechanisms underlying this privilege remain poorly understood. The phenomenon of testicular immune privilege emerged as early as 1767 when John Hunter transplanted a cock testis into the belly of a hen and subsequently recovered a testis of normal structure from the hen.[18] The term immune privilege is often used synonymously with BTB. In fact, it is often stated that immune privilege in the testis is because of the BTB. However, immune privilege in the testis is more complex, and the *whole testis, not just the adluminal compartment, is immune-privileged.*[12] The immunoprotective ability of the space outside of the Sertoli cell barrier was demonstrated by several studies in which allogeneic or xenogeneic tissues, such as fragments of skin, pancreas, or parathyroid, transplanted into the testis interstitium, survived for an extended duration of time.[19] Transplantation of allogeneic or xenogeneic pancreatic islets within the interstitial space of the testis not only resulted in prolonged graft survival compared to nonimmune privileged sites, kidney, and liver, but also normalized blood glucose levels in diabetic rodents and rhesus monkeys.[18, 19] If the whole testis is immunologically privileged, what role, if any, does the Sertoli cell (SC) barrier play in testicular immune privilege and creating an immunological barrier to protect autoan-

tigenic germ cells? It has been proposed that the SC barrier acts as an immunological barrier (i) by sequestering germ cell antigens and (ii) preventing entry of leukocytes and antibodies into the adluminal compartment of the testis.[20] This is partially true in that antibodies and immune cell entry is hindered by the SC barrier under normal conditions. However, in the testis, although most of the autoantigenic germ cells are sequestered behind the SC barrier, preleptotene spermatocytes and spermatogonia are located within the basal compartment of the seminiferous epithelium, and express antigens that can evoke an immune response.[12] Therefore, the idea of complete sequestration of testis autoantigens as the chief basis of immunologic unresponsiveness no longer holds.[19] Instead, testicular *immune privilege also involves local production of anti inflammatory cytokines and immunosuppressive factors by cells within the testis, such as Sertoli cells, Leydig cells, testicular macrophages, and regulatory T cells, which control the immune response.*[20] The importance of Sertoli cells in creating an immune privileged environment has been demonstrated by transplantation experiments, where Sertoli cells placed in ectopic sites were able to survive and protect cografted allogeneic and xenogeneic cells.[21, 22] In addition to the SC barrier, studies have shown that Sertoli cells per se may play a critical role in maintaining the testis as an immune privileged organ by secreting immunosuppressive molecules to block immune response to transiently expressed autoantigens in developing germ cells during spermatogenesis, as demonstrated in cotransplantation experiments.[1]

Overall, the immunological component of the SC barrier is an important part of testicular immune-privilege. By preventing antibodies and immune cells from entering into the tubule, sequestering the majority of the autoantigens and allowing for controlled antigen exposure/presentation, this could decrease the immune response to the germ cells. Combined with the other immunomodulatory properties of the testis, this leads to creation of an effective immune-privileged environment throughout the testis that, under normal circumstances, protects the autoantigenic germ cells from immunological destruction.[23]

References

1. Cheng CV, Mruk DD. The Blood–testis barrier and its implications for male contraception. Pharmacol Rev. 2012; 64: 16–64.
2. Sharpe RM. Hormones and testis development and the possible adverse effects of environmental chemicals. Toxicology Letters 2001; 120: 221–32
3. Sharpe RM. Regulation of spermatogenesis. In The Physiology of Reproduction. 2nd Edn. pp 1363–1434. Eds, E Knobil and JD Neill. Raven Press, New York, 1994.
4. Sharpe RM, McKinnell C, Kivlin C, et al. Review Proliferation and functional maturation of Sertoli cells, and their relevance to disorders of testis function in adulthood. Reproduction. 2003; 125: 769–84.
5. Orth JM, Gunsalus GM, Lamperti AA. Evidence from Sertoli cell-depleted rats indicates that spermatid numbers in adults depends on numbers of Sertoli cells produced during perinatal development Endocrinology. 1988; 122: 787–94.
6. Bellieve AR, Cavicchaia JC, Millette CF, et al. Spermatogenic cells of prepubertal mouse. Published Online: 1 July, 1977 | Supp Info: http://doi.org/10.1083/jcb.74.1.68
7. deRooij DG. The spermatogonial stem cell niche. Microsc Res Tech. 2009; 72: 580–85.
8. Russell LD, Peterson RN. Sertoli cell junctions: morphological and functional correlates. Int Rev Cytol. 1985; 94: 177–211.
9. Li N, Wong CKC, Cheng CY. Plastins regulate ectoplasmic specialization via its actin bundling activity on microfilaments in the rat testis. J Androl. 2016 Sep-Oct; 18(5): 716–722. Published online 2015 Nov 20. doi: [10.4103/1008-682X.166583
10. Mruk DD, Cheng CY. The mammalian blood-Testis barrier: Its biology and regulation. Endocrine Reviews. 2015; 36: 564–91.
11. Setchell BP, Voglmayr JK, Waits GM. A blood-testis barrier restricting passage from blood into rete testis fluid but not in lymph. J Physiol. 1969; 200: 73–85.
12. Mital P, Hinton BT, Dufour JM. The blood–testis and blood–epididymis barriers are more than their tight junctions. Biol Reprod. 2011; 84: 851–58.
13. Setchell BP. Blood-testis barrier, transport proteins and spermatogenesis. Adv Exp Med Biol. 2008; 836: 212–33.
14. Klein DM, Cherrington NJ. Organic and inorganic transporters of the testis: A review. Spermatogenesis. Published online 2015 Jan 7. doi: [10.4161/21565562.2014.979653]
15. Yan HNN, Cheng CY. Blood–testis barrier dynamics are regulated by an engagement-disengagement mechanism between tight and adherens junctions via peripheral adaptors. PNAS. 2005; 102: 11722–27.
16. Smith BE, Braun RE. Germ cell migration across Sertoli cell tight junctions. Science. 2012; 338: 798–802.
17. Yan HHN, Mruk DD, Lee WM, et al. Blood–testis barrier dynamics are regulated by testosterone and cytokines *via* their differential effects on the kinetics of protein endocytosis and recycling in Sertoli cells. FASEB J. 2008; 22: 1945–59.
18. Setchell BP. The testis and tissue transplantation: historical aspects. J Reprod Immunol. 1990; 18: 1–8.
19. Barker CF, Billingham RE. Immunologically privileged sites. AdvImmunol. 1977; 25: 1–24.
20. Dufour JM, Rajotte RV, Korbutt GS, et al. Harnessing the immunomodulatory properties of Sertoli cells to enable xenotransplantation in type 1 diabetes. Immunol Invest 2003; 32: 275–97.
21. Yule TD, Mahi-Brown CA, Tung KS. The role of testicular autoantigens and influence of lymphokines in testicular autoimmune disease. J ReprodImmunol. 1990; 18: 89–103.
22. Fijak M, Meinhardt A. The testis in immune privilege. Immunol Rev. 2006; 213: 66–81.
23. Mital P, Kaur G, Dufour JM. Immunoprotective Sertoli cells: making allogeneic and xenogeneic transplantation feasible. Reproduction. 2010; 139: 495–504.

Chapter 26

Kisspeptins

INTRODUCTION

Reproduction is under a complex neuroendocrine control. The regulatory factors mainly control the hypothalamic-pituitary-gonad axis. It was in 1906 that, secretions from the ovaries were shown to produce estrus (cyclic sexual activity in the rat) and the term "estrogen" was born. It was derived from the Greek *oistros* (mad desire) and *gennan* (to produce), but the hormone estrogen was isolated only in 1929. Same year the hormone progesterone was isolated from corpus luteum of rabbit. By 1935, testosterone had also been isolated from the testes. It is remarkable that with in six years, the chief sex hormones had been identified and isolated.

The first experimental evidence, suggesting that the pituitary has a role in regulating the gonads, stems from the experiments in 1910 that showed that pituitary ablation resulted in atrophy of the genital organs in adult dogs, and a persistence of infantilism and sexual inadequacy in puppies.[1] Two years later, it was demonstrated that sectioning of the pituitary stalk affected the genital organs, and therefore hypothesized that pituitary extracts may affect the gonads. In 1926, it was shown that daily implants of fresh anterior pituitary gland tissue from mice, rats, cats, rabbits and guinea pigs into immature male and female mice and rats rapidly induced precocious sexual maturity, marked enlargement of the ovaries and "superovulation".[1] These pioneering experiments revealed that ovarian function is regulated by the pituitary. By 1930, the pituitary–gonadal relationship was established as we know it today.[1] This hypothesis was confirmed a year later with the extraction of two different hormones from the pituitary, one of which acted as a follicle-stimulating factor and one as a luteinizing factor.[2] The role of hypothalamus in reproductive function was initially discovered by Harris.[3] in 1948, who observed that electrical stimulation of the medial basal hypothalamus induced ovulation without any coitus. Only in 1971, GnRH was isolated.[4] Thus, by 1971, the fundamentals of hypothalamus-pituitary-gonad axis (Fig. 26.1) had been worked out. After that, the chief discovery in the knowledge of reproductive physiology was the role of kisspeptins in the regulation of onset of puberty.

Identification of kisspeptins is now considered as one of the major breakthroughs in reproductive biology since the isolation of GnRH back in the early 1970s. The first evidence about the reproductive roles of kisspeptins dates back to 2003, when

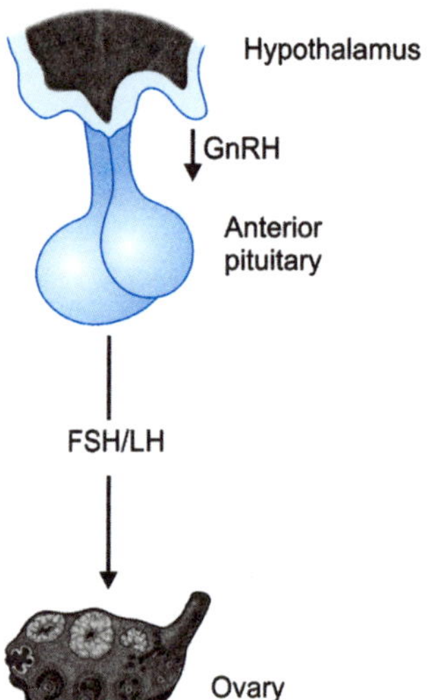

Fig. 26.1: The hypothalamic-pituitary-gonad axis in females

inactivating mutations of its receptor were described in patients with hypogonadotropic hypogonadism, a pathological condition of impuberism and infertility of central origin. Kisspeptin-expressing neurons reside in the anteroventral periventricular nucleus and the arcuate nucleus, among others, and send projections into the medial preoptic area, where there is an abundance of GnRH cell bodies. Kisspeptin appears to act directly on GnRH neurons to stimulate the secretion of GnRH.[5, 6]

KISSPEPTIN TERMINOLOGY

Kisspeptins are a family of structurally related peptides, encoded by the *KISS1/Kiss1* gene, that act through binding and subsequent activation of the G protein-coupled receptor GPR54. **KISS1 and Kiss1 refer to human and non-human genes for kisspeptins.** The gene products of KISS1 and Kiss1 are collectively known as kisspeptins. *KISS1 R* and *Kiss1 r* refer to human and non-human receptors of kisspeptins, respectively. Kisspeptins were given this name because the peptides were discovered by Dr Danny Welch's lab in Hershey, Pennsylvania in 1996. *"Hershey's Kisses"* is a brand of chocolate manufactured by The Hershey Company.

1. Role of Kisspeptin in Regulation of Puberty

Puberty is defined as the transient period between childhood and adulthood during which reproductive function is attained. During this period, the secondary sexual characteristics appear, the adolescent growth spurt occurs, the gonads start to produce mature gametes (sperm or oocytes) capable of fertilization, and major psychological changes occur. In the 1980s, it was known that puberty begins when the pulsatile release of LH releasing hormone (GnRH or LHRH) increases. The hypothalamic decapeptide, GnRH, stimulates pituitary release of LH and FSH, which are essential for the production of mature gametes and gonadal steroid secretion. Thus, the question of "what triggers the onset of puberty?" becomes rather the question of "what triggers the pubertal increase in GnRH release?"[7] By the end of 20th century, it became apparent that during the prepubertal period, an inhibitory neuronal system suppresses GnRH release. During the subsequent maturation of the hypothalamus, this prepubertal inhibition is removed, allowing the adult pattern of pulsatile GnRH release. Initially GABA was identified as the inhibitory neurotransmitter. However, during the last decade, increased secretion of **kisspeptins** emerged as the chief regulator of onset of puberty.

a. Hormonal Changes at the Onset of and during Puberty

Before the onset of puberty, plasma LH and FSH levels are low. In both boys and girls, preceding the physical signs of puberty, LH and FSH levels become elevated, pulsatility of these hormones becomes more pronounced, and the nocturnal increase in gonadotropin release is enhanced (Fig. 26.2). In rhesus monkeys, the first hormonal sign of puberty in female appears several months earlier than menarche. During the early

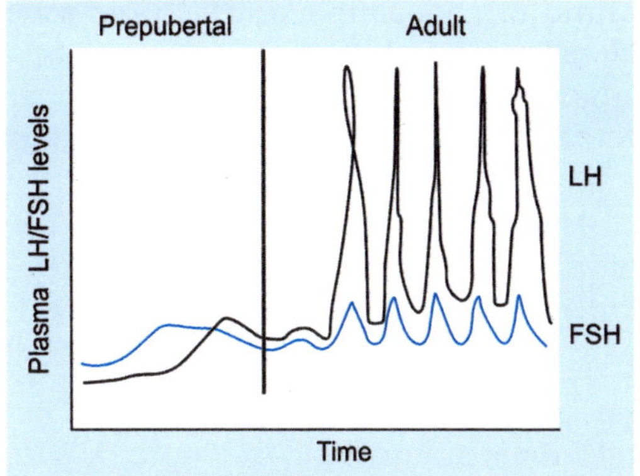

Fig. 26.2: Pulsatile plasma LH/FSH levels after puberty

pubertal period, mean FSH levels increase, followed by an increase in mean LH levels. These changes in the LH release pattern during puberty are the result of developmental changes in pulsatile GnRH release. The pulse frequency of GnRH release increases at the onset of puberty and remains elevated throughout the pubertal period.[8, 9]

b. Kisspeptin Neurons: Gatekeepers of Puberty Onset

A key event in the onset of puberty is the progressive increase of the neurosecretory activity of GnRH neurons, which in turn, maximally stimulates gonadotropin secretion to allow complete gonadal maturation and function thereafter. Such pubertal changes in the pattern of GnRH secretion are the result of concerted modifications in trans-synaptic and glial inputs to the GnRH neuronal network as well as plastic changes in GnRH neurons themselves. Among these maturational events, functional and structural changes in neuronal afferents to GnRH neurons appear crucial as ultimate triggers of puberty. It could possibly be through the increase in excitatory inputs and/or decrease in inhibitory signals projecting to GnRH neurons. The nature of such excitatory and inhibitory neuronal transmitters has been partially elucidated in recent years.[10] Among the different trans-synaptic

regulators of GnRH neurons, kisspeptins have drawn substantial attention as essential gatekeepers of puberty onset and reproductive function.

To study the consequences of increased kisspeptin secretion during puberty, studies were undertaken to look at the effect of repetitive administration of exogenous kisspeptin to juvenile female rats and primates. The result would answer the crucial question: Are kisspeptin alone sufficient to kick start the onset of puberty? In both studies, repetitive kisspeptin administration was able to advance puberty. In both species, gonadotrophins were elevated to adult levels. Thus, it was revealed in both cases to be due to kisspeptin stimulation of GnRH secretion, as GnRH antagonists completely abolished the advancement of puberty.[8]

The studies discussed above convincingly demonstrate that kisspeptins are key regulators of puberty in mammals. There seems to be a programmed increase in KISS1 mRNA and increase in receptor sensitivity to kisspeptin, possibly due to an increase in receptors at the GnRH secreting neurons. Activation of the kisspeptin system facilitates increased pulsatile release of GnRH, awakening the reproductive axis and bringing about pubertal maturation.

That the maturation of kisspeptin neurons is responsible for onset of puberty is supported by the following observations.

1. Animal studies demonstrated an increase in KISS1 neurons at puberty.

2. At hypothalamus, it was found that over three-fourths of GnRH neurons coexpress the receptor for kisspeptin, GPR54, in their RNA.

3. Kisspeptin was also able to bring about the release of GnRH both *ex vivo* and *in vivo* in rat and sheep.

4. In humans, it was shown that females at the beginning stages of puberty had much higher kisspeptin levels than those females of the same age who had yet to begin puberty.

5. Exogenous kisspeptins induced precocious puberty in rodents and monkeys, whereas kisspeptin antagonists delayed puberty in these animals.
6. Inactivating mutations in KISS1/Kiss1 or their receptors (KISS1R/Kiss1r) led to pubertal failure.

It can be concluded that by activating GnRH neurons in the hypothalamus, kisspeptin causes GnRH release which leads to the release of FSH and LH. Maturation of Kiss1 neurons seems to be essential for the normal timing of puberty. This phenomenon includes not only the increase of kisspeptin tone but also of the increased sensitivity of GnRH neurons to the stimulatory effects of kisspeptins. In addition, there is a rise in the number of kisspeptin neurons and their projections to GnRH neurons during puberty[8–10] (Fig. 26.3).

2. Role for Kisspeptin as a Link Between Nutrition and Puberty

Many studies provided evidence that girls in the United States, are starting puberty at a younger age than earlier studies had found. Because nutritional status is known to affect timing of puberty and there is a clear trend for increasing obesity in US children during the past 25 years, it was hypothesized that the earlier onset of puberty could be attributable to the increasing prevalence of obesity in young girls.[11]

The signals regulating body weight and energy expenditure have been extensively studied in recent years. Leptin is a peptide hormone secreted by adipocytes. Deficiency of leptin results in delayed puberty and hypogonadotropic hypogonadism in mice and humans. Furthermore, leptin administration reverses the infertility associated with leptin deficiency.[12] On the basis of such observations, it was hypothesized that leptin may constitute a link between nutrition and onset of puberty. However, GnRH neurons lack receptors for many of the major metabolic signaling peptides, including insulin and leptin.[13]

Now, kisspeptin is implicated as an intermediary between leptin signaling and GnRH function. Kisspeptin neurons express the leptin receptor, and *Ob/Ob* mice, genetically deficient in leptin, have reduced ARC levels of kisspeptin mRNA compared with controls. Furthermore, kisspeptin expression is increased following exogenous leptin administration. Fasting has been shown to reduce hypothalamic kisspeptin mRNA and

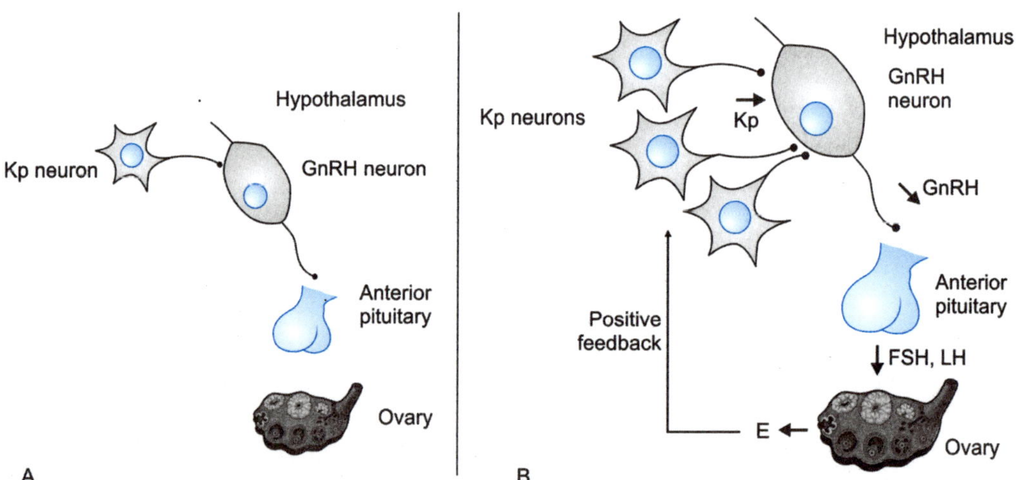

Fig. 26.3: The role of kisspeptin in the hypothalamus-pituitary-ovary axis. Prepubertal (A) and peripubertal (B)

delay the onset of puberty in rats. In addition, central administration of kisspeptin to chronically undernourished prepubertal rats restored parameters of delayed puberty. In brief, now it is believed that, via action of leptin and kisspeptin, obesity results in early onset of puberty, whereas undernutrition delays the onset of puberty[14] (Fig. 26.4).

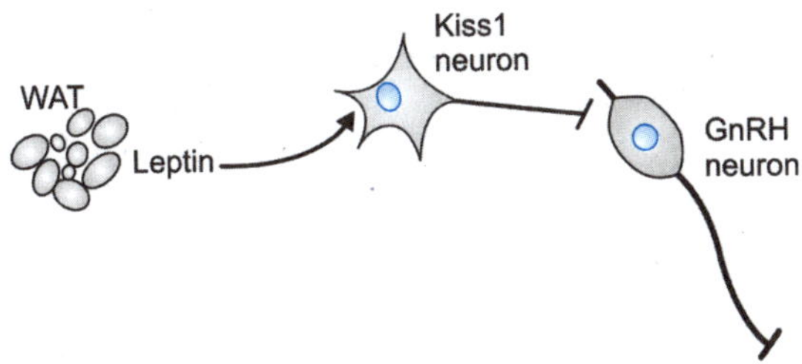

Fig. 26.4: Mechanism of effect of body weight on the onset of puberty

3. Role of Kisspeptins in Adults

i. **Stimulator of GnRH secretion:** Besides their role in the regulation of puberty onset, kisspeptins remain one of the strongest stimulators of GnRH secretion in adults. A large number of studies have demonstrated that injection of kisspeptins can stimulate the secretion of GnRH and the pituitary gonadotropins LH and FSH in adult rats (Fig. 26.5). In fact, adult mice respond to lower doses of kisspeptin than peripubertal rats. It appears that GnRH neurons become

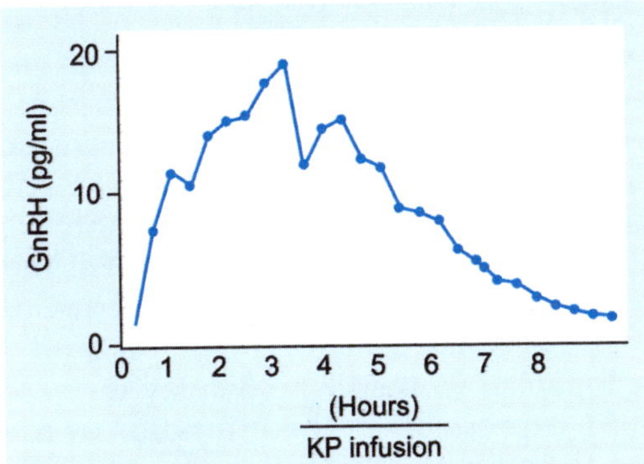

Fig. 26.5: Effect of kisspeptin administration on levels of GnRH (Adapted from PNAS 2005;102:1761–66)

developmentally activated by kisspeptin over the course of puberty. An LH peak has been successfully induced in all species studied (rodents, sheep, monkeys, humans) irrespective of the administration route (intracerebroventricular, intravenous, subcutaneous) even at very low doses of kisspeptin. Several elements point to GnRH neurons as the main sites of action of kisspeptin. First, GnRH neurons express the kisspeptin receptor KISS1R; second, when animals are pretreated with GnRH antagonists, kisspeptin-induced LH and FSH release is blocked. Furthermore, *in vitro*, kisspeptins induce GnRH secretion by rat hypothalamic explant.[15–17]

ii. **Feedback control of kisspeptin neurons by sex steroids:** Changes in the frequency and amplitude of pulsatile GnRH release, and consequent pulsatile release of the gonadotropins LH and FSH from the anterior pituitary gland, are essential for the appropriate control of fertility. In turn, LH and FSH regulate the gonadal function including sex steroid hormone production and gamete development at the level of the gonads. In a feedback loop, steroid hormones act through an afferent neuronal network to provide critical information to GnRH neurons, regulating their pattern of activity. Identification of the neuronal structures involved in pulsatile release of GnRH, and the afferent pathways through which gonadal hormones regulate pulsatile GnRH release, have been unresolved questions in reproductive neuroendocrinology.[17] Now, it has been discovered that, in males, testosterone has only negative feedback control over GnRH neurons via kisspeptin neurons. However, in females, the feedback control by sex hormones is both negative and positive in different phases of ovarian cycle (Fig. 26.6). This complex feedback control of estrogens and progesterone has been explained by what is known as *"KNDy hypothesis."*

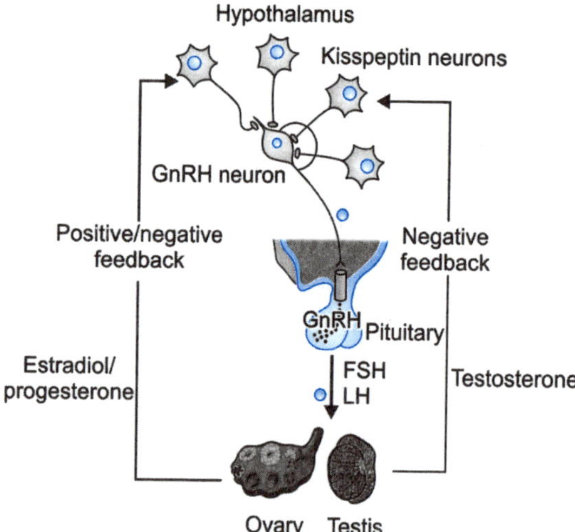

Fig. 26.6: Feedback control of sex steroids on kisspeptin neurons

KNDy HYPOTHESIS

In arcuate nucleus of the hypothalamus, a group of neurons are known as the kisspeptin/neurokinin B/dynorphin (KNDy) neurons. A distinguishing feature of these kisspeptin secreting neurons is that besides kisspeptin, these neurons release two additional neuropeptides, neurokinin B (NKB) and dynorphin, each of which play an important role in steroid feedback control of GnRH and LH secretion.

The hypothesis that KNDy neurons form a critical component of the GnRH pulse generator was initially based on the co-localization of three neuropeptides with

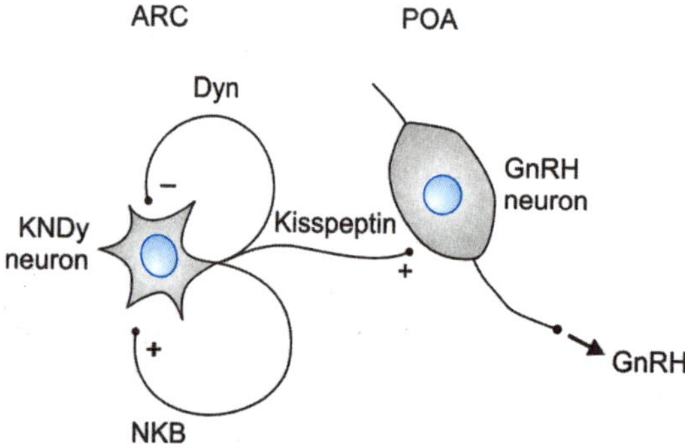

Fig. 26.7: KNDy neurons co-secrete kisspeptins, neurokinin B (NKB) as well as dynorphin (Dyn)

opposing effects on GnRH/LH release, *kisspeptin and NKB being stimulatory* and *dynorphin inhibitory*. In brief, the so-called "KNDy hypothesis" stated that *NKB was the signal responsible for pulse onset by triggering activation among KNDy neurons, kisspeptin served as the output signal from KNDy neurons driving GnRH secretion, and dynorphin served as the signal terminating each pulse.*[17]

Kisspeptin/ KNDy Neurons: Mediators of Positive and Negative Feedback Control on GnRH Neurons by Gonadal Steroids

A crucial role for sex steroids in the regulation of GnRH neurons and/or anterior pituitary gonadotrophs in humans was initially proposed as serial blood sampling and gonadotropin assays in women through phases of menstrual cycle showed an uneven distribution, with a clear mid-cycle surge in LH and FSH. Two mechanisms were proposed to mediate this effect: First, GnRH secretion is altered in response to the steroid milieu; second, sensitivity of the gonadotrophs to a GnRH input is sex-steroid dependent, although the exact mechanism remained controversial.[18] Now it has been shown that kisspeptin neurons have an important role in the production of positive and negative feedback control of sex steroids in females.

Mediator of Negative Feedback Control

The ARC kisspeptin neurons are implicated in the negative feedback control by sex steroids. This neuron population has been demonstrated by experiments in which ovariectomy in rodents led to increased concentration of *kiss1* mRNA in the ARC and this increase could be prevented by estrogen or testosterone replacement therapy. This negative feedback control has also been demonstrated in primates and in menopausal women who display increased *KISS1* expression in the infundibular nucleus (human equivalent to the ARC).[19] A remarkable feature of Kiss1 neurons is that virtually all of them express estrogen receptors α (ERα), while only a modest fraction expresses ER β. This is of great physiological relevance, especially considering the *dominant roles of ERα in the negative feedback control of gonadotropins but the conspicuous lack of this receptor in GnRH neurons*. All these researches support the relevance of the ARC ERα-Kiss1 pathway in the negative feedback control of gonadotropins (Fig. 26.8).

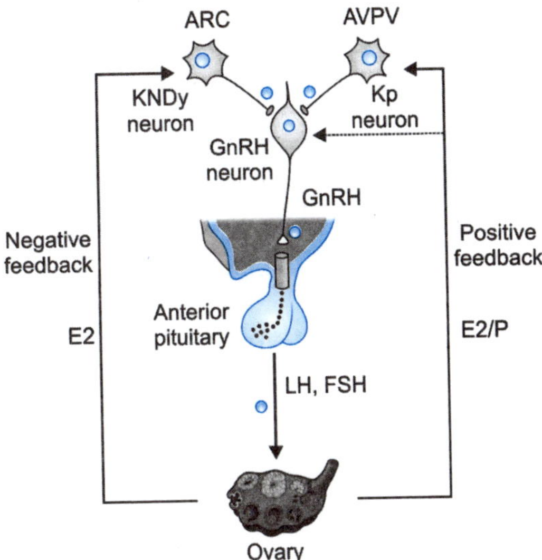

Fig. 26.8: Differential negative feedback and positive feedback regulation of ARC *vs* AVPV Kiss1 neurons in the control of GnRH (for details, *see* text)

Mediators of Positive Feedback Control

While ARC kisspeptin neurons are implicated in negative feedback control of GnRH secretion, the anteroventral periventricular (AVPV) nueleus kisspeptin neurons are implicated in the positive feedback control of GnRH secretion by estrogens. Expression and functional studies in the context of sex steroid regulation of *Kiss1* expression described above also revealed the existence of a second population of Kiss1 neurons in AVPV, displaying opposite responses to sex steroids. The second set of Kiss1 neurons seem to be ideally suited for mediating the positive feedback effects of estrogen in the generation of the preovulatory surge of gonadotropins (Fig. 26.8). Contrary to the ARC, neurons at the AVPV displayed decreased *Kiss1mRNA* levels after elimination of gonadal steroids and increased expression following sex steroid replacement. This pattern of responses was initially detected in both male and female mice; however, responses in the female were much more prominent.

Using antagonists of the alpha estrogen receptor (ERα), it has been demonstrated that this positive feedback control is also mediated via the ERα and not the second type of estrogen receptors ERβ. But as mentioned earlier, GnRH neurons express ERβ and not ERα, suggesting there would be an intermediary pathway relaying the estrogen action on GnRH neurons.[19] Kisspeptin neurons in the anteroventral periventricular nucleus (AVPV) neurons are more numerous in females than males, express ERα, and project on GnRH neurons, making them ideal candidates for this intermediary pathway.[20] It has been shown that estrogens induce increased expression of kisspeptin neurons in the AVPV, that the same neurons are activated during the preovulatory period, and that blocking endogenous kisspeptins in the preoptic area abolishes the LH peak in the rat.[21]

References

1. Lunenfeld B. Historical perspectives in gonadotrophin therapy. Human Reproduction Update, 2004; 10: 453–67.
2. Fevold SL, Hisaw FL, Leonard SL. The Gonad-stimulating and the luteinizing hormones of the anterior lobe of the hypophysis. Am J Physiol 1931; 97: 291–301.
3. Harris GW. Electrical stimulation of the hypothalamus and the mechanism of neural control of the adenohypophysis. Jour Physiol. 1948; 107, 418–29.
4. Amoss M, Burgus R, Blackwell R, et al. Purification, amino acid composition and N-terminus of the hypothalamic luteinizing hormone releasing factor (LRF) of ovine origin. Biochemical and Biophysics Communication. 1971; 44, 205–210.
5. Tng EL. Kisspeptin signalling and its roles in humans. Singapore Med J 2015; 56: 649150656

6. Roseweir AK, Millar RP. Human Reproduction Update, Volume 15, Issue 2, 1 March 2009, https://doi.org/10.1093/humupd/dmn058

7. Terasawa E, Nass TE, Yeoman RR, et al. Hypothalamic control of puberty. In: Norman RL (ed). Neuroendocrine Aspects of Reproduction. Academic Press, New York. 1983, p149.

8. Plant TM, Ramaswamy S, Dipietro MJ. Repetitive activation of hypothalamic G protein-coupled receptor 54 with intravenous pulses of kisspeptin in the juvenile monkey (Macacamulatta) elicits a sustained train of gonadotropin-releasing hormone discharges. Endocrinology. 2006; 147: 1007–13.

9. Pinilla L, Aguilar E, Dieguez C, et. al. Kisspeptins and reproduction: physiological roles and regulatory mechanisms. Physiol Rev. 2012; 92: 1235–316.

10. Chehab FF, Lim ME, Lu R. Correction of the sterility defect in homozygous obese female mice by treatment with the human recombinant leptin. Nat Genet. 1996; 12: 318–20.

11. Kaplowitz PB, Slora EJ, Wasserman RC, et al. Earlier Onset of Puberty in Girls: Relation to Increased Body Mass Index and Race. Pediatrics. 2001; 108: 347–53.

12. Ojeda SR, Dubay C, LomnicziA, et al. Gene networks and the neuroendocrine regulation of puberty. Mol Cell Endocrinol. 2010; 324: 3–11.

13. Clarke H, Dhillo WS, Jayasena CN. Comprehensive Review on Kisspeptin and its Role in Reproductive Disorders. Endocrinol Metab (Seoul). 2015; 30: 124–41.

14. Chehab FF. Leptin and reproduction: past milestones, present undertakings, and future endeavors. Journal of Endocrinology. 2014; 223: T37–48.

15. Oakley AE, Clifton D.K., Steiner R.A. Kisspeptin signaling in the brain. *Endocr Rev.* 2009; 30: 713.

16. Moore AM, Coolen LM, Porter TD, et al. KNDy Cells Revisited. *Endocrinology.* 2018; 159: 3219–34.

17. Sonigo C, Binart N. Overview of the impact of kisspeptin on reproductive function. Annalesd' Endocrinologie 2012; 73: 448–58.

18. Marques P, Skorupskaite K, George JT, et al. Physiology of GNRH and Gonadotropin Secretion. 2018 Jun 19. In: De Groot LJ, Chrousos G, Dungan K, et al., editors. Endotext[Internet]. South Dartmouth (MA): MDText. com, Inc.; 2000-. Available from: https://www.ncbi.nlm.nih.gov/books/NBK279070/

19. Hameed S., Jayasena C.N., Dhillo W.S. Kisspeptin and fertility J Endocrinol. 2011; 208: 97–105.

20. Roa J, Castellano JM, Navarro VM, et al Kisspeptins and the control of gonadotropin secretion in male and female rodents. Peptides. 2009; 30: 5715066.

21. Shibata M, Friedman RL, Ramaswamy S, et al. Evidence that downregulation of hypothalamic KISS1 expression is involved in the negative feedback action of testosterone to regulate luteinising hormone secretion in the adult male rhesus monkey (Macacamulatta). *J Neuroendocrinol* 2007; 19: 432–38.

Chapter 27

Parturition

INTRODUCTION

Historically, the central role of the hormone progesterone, as necessary to maintain pregnancy and block the onset of labor, has been recognized for more than 70 years. This "progesterone block" hypothesis remained a cornerstone of parturition physiology. Subsequent *in vivo* studies in sheep by Liggins and colleagues in the 1960s established activation of the fetal hypothalamic–pituitary–adrenal axis as the driver of withdrawing progesterone (removing the block) and initiating parturition. Findings in the sheep, however, proved problematic to extrapolate to human parturition, as systemic progesterone withdrawal, a fall in circulating serum progesterone, does not occur in women before the onset of labor[1] (Fig. 27.1).

However, certain key events in the parturition process can be recognized in all the mammalian species. Pivotal among these are: (1) Transformation of the uterine smooth muscle (myometrium) from a quiescent to a highly excitable state; it starts contracting forcefully and rhythmically, and (2) remodeling of the extracellular matrix of the cervical stroma such that the walls of the cervix distend in response to the contractions of labor to open the cervical canal and allow the conceptus to be moved into the vagina. The synchronization of these events, referred to as active labor, occurs over a relatively

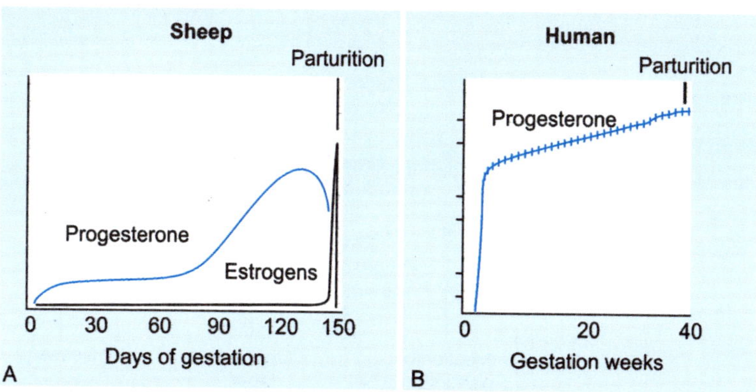

Fig. 27.1: Plasma progesterone levels during pregnancy and at term in sheep (A), and in humans (B)

short time and usually at term. Parturition occurs at a time in the course of pregnancy which is appropriate for adequate fetal maturity for survival outside the uterus.[2]

TIMING OF PARTURITION

The success of pregnancy is dependent on the appropriate timing of parturition. Ideally, birth should occur at a stage in gestation when the fetus is physiologically prepared for life outside of the uterus. Full term for human pregnancy is at the 40th completed week of gestation (calculated from the first day of the mother's last menstrual period). At this time, the probability of the fetus and mother surviving the parturition process is at its apex. The risk for neonatal and/or maternal morbidity and mortality increasing exponentially with the deviation (whether preterm or post-term) from term at which birth occurs.[2]

The last few hours of human pregnancy are characterized by uterine contractions that effect cervical dilatation and expulsion of the fetus out of the uterus. Long before these forceful, painful contractions, there are extensive preparations in both the uterus and cervix. During the first 36 to 38 weeks of normal gestation, the myometrium is in an unresponsive state. Following this prolonged uterine quiescence, there is a transitional phase during which myometrial contractile unresponsiveness is replaced by activation, and the cervix undergoes ripening, effacement, and loss of structural integrity (preparation for labor). This phase is followed by the stage of active labor and finally the stage of uterine involution (parturient recovery).[3] The four stages of pregnancy and labor are depicted in Fig. 27.2.

Stage of Uterine Quiescence

It is generally accepted that progesterone plays an important role in the maintenance of uterine quiescence until the perinatal period, when estrogen action dominates.[4] A successful pregnancy requires near complete relaxation of the uterus for more than ninety-five percent of gestation, overcoming the inherent tendency of the myometrium to contract with stretch. This active and highly regulated process is called myometrial quiescence. It requires not only the near absence of myometrial contractions, but also its refractoriness to contractile agents. Plasma progesterone levels rise progressively human pregnancy (Fig. 27.1). At term, the plasma progesterone levels are approximately 10-times the level at the time of ovulation. However, the molecular mechanisms by which progesterone reduces uterine contractility in human gestation are not completely known. The resting membrane potential and corresponding excitability of uterine myocytes, and the frequency and duration of action potentials

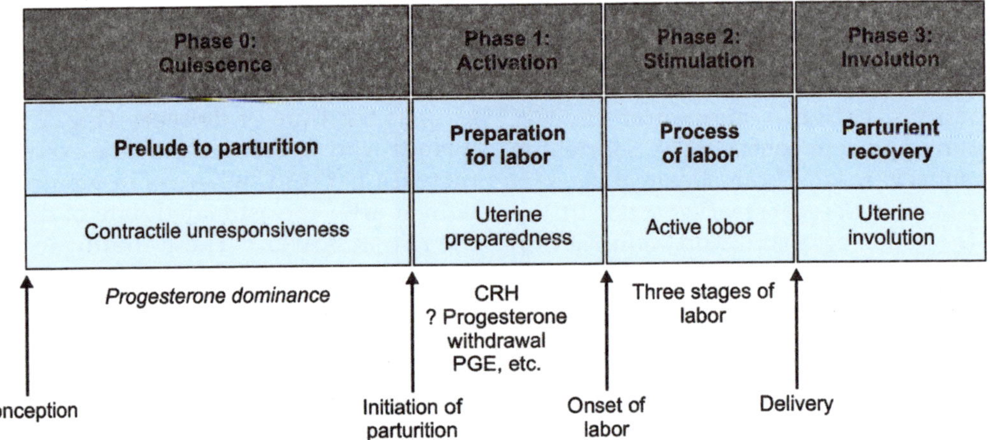

Fig. 27.2: Four stages of pregnancy and labor

are differentially influenced by the enhancing effects of estrogen and suppressive actions of progesterone.[5] In one experiment, membrane potential of uterine smooth muscle increased from −35 to −65 under the effect of progesterone.[6] Progesterone increases expression of calcium- and voltage-gated K[+] channels; allowing efflux of potassium, thereby increasing the potential difference across the cell membrane and making it less likely to depolarize. Progesterone appears to inhibit uterine excitability by a number of other mechanisms as well, including inhibiting the synthesis of prostaglandins and inhibiting the expression of contraction-associated proteins genes and reduction in cell signal components that lead to increased intracellular Ca^{2+} concentrations in response to contractile stimuli.[7]

MECHANISMS OF ACTIVATION OF MYOMETRIUM AT TERM

An important event in labor is the expression of a group of proteins termed "contraction-associated proteins. There are three types of contraction-associated proteins: Those that enhance the interactions between the actin and myosin proteins that thereby increase myometrial contractility, those that increase the excitability of individual myometrial cells, and those that promote the intercellular connectivity that permits the development of synchronous contractions.

a. Proteins Increasing Myometrial Contractility

Interactions between actin and myosin determine myocyte contractility. Shortening of the muscle results from *sliding of the actin filaments over the myosin filaments*. In the smooth muscle, excitation contraction coupling is mediated by voltage-gated Ca^{2+} channels which transfer ionic calcium from ECF into smooth muscle. Increased presence of voltage-dependent calcium channels (VDCC) in gestational uteri indicate that progression toward term is accompanied by increased VDCC activity.[8]

b. Proteins Increasing Myocyte Excitability

At the time of labor, changes in the distribution and function of the K[+] channels lower the intensity of the stimulus required to depolarize myocytes and to produce the associated influx of calcium that generates contraction.[7]

c. Proteins Promoting Intercellular Connectivity

A critical aspect of myometrial activity at labor is the development of synchrony. Synchronous activity of myometrial cells results in the powerful contractions needed to expel the fetus. The uterus lacks a pacemaker that regulates the contractions. However, as parturition progresses, there is increasing synchronization of the electrical activity of the uterus. At the cellular level, this synchrony is achieved by electrical conduction gap junctions (*refer to* Fig. 12.8). Recent ultrastructural studies of the myometrium have shown that gap junctions develop in large numbers between smooth muscle cells at term.[9]

NEUROHUMORAL CONTROL OF PARTURITION

A. Role of Corticotropin-releasing Hormone (CRH)

i. Maternal CRH

Several studies have shown an association between levels of maternal plasma CRH, which is of placental origin, and the timing of birth.[10] Maternal plasma CRH levels increase exponentially as pregnancy advances, peaking at the time of delivery (Fig. 27.3). In women who deliver preterm, the exponential increase is rapid, whereas in women who deliver after the estimated date of delivery, the rise is slower.[11] These findings suggest that a placental clock determines the timing of delivery.

Production of CRH by the placenta is restricted to primates. Humans also produce a circulating binding protein for CRH (CRHBP). At the end of pregnancy, CRHBP levels fall, thereby increasing the bioavailability of CRH.

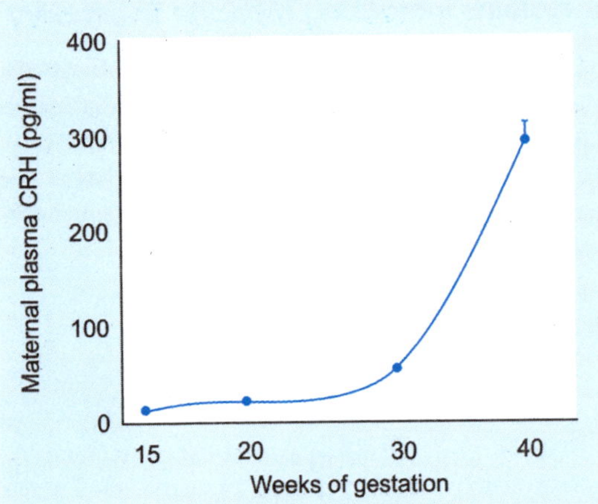

Fig. 27.3: Maternal plasma CRH levels in second and third trimesters of human pregnancy

Glucocorticoids stimulate expression of the *CRH* gene and production of CRH by the placenta.[12] In turn, CRH stimulates the pituitary to produce corticotropin, which causes the adrenal cortex to release cortisol. This arrangement permits a positive feed-forward system that explains the progressively increasing concentration of estrogens observed in human pregnancy, as explained below.[13]

CRH is secreted from the placenta predominantly into the maternal blood, but it also enters the fetal circulation. In the mother, CRH receptors are present in the pituitary, the myometrium, and probably the adrenal glands. In the fetus, there are CRH receptors in the pituitary, the adrenal glands, and perhaps the lungs. Rising levels of CRH can therefore act at multiple sites in mother and fetus to initiate the changes associated with parturition. Increased placental CRH levels drive the rise in maternal corticotropin and cortisol levels as gestation advances. The increased levels of CRH and corticotropin promote the production of cortisol and dehydroepiandrosterone sulfate (DHEAS) by the maternal adrenal glands; the increased cortisol stimulates further placental release of CRH, and DHEAS provides a substrate for placental estrogen synthesis[13] (Fig. 27.4).

ii. Fetal CRH

Placental CRH is also released into the fetus, and although the concentrations are lower in the fetal circulation than in the maternal circulation, they still rise with advancing gestation.[14] As mentioned above, in the fetus, there are CRH receptors in the pituitary, the adrenal glands, and perhaps the lungs. Stimulation of the fetal pituitary by CRH increases corticotropin production and, consequently, the synthesis of cortisol by the fetal adrenal gland and maturation of the fetal lungs. In turn, the rising cortisol concentrations in the fetus further stimulate placental CRH production. The maturation of the fetal lungs as a result of increasing cortisol concentrations is

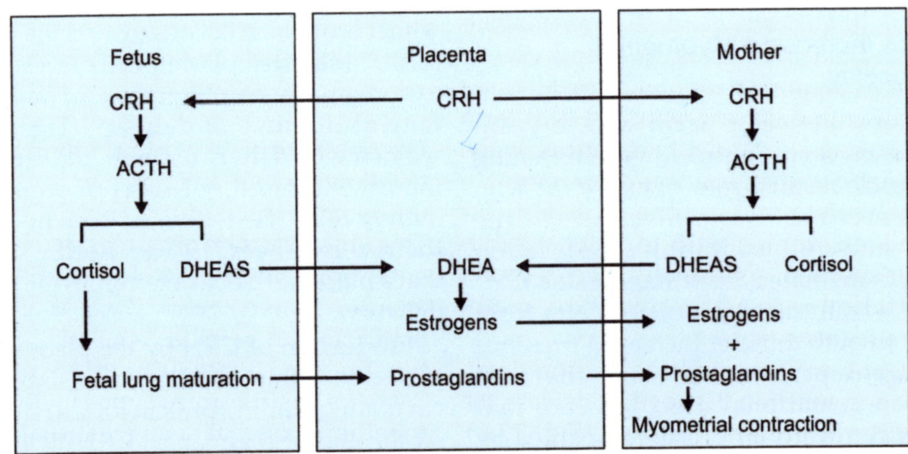

Fig. 27.4: Role of placental CRH in parturition

associated with increased production of surfactant protein A and phospholipids, both of which have proinflammatory actions and may stimulate myometrial contractility through increased production of prostaglandins by fetal membranes and the myometrium itself[13] (Fig. 27.4). CRH stimulation of fetal adrenal zone cells, which lack 3β-hydroxysteroid dehydrogenase, preferentially causes placental formation of DHEA, the precursor of estrogen.[14] The fetal zone of the adrenal glands involutes rapidly after delivery of the placenta, indicating that placental factors, such as CRH, maintain the fetal zone. Thus, CRH may stimulate adrenal steroidogenesis, thereby providing the substrate for the placental production of estrogens, which favor parturition by inducing contraction (Fig. 27.4).

In summary, it appears that *positive feedforward systems* in the mother and fetus drive an exponential increase in placental CRH production as gestation advances. Increased placental CRH production which, in turn, increases fetal cortisol concentrations and fetal lung maturation and synthesis of prostaglandins, that precipitates labor and delivery (Fig. 27.4). These pathways, each capable of stimulating parturition, make the mechanism of labor robust.[13]

B. Role of Estrogens

After the first three months of pregnancy, placenta becomes the chief site of synthesis and secretion of estrogens. Plasma concentrations of estrone, estradiol, and estriol increase as human pregnancy progresses (Fig. 27.5) with daily excretion rates at term approximating 2, 1, and 40 mg, respectively. As explained above, fetus and placenta are intimately involved in the synthesis of estrogens. In women with threatened first-trimester abortion, abnormally high estradiol concentrations are associated with a subsequent pregnancy loss.

Estrogens are essential for uterine development and function. Estrogens increase the excitability of pregnant uterus (Fig. 27.6).

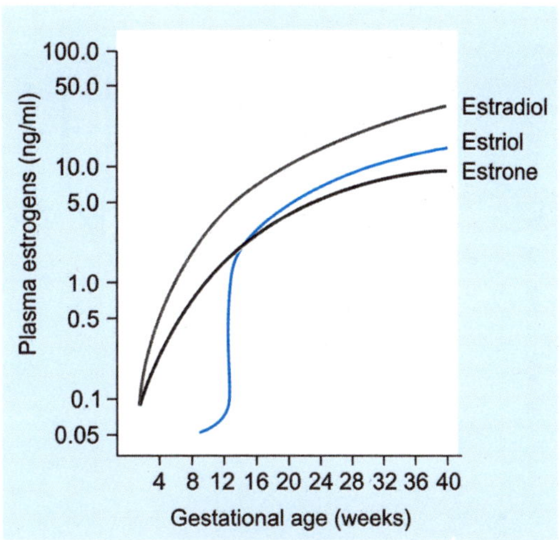

Fig. 27.5: Plasma estrogens in pregnancy

Various mechanisms seem to be involved for this response. Estrogens are responsible for the synthesis of the contractile proteins and the regulatory enzymes necessary for uterine contractility. Estrogens increase the concentration of receptors for oxytocin, which modulates membrane calcium channels. Estrogens increase gap junction formation in the myometrium. This allows for coordinated uterine contractions. Estrogen also stimulates the production of prostaglandins F_{2a} and E_2, which stimulate uterine contractions. Estrogens control cervical ripening also. In women, although plasma estradiol levels continue to remain high during pregnancy, it does not have a sharp predelivery increase as in sheep (Fig. 27.1), where it is responsible for the onset of labor.

C. Role of Progesterone Withdrawal

Progesterone plays a critical role in pregnancy by allowing implantation and, subsequently, the maintenance of myometrial relaxation. The administration of a progesterone receptor antagonist such as RU486 readily induces abortion if given before 7 weeks of gestation. Similarly, the surgical removal of the corpus luteum, the source of progesterone, before 7 weeks results in pregnancy loss. Taken together, these data suggest that adequate

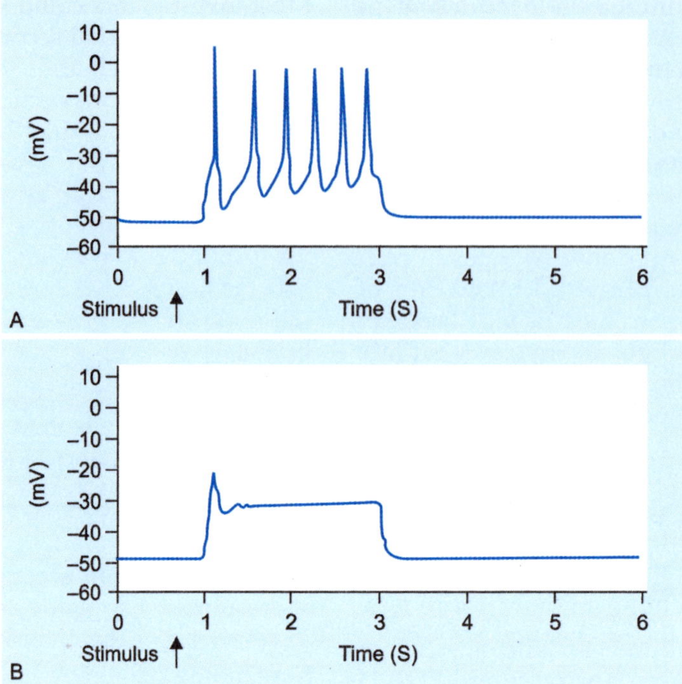

Fig. 27.6: Action potential of a uterine muscle fiber of a pregnant rat in the presence (A) or absence (B) of estradiol

production of progesterone by the corpus luteum is critical to the maintenance of early pregnancy until the placenta takes over this function at approximately 7 to 9 weeks of gestation.[16] In many mammals (e.g. sheep and cattle), a drop in circulating progesterone levels precipitates parturition (Fig. 27.1). However, a characteristic of human pregnancy is that the level of circulating progesterone does not fall prior to the onset of labor[17] (Fig. 27.7). However, circulating hormone levels do not necessarily reflect tissue levels, and there is increasing evidence from both *in vitro* and *in vivo* studies that the spontaneous onset of labor at term may be preceded by a physiologic (functional) withdrawal of progesterone activity at the level of the uterus.[16] A search for mechanisms

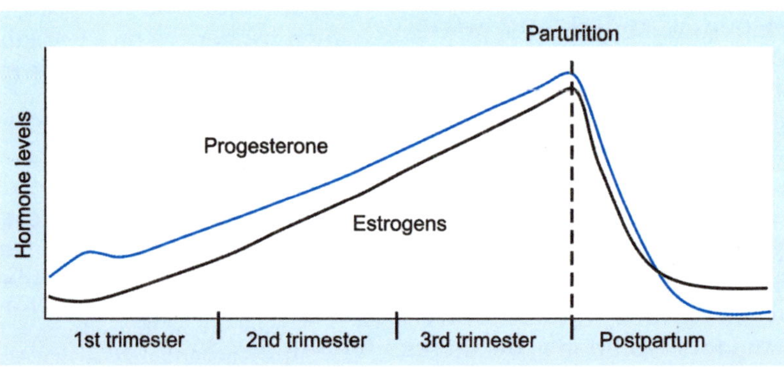

Fig. 27.7: Concentrations of estrogens and progesterone during pregnancy in human female. The hormonal level decline only *after* the onset of parturition

that could account for a *functional withdrawal of progesterone* has identified several forms of the progesterone receptors. With the onset of labor, the progesterone receptors function changes in a way that could constitute a mechanism of progesterone withdrawal.[17, 18]

D. Role of Prostaglandins

Besides progesterone, prostacyclins, the inhibitory prostaglandins, present throughout early pregnancy, are also responsible for uterine quiescence during pregnancy. There is strong evidence that prostaglandins are of major importance in mechanism of labor in women.[19] Prostaglandins are produced in the placenta and fetal membranes. Prostaglandin levels are increased before and during labor in the uterus and membranes.[20] $PGF_{2\alpha}$ is produced primarily by the maternal decidua and acts on the myometrium to upregulate oxytocin receptors and gap junctions, thereby promoting uterine contractions. PGE_2 is primarily of fetoplacental origin and is likely to be more important in promoting cervical ripening (maturation) associated with collagen degradation and dilation of cervical small blood vessels and spontaneous rupture of the fetal membranes. Prostaglandins production is decreased by progesterone and increased by estrogens[21] (Fig. 27.4).

E. Role of Oxytocin

More than a century ago, Dale noted the ability of posterior pituitary extract to stimulate uterine contractions. By the 1940s, oxytocin was introduced into clinical practice to stimulate labor as well as to treat postpartum hemorrhage.[22] After the demonstration of the potent uterine contractile properties of oxytocin, many investigators attempted to establish oxytocin as the major factor in the initiation of labor. However, most of the results failed to support such a hypothesis and suggested that the importance of oxytocin was confined to the expulsive stage of labor and for involution of the uterus. It was reported that serum oxytocin concentrations remain low throughout pregnancy.[23]

Most investigators could not demonstrate an increase in maternal serum oxytocin concentrations until well after the onset of labor.[24] Peak concentrations occur at the time of fetal expulsion.[22] Thus, till 1992, the majority of evidence did not support a major role for oxytocin in the physiological onset of labor. Most investigators concluded that the role of oxytocin in labor was, at most, facilitatory and that the hormone had perhaps a more important role in the prevention of postpartum hemorrhage.

Recent findings suggest that the uterus itself is a major site of oxytocin production during late pregnancy. In the rat, oxytocin is synthetized in the epithelial cell layer of the endometrium. Endometrial oxytocin mRNA increases more than 150-fold during gestation and at term, exceeds hypothalamic oxytocin m RNA by 70-fold. Expression of the oxytocin gene has also been demonstrated in the human uterus, with the presence of oxytocin mRNA in decidua, and to a lesser extent in amnion and chorion. Estradiol stimulates the expression of oxytocin mRNA by choriodecidual tissue *in vitro*. Thus, oxytocin may act primarily as a local mediator of uterine contraction during parturition. Endogenous oxytocin may exert a paracrine control of myometrial contractions both directly by interacting with myometrial oxytocin receptors, and indirectly by stimulating the production of prostaglandins $F_{2\alpha}$ ($PGF_{2\alpha}$). The rise in uterine oxytocin production, in concert with the rise in uterine oxytocin receptors synthesis may be involved in mechanisms which trigger the onset of labor. Circulating oxytocin, however, still has a role in later stage of labor.[25]

References

1. Liggins GC, Fairclough RJ, Grieves SA, et al. The mechanism of initiation of parturition in the ewe, Rec. Progr. Hormone Res. 1973; 29: 111–59.
2. Plant TM, Zeleznik AJ. Knobil and Neill's Physiology of Reproduction (Fourth Edition), 2015. Amsterdam: Elsevier Academic Press.
3. Corton MM. Williams Obstetrics, 24e2014 by McGraw-Hill.

4. Casey ML, MacDonald PC. The endocrinology of human parturition. Ann N Y Acad Sci. 1997; 828: 273–84.

5. Kuriyama H, Suzuki H. Changes in electrical properties of rat myometrium during gestation and following hormonal treatments. J Physiol [Lond]. 1976; 260: 315–33.

6. Goto M, Csapo A. The effect of ovarian steroids on the membrane potential of uterine muscle. J Physiol. 1959; 43: 455–66.

7. Soloff MS, Jeng Y, Izban MG. Effects of progesterone treatment on expression of genes involved in uterine quiescence. Reprod Sci. 2011; 18: 781–97.

8. Collins PL, Moore JJ, Lundgren DW, et al. Gestational changes in uterine L-type calcium channel function and expression in guinea pig. Biology of Reproduction. 2000; 63: 1262–70.

9. Garfield RE, Kannan MS, Daniel EE. Gap junction formation in myometrium. Control by estrogens, progesterone, and prostaglandins. Am J Physiol. 1980; 238: C81–90.

10. Wadhwa PD, Porto M, Garite TJ, et al. Maternal corticotropin-releasing hormone levels in the early third trimester predict length of gestation in human pregnancy. Am J Obstet Gynecol. 1998; 179: 1079–85.

11. Warren WB, Patrick SL, Goland RS. Elevated maternal plasma corticotropin-releasing hormone levels in pregnancies complicated by preterm labor. Am J Obstet Gynecol. 1992; 166: 1198–1204.

12. Robinson BG, Emanuel RL, Frim DM, et al. Glucocorticoid stimulates expression of corticotropin-releasing hormone gene in human placenta. Proc Natl Acad Sci USA. 1988; 85: 5244–48.

13. Smith R. Parturition. N Engl J Med. 2007; 2007; 356: 271–83.

14. Nodwell A, Carmichael L, Fraser M, et al. Placental release of corticotrophin-releasing hormone across the umbilical circulation of the human newborn. Placenta. 1999; 20: 197–202.

15. Weiss G. Endocrinology of Parturition. J Clin Endocrinol and Metab.2000; 85: 4421–25.

16. Snegovskikh V, Park JS, Norwitz ER. Endocrinology of Parturition. Endocrinol Metab Clin N Am. 2006; 35: 173–91.

17. Mazouni C, Provensal M, Porcu G, et al. Termination of pregnancy in patients with previous cesarean section. Contraception 2006; 73: 244–48.

18. Condon JC, Jeyasuria P, Faust JM, et al. A decline in the levels of progesterone receptor coactivators in the pregnant uterus at term may antagonize progesterone receptor function and contribute to the initiation of parturition. Proc Natl AcadSci USA. 2003; 100: 9518–23.

19. Gibb W. The role of prostaglandins in human parturition. Ann Med. 1998; 30: 235–41.

20. Skinner KA, Challis JR. Change is the synthesis and metabolism of prostaglandins by human fetal membranes and decidua in labor. Am J Obstet Gynecol. 1985; 151: 519–23.

21. Kota SK, Gayatri K, Jammula S, et al. Endocrinology of parturition. Indian J Endocrinol Metab. 2013; 17: 50–59.

22. Mitchell BF, Fang X, Wong S. Oxytocin: a paracrine hormone in the regulation of parturition? (review). Reviews of Reproduction. 1998; 3, 113–22.

23. Thornton S, Davison J and Baylis PH. Plasma oxytocin during the first and second stages of spontaneous human labour. Acta Endocrinologica. 1992; 126 425–29.

24. Hirst JJ, Chibbar R, Mitchell BF. Role of oxytocin in the regulation of uterine activity during pregnancy and in the initiation of labor Seminars in Reproductive Endocrinology. 1993; 11: 219–33.

25. Blanks AM, Vatish M, Allen MJ, et al. Paracrine oxytocin and estradiol demonstrate a spatial Increase in human intrauterine tissues with labor. J clin Endocrinol Metab. 2003; 88: 3392–3400.

Chapter 28

Prolactin

INTRODUCTION

Prolactin is a polypeptide hormone that is synthesized in and secreted from lactotrophs in the anterior pituitary gland. The hormone was discovered in Cattle around 1930 and confirmed in humans as late as 1970. The hormone was given its name based on the fact that an extract of bovine pituitary gland would promote lactation in rabbits. Prolactin (PRL) is best known for the stimulation of the growth and function of the mammary gland. In the primed breast, PRL is required to initiate and maintain lactation. Serum PRL levels are about 10–25 μg/L in women. Even in men, plasma prolactin level is almost in the same range (10–20 μg/L).

During pregnancy and lactation, lacto-trophs constitute as many as 50% of all the cells in anterior pituitary. It is difficult to believe that so many cells have only mammo-tropic function. Moreover, if lactotrophs are only concerned with mammary gland function, their presence in large numbers in anterior pituitary in males cannot be accounted for. Therefore, nowadays, it is believed that despite its rather simple and one-dimensional name, prolactin does much more than simply *PROmote LACTation*. It is now recognised as a pleiotropic hormone with the wide range of physiological actions.[1]

Lactotrophs are acidophils, staining bright red with erythrosine. The animal models in which they could be best studied were lactating or pregnant animals because their pituitaries showed enlarged, bright red lacto-trophs. Lactotrophs produce the polypeptide hormone prolactin, which has 199-amino acids. Initially described by light microscopy using conventional staining techniques, lactotrophs were subsequently identified unequivocally by immunocytochemistry in the anterior pituitary gland. In normal animals, lactotrophs represent approximately 15–20% of the pituitary cell population, with more lactotrophs in females. During pre-gnancy-lactation lactotrophs may constitute as many as 50% of pituitary cell population.[2]

NEURAL CONTROL OF PROLACTIN SECRETION

The well-accepted view is that lactotrophs have spontaneously high secretory activity and pituitary prolactin secretion is under a tonic and predominant inhibitory control exerted by the hypothalamus. This view is based on the following observations: (1) Surgical disconnection of the pituitary gland and the medial basal hypothalamus (median eminence lesion) or pituitary stalk section results in gradual increase in plasma prolactin levels that reach a plateau within a week after these surgeries, and (2) prolactin secretion occurs at a high spontaneous rate when (i) the anterior lobe is transplanted to a site that has no vascular or neural connection to the hypothalamus (e.g. under the kidney capsule) or (ii) when pituitary cells are cultured *in vitro*.[3]

The neural control of prolactin secretion, and indeed the whole hypothalamopituitary prolactin axis is different from other hypothalamopituitary hormonal systems. First, the hypothalamic regulation on prolactin secretion is predominantly inhibitory, as opposed to stimulatory in other anterior pituitary hormones. Second, it also involves a catecholamine neurotransmitter, dopamine, rather than the more typical peptide hypothalamic hormones involved in regulating all other anterior pituitary hormones. Third, prolactin is also the only anterior pituitary hormone that does not have an endocrine target tissue, and therefore lacks a classical long-loop hormonal feedback system. It is regulated, instead, by a short loop feedback whereby prolactin itself stimulates the secretion of the inhibitory factor, dopamine (Fig. 28.1). Hypothalamic dopamine is released by tuberoinfundibular dopamine (TIDA) neurons in the arcuate nucleus of hypothalamus.

Even after the clear demonstration of inhibitory regulation of prolactin secretion in the 1950s, identification of the inhibitory hormone mediating this action remained problematic. All hypothalamic hypophysiotropic hormones identified had been peptides, and the expectation was that 'prolactin-inhibitory factor' would also be a peptide. Initial clues that this might not be the case came from observations that drugs such as reserpine, which depletes endogenous catecholamines, induced pseudopregnancy in rats[4], indicative of elevated prolactin levels in the blood. Based on the evidence of dopaminergic nerve terminals in the median eminence, McLeod[5] proposed that dopamine may be released into the pituitary portal system, and thereby acting as a hypothalamic inhibitory hormone. He demonstrated that dopaminergic agonists were effective at suppressing prolactin secretion *in vivo*, and more importantly, that dopamine could inhibit prolactin secretion from isolated pituitary glands.[5] Dopamine was subsequently detected in the pituitary portal blood and it was reported that variations of levels of dopamine in the portal blood accounted for changes in prolactin secretion in various physiological conditions.[6] Soon, dopamine receptors were identified on lactotrophs in the anterior pituitary. The observation that mice lacking the dopamine D2 receptor are hyperprolactinemic, clearly demonstrated the critical role of dopamine in suppressing endogenous prolactin secretion.[7]

PATTERNS OF PROLACTIN SECRETION IN DIFFERENT REPRODUCTIVE STATES

1. Pregnancy

During pregnancy, there is a progressive increase in plasma prolactin level (Fig. 28.2).

During the later months of pregnancy, the neuroendocrine control of prolactin secretion is markedly altered to allow a state of hyperprolactinemia to develop. Prolactin secretion is normally tightly regulated by a short-loop feedback mechanism. In non-pregnant state, any increase in plasma prolactin level stimulates activity of tuberoinfundibular dopamine (TIDA) neurons to increase dopamine secretion into the pituitary portal blood. Dopamine

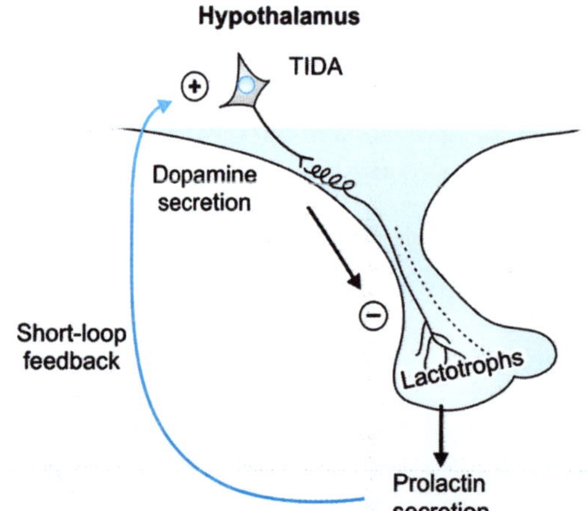

Fig. 28.1: Short-loop feedback control on secretion of prolactin

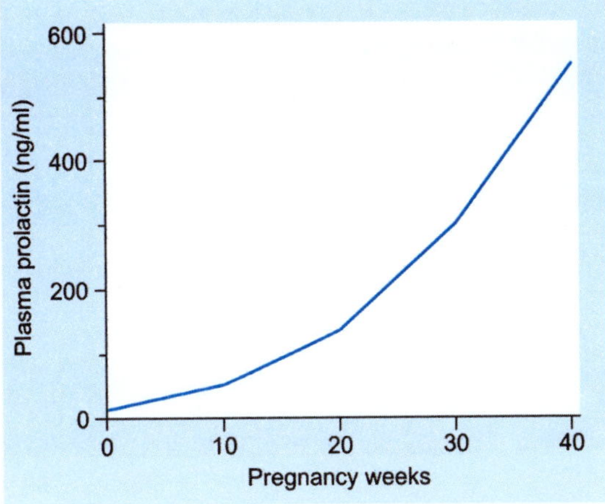

Fig. 28.2: Plasma prolactin level during pregnancy

inhibits prolactin secretion, thus reducing prolactin concentrations in the circulation back to the normal low level (Fig. 28.1). Activation of this feedback secretion by placental lactogen during pregnancy maintains relatively low levels of prolactin secretion during early and mid-pregnancy. Despite the continued presence of placental lactogen, however, dopamine secretion from TIDA neurons is reduced during late pregnancy. The TIDA neurons become completely unresponsive to endogenous or exogenous prolactin at this time, allowing a large nocturnal surge of prolactin to occur from the maternal pituitary gland in later months of pregnancy (Fig. 28.3). The loss of response to prolactin is an important maternal adaptation to pregnancy, allowing the prolonged period of hyperprolactinemia required for mammary gland development and function and for maternal behavior immediately after parturition, and possibly also contributing to a range of other adaptive responses in the mother.[8]

Role of Estrogens

In the pituitary gland, estradiol is a major stimulator of prolactin secretion, principally through (i) a classical genomic regulation of prolactin gene expression, (ii) by increasing the number of lactotrophs (direct actions on lactotrophs) and (iii) by modifying lactotroph

responsiveness to other regulators.[9, 10] The rising levels of estradiol during pregnancy are critical to promoting prolactin secretion, particularly during late pregnancy.

The TIDA neurons also express receptors for both estradiol and progesterone[11], and gonadal steroids regulate prolactin secretion indirectly through actions on these neurons. The predominant direct action of estradiol on TIDA neurons is one of inhibition, reducing secretion of dopamine into the portal blood, thereby facilitating prolactin secretion. Thus, estradiol acts at multiple levels to both directly and indirectly regulate prolactin synthesis and secretion (Fig. 28.2).

2. Lactation

The varied effects of prolactin on the mammary gland include growth and development of the mammary gland (mammogenesis), synthesis of milk (lactogenesis), and maintenance of milk secretion (galactopoiesis). It has long been accepted that prolactin is involved in the development of the mammary gland. Targeted disruption of the prolactin gene (prolactin knockout) or knockout of the prolactin receptor in mice results in abnormal mammogenesis characterized by complete absence of lobuloalveolar units in adult homozygous females. However,

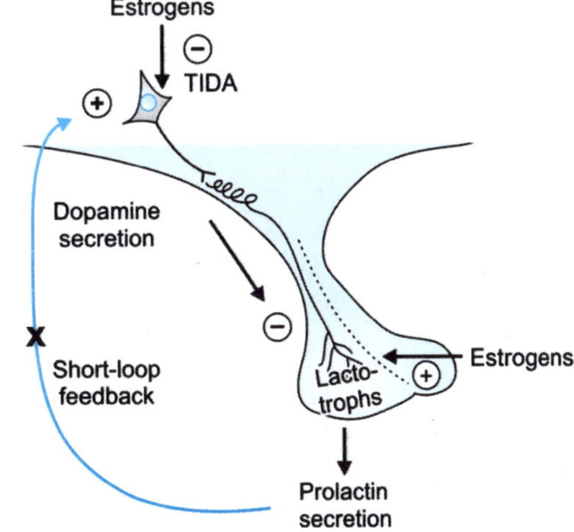

Fig. 28.3: Mechanisms of increased prolactin secretion during pregnancy

prolactin knockout heterozygous mice appear to have nearly normal mammogenesis.[12] Lactogenesis clearly requires pituitary prolactin, since hypophysectomy during pregnancy prevents subsequent lactation.

While hypophysectomy of rats and mice stops lactation, the replacement cocktail should minimally consist of prolactin and glucocorticoid or adrenocortiotropin to maintain sufficient nurturing of the pups. Addition of growth hormone permits the maintenance of maximal lactation. Thus, prolactin is merely one of many hormones and growth factors that are required for the mammary gland development during pregnancy and lactation. A great deal of evidence from hypophysectomized-ovariectomized-adrenalectomized rodents suggests that the mammary gland's lobuloalveolar growth and development *in vivo* after puberty requires prolactin, estrogen, progesterone, and glucocorticoids. During pregnancy, the extensive branching of the ducts and development of the alveoli is a function of progesterone and prolactin or placental lactogen.[13] There is evidence that insulin, growth hormone, thyroid hormone, parathyroid hormone, calcitonin, several growth factors, and even oxytocin may also play a role in galactopoiesis in various mammals. In the process of lactogenesis, prolactin stimulates uptake of some amino acids, the synthesis of the milk proteins casein and α-lactalbumin, uptake of glucose, and synthesis of lactose as well as milk fats.[13]

The best-known physiological stimulus affecting prolactin secretion is the suckling stimulus applied by the nursing young. This has been characterized as a classical neuro-endocrine reflex. In rats, blood prolactin concentrations begin to rise within 1–3 min of initiation of nursing, peak within 10 min, are sustained at a constant level as long as nursing continues, and fall when nursing is termi-nated (Fig. 28.4). The expression of prolactin mRNA in the pituitary gland follows the same pattern. Cessation of the suckling stimulus results in termination of prolactin secretion, and the rate of decrease in blood prolactin

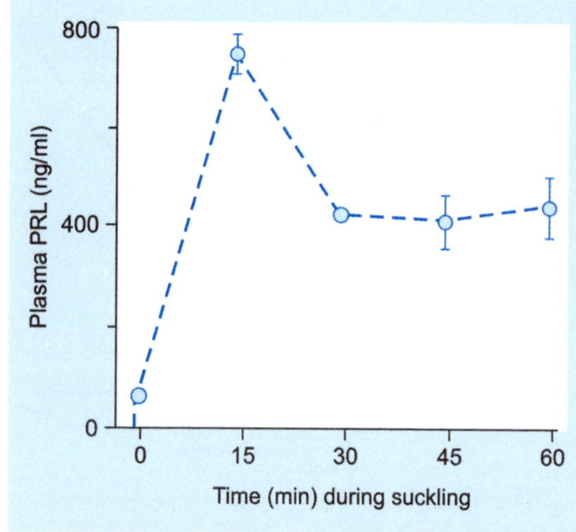

Fig. 28.4: Effect of suckling on plasma prolactin (PRL) levels in the rat

levels is proportional to the metabolic clearance rate of the hormone.

Mechanism of Increased Prolactin Secretion During Lactation

In order to support a period of hyper-prolactinemia during lactation, and thereby promote the milk production, there is an apparent loss of sensitivity of the short-loop feedback system during *late pregnancy and lactation*.[8] This is a remarkable example of adaptive plasticity within a neuroendocrine control network, allowing a sustained period of high prolactin secretion to be maintained unencumbered by a regulatory feedback pathway. The mechanisms mediating this adaptive response are only recently becoming elucidated. Hypothalamic dopamine neurons apparently no longer release dopamine in response to prolactin, rendering the short-loop feedback system functionally inactive. This adaptation persists into lactation, and dopamine secretion remains low throughout this period of elevated prolactin secretion (Fig. 28.5). This is an important adaptation, because prolactin is required for milk pro-duction and maternal behavior at this time.[1]

There is a highly coordinated release of prolactin during lactation, caused by the suckling stimulus. It is interesting to note that

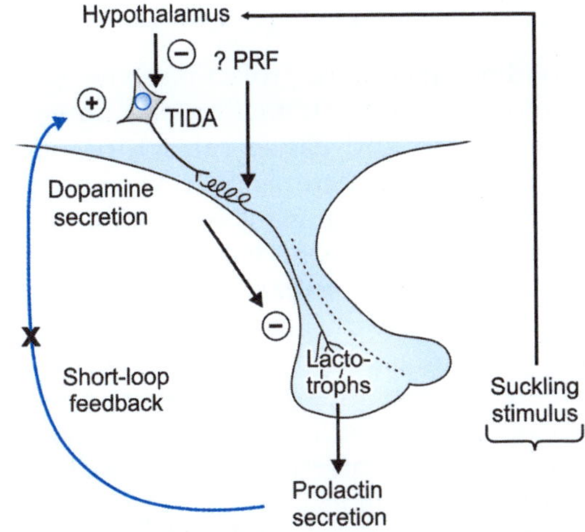

Fig. 28.5: Mechanisms of increased secretion of prolactin during lactation

while the suckling-induced prolactin secretion could be consistent with the 'dopamine withdrawal' or 'disinhibition' model of prolactin secretion, previously discussed, the chronic reduction in dopamine output seen during lactation complicates this interpretation. If dopamine production is already lost, how can suckling cause an acute increase in prolactin secretion through dopamine withdrawal? Thus, there is a possibility that a *suckling-induced 'prolactin-releasing factor'* (PRF) may be involved in stimulating prolactin secretion at this time.[14] It seems, besides dopamine, other regulatory factors must be involved in the physiological regulation of prolactin secretion, particularly during lactation when it is most required. Whether one or more of these factors becomes the long sought 'prolactin-releasing factor' predicted by Harris remains an exciting area needing further investigation[1] (Fig. 28.5).

Metabolic Functions of Prolactin

Metabolic functions of prolactin are the best example of its pleiotropic role, i.e. multiple seemingly diverse actions. Prolactin receptors are expressed on multiple tissues, including adipose tissue, liver, pancreas and the brain. It appears to play a broad role in both pancreatic and adipose development. In adipose tissue, prolactin is essential in adipogenesis and adipocyte differentiation, as well as modulating lipid metabolism. It also regulates the secretion of several adipokines, including stimulation of leptin and inhibition of adiponectin production.[15] In the pancreas, it promotes growth of islets during development and increases insulin expression and glucose-stimulated insulin secretion.[16]

The potential relevance of PRL in body weight homeostasis has been highlighted by different reports showing that up to 58% of men and 18% of women with PRL-secreting adenomas at initial presentation were overweight.[17] Physiological hyperprolactinemia, both lactational and clinical, leads to increased leptin levels, a surrogate marker of adipocyte cell expansion. Thus, hyperprolactinemia has emerged as one of the obesogenic factors and a potential contributor to the development of dyslipidemia and insulin resistance. Moreover, experience accumulated over many years has shown that D_2-agonist therapy is effective and safe for the reduction of body weight in patients with hyperprolactinemia-associated obesity.[18]

Metabolic actions of prolactin in both adipose tissue and pancreas are likely to be profoundly important during pregnancy and lactation. During lactation, lipid metabolism is altered, with lipid mobilization from stores and utilization in mammary gland promoted by prolactin. Adaptive changes in glucose homeostasis are also important in pregnancy. Maternal tissues develop insulin resistance to preferentially direct glucose to the fetal/placental compartment. Moreover to ensure the maternal tissues continue to receive the nutrients required, there is increased maternal insulin secretion, and glucose-stimulated insulin secretion increases. To adapt to this altered demand, there is significant proliferation of β cells in the islets, enhanced insulin synthesis and decreased threshold for glucose-stimulated insulin secretion, with prolactin playing a critical adaptive role in promoting these changes. Failure of this adaptive response results in gestational diabetes.[19] According to Stuebe and Rich-Edward[20], maternal metabolic alterations in pregnancy end not with birth, but with weaning. Data from animal models and human studies suggest that lactation is associated with favorable changes in adiposity, glycemic control, and lipid homeostasis that persist long after weaning (Fig. 28.6).

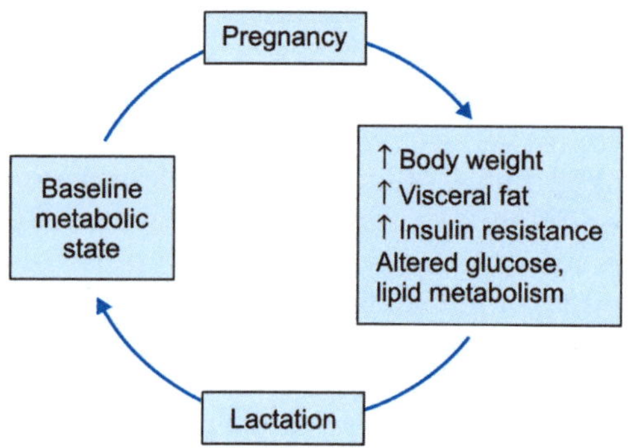

Fig. 28.6: Metabolic changes in pregnancy and their reversal after lactation

These peripheral actions of prolactin on metabolism are complemented by CNS actions of prolactin to promote appetite and potentially regulate glucose homeostasis. Systemic prolactin administration increases food intake in a variety of species. Thus, the elevated prolactin secretion is likely to contribute to the rapid increase in food intake during pregnancy and the extreme hyperphagia of lactation.[21] Prolactin also induces *functional* leptin resistance, which would contribute to increased food intake, potentially mediating the well-established leptin resistance of pregnancy and lactation.[22]

Thus, seemingly diverse metabolic actions of prolactin different tissues can be unified into a single adaptive function, which is metabolic adaptation to pregnancy, increasing energy availability to the mother and offspring.[1] Should hyperprolactinemia occur at an inappropriate time, e.g. due to pituitary adenomas, one might predict this could contribute to metabolic disorders. There is some evidence for this. As mentioned earlier, many patients with hyperprolactinemia are prone to excessive weight gain and normalization of prolactin levels using dopamine agonists is associated with weight loss.[23]

Role of Prolactin in Orgasm

Another function of prolactin seems to be to provide sexual satisfaction during sexual intercourse and cause refractory period afterwards. The neuroendocrine response to sexual activity in humans is characterized by a pronounced orgasm-dependent increase in plasma levels of prolactin both in males as well as females[24] (Fig. 28.7). Leeners et al estimated plasma prolactin levels in women before and just after sexual intercourse. They observed that women who had orgasm during sexual intercourse showed elevated plasma prolactin levels; plasma prolactin did not rise in women who did not have orgasm.[25]

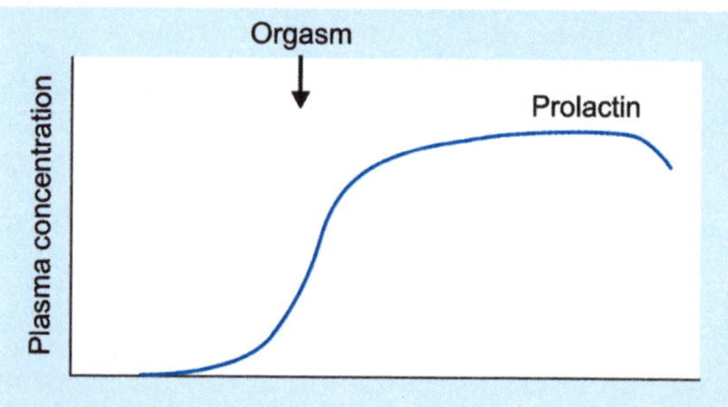

Fig. 28.7: Plasma prolactin level during and after sexual intercourse

Hyperprolactinemia and Infertility

Hyperprolactinemia causes infertility in both males and females. The mechanism by which prolactin inhibits the reproductive axis is not clear, but evidence suggests that prolactin impacts fertility through actions on GnRH neurons. In humans, hyperprolactinemia is associated with a marked reduction in both the frequency and amplitude of LH pulses indicative of a decrease in GnRH secretion. Suppression of LH pulsatility can be reversed by reducing serum prolactin concentrations to normal.[26] Similarly, prolactin suppresses both the frequency and amplitude of LH pulses in male and female *rats* and measurements of GnRH secretion into the portal blood have revealed prolactin-induced suppression of GnRH release.[27] Thus, in animal models, as in clinical studies, the primary cause of infertility appears to be the suppression of the activity of GnRH neurons. This effect of prolactin is unlikely to be mediated directly by an *action on GnRH neurons, as these neurons do not express the prolactin receptors.*[28] Therefore, prolactin-induced inhibition of GnRH neurons must involve prolactin-sensitive afferents to these cells. As such, kisspeptin neurons have emerged as the most likely intermediate regulators.

Kisspeptin neurons are now recognised as critical parts of the circuit regulating activity of the GnRH neurons that control fertility.[29] Kisspeptin neurons may have an important role in mediating the suppressive effect of prolactin on fertility. Kisspeptin is the most potent stimulator of GnRH neuronal activity yet identified and *prolactin receptors are expressed in the majority of kisspeptin neurons.*[28] These observations are consistent with the hypothesis that prolactin-induced suppression of GnRH secretion is mediated by an inhibition of kisspeptin neurons (Fig. 28.8).

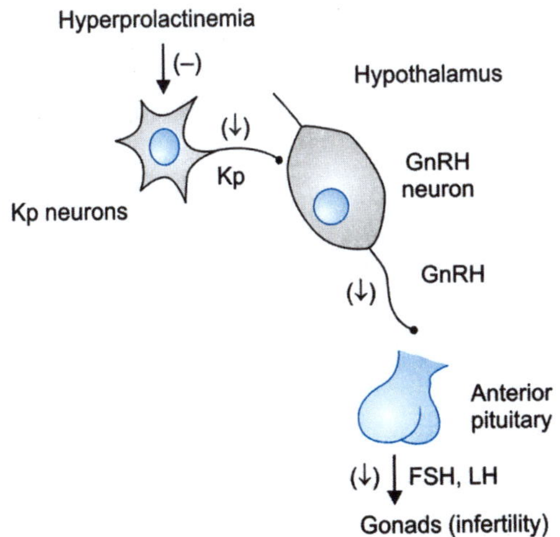

Fig. 28.8: Mechanism of infertility in hyperprolactinemia

What does Prolactin do in the Male?

The hypothesis, that most prolactin actions in the body can be tied to the adaptation to pregnancy and lactation, does not explain the possible role of prolactin in the male other than in orgasm. Compared to 20% in females, lactotrophs constitute approximately 15% of cell population in the anterior pituitary gland, suggesting that prolactin has some functions in males. But knockout studies have not identified any essential function of prolactin in the male. While there is no male equivalent of lactation, metabolic actions of prolactin in females can also be observed in males. For example, many of the metabolic functions of prolactin can be observed in males, but whether prolactin levels are ever sufficient for these effects to be of physiological significance is uncertain.[1] Perhaps the most consistent stimulus for prolactin secretion in males is stress, but the functional consequences of this response are not well-understood.

References

1. Grattan DR. 60 years of neuroendocrinology: The hypothalamo-prolactin axis. J Endocrinol. 2015; 226: T101–22.
2. Freeman ME, Kanyicsaka, BL, Lerant, A, et al. Prolactin: Structure, function, and regulation of secretion. Physiol Rev. 2000; 80; 1523–1631.
3. Neill JD. Neuroendocrine control of prolactin secretion. In Frontiers in neuroendocrinology. Eds, Martini L & Ganong WF. New York, Raven, 1980; p 129–155.
4. Barraclough CA, Sawyer CH. Induction of pseudopregnancy in the rat by reserpine and chlorpromazine. Endocrinology. 1959; 65: 563–71.
5. McLeod RM, Fontham EH, Lehmeyer JE. Prolactin and growth hormone production as influenced by catecholamines and agents that affect brain catecholamines. Neuroendocrinology. 1970; 6: 283–94.
6. Ben-Jonathan N, Neill MA, Arbogast LA, et al. Dopamine in hypophysial portal blood: relationship to circulating prolactin in pregnant and lactating rats. Endocrinology. 1980; 106: 690–96.
7. Kelly MA, Rubinstein M, Asa SL, et al. Pituitary lactotroph hyperplasia and chronic hyperprolactinemia in dopamine D_2 receptor-deficient mice. Neuron. 1997; 19: 103–13.
8. Grattan DR, Steyn FJ, Kokay IC. Pregnancy-induced adaptation in the neuroendocrine control of prolactin secretion. Jour Neuroendocrinol. 2008; 20: 497–507.
9. Kansra S, Chen S, Bangaru ML, et al. Selective estrogen receptor down-regulator and selective estrogen receptor modulators differentially regulate lactotroph proliferation. PLoS.ONE. 2010; 5: e10060 doi:10.1371/journal.pone. 0010060 [PMC free article]

10. Nolan LA, Levy A. The trophic effects of oestrogen on male rat anterior pituitary lactotrophs. Journal of Neuroendocrinology. 2009; 21: 457–64.

11. Steyn FJ, Anderson GM, Grattan DR. Expression of ovarian steroid hormone receptors in tuberoinfundibular dopaminergic neurons during pregnancy and lactation. Journal of Neuroendocrinology. 2007; 19: 788–93.

12. Horsemen N, Zhao W, Montechcino E, et al. Defective mammopoiesis but normal hematopoiesis in mice with targeted disruption of prolactin gene. EMBO J. 1997; 16: 627–49.

13. Imagawa W, Yang J, Guzman RC, et al. Control of mammary gland growth and development. In: Physiology of reproduction, eds. Knobil E and Neill JD. New York, Raven, 1994; p. 1033–1063.

14. Boyd E, Spencer E, Jackson IMD, et al. Prolactin-releasing Factor (PRF) in porcine hypothalamic extract distinct From TRH 1. Endocrinology. 1976; 99: 861–71.

15. Carre N, Binart N. Prolactin and adipose tissue. Biochimie. 2014; 97: 16–21.

16. Brelje TC, Stout LE, Bhagroo NV, et al. Distinctive roles for prolactin and growth hormone in the activation of signal transducer and activator of transcription 5 in pancreatic islets of Langerhans. Endocrinology. 2004; 145: 4162–75.

17. Doknic M, Pekic S, Zarkovic M, et al. Dopaminergic tone and obesity: an insight from prolactinomas treated with bromocriptine. Eur J Endocrinol. 2002; 147: 77–84.

18. Tovar S, Dieguez C. Prolactin and energy homeostasis: Pathophysiological mechanisms and therapeutic considerations. Endocrinology. 2014; 155: 659–62.

19. Ramos-Roman MA. Prolactin and lactation as modifiers of diabetes risk in gestational diabetes. Hormone and Metabolic Research. 2011; 43: 593–600.

20. Stuebe AM, Rich-Edwards JW. The reset hypothesis: lactation and maternal metabolism. *Am J Perinatol.* 2008; 2: 81–8.

21. Woodside B, Budin R, Wellman MK, et al. Many mouths to feed: the control of food intake during lactation. Frontiers in Neuroendocrinology. 2012; 33: 301–14.

22. Augustine RA, Grattan DR. Induction of central leptin resistance in hyperphagic pseudopregnant rats by chronic prolactininfusion. Endocrinology. 2008; 149: 1049–55.

23. Galluzzi F, Salti R, Stagi S, et al. Reversible weight gain and prolactin levels–long-term follow-up in childhood. Journal of Pediatric Endocrinology and Metabolism. 2005; 18: 921–24.

24. Krüger TH, Haake P, Chereath D, et al. Specificity of the neuroendocrine response to orgasm during sexual arousal in men. J Endocrinol. 2003 Apr; 177(1): 57–64.

25. Leeners B, Kruger TH, Brody S, et al. The quality of sexual experience in women correlates with post-orgasmic prolactin surges: results from an experimental prototype study. J Sex Med. 2013 May; 10(5): 1313–19.

26. Moult PJ, Rees LH, Besser GM. Pulsatile gonadotrophin secretion in hyperprolactinaemic amenorrhoea and the response to bromocriptine therapy. Clinical Endocrinology. 1982; 16: 153–62.

27. Park SK, Keenan MW, Selmanoff M. Graded hyperprolactinemia first suppresses LH pulse frequency and then pulse amplitude in castrated male rats. Neuroendocrinology. 1993; 58: 448–53.

28. Kokay IC, Petersen SL, Grattan DR. Identification of prolactin-sensitive GABA and kisspeptin neurons in regions of the rat hypothalamus involved in the control of fertility. Endocrinology. 2011; 152: 526–35.

29. Pinilla L, Aguilar E, Dieguez C, et al. Kisspeptins and reproduction: physiological roles and regulatory mechanisms. Physiological Reviews. 2012; 92: 1235–316.

Chapter 29

Juxtaglomerular Apparatus

INTRODUCTION

More than a century ago, Golgi in 1889 described the essential components of the anatomical unit which is known today as the *juxtaglomerular apparatus* (JGA). He showed that the ascending limb of Henle's loop regularly returns to the original renal corpuscle and attaches itself precisely to the point where the afferent arteriole enters and the efferent arteriole exits the glomerular hilus. In 1898, the pressor-active principle of the kidney was discovered and named renin. The discovery of renin was based on the experiment in which injection of extracts of donor rabbit kidneys into the jugular vein of recipient rabbits consistently increased BP in the latter. In further experiments, it was shown that this pressor substance was located in the renal cortex and that the pressure response did not require an intact nervous system. Intriguingly, these observations were forgotten for many years. The possibility that the kidney may release a pressor substance was revived by Goldblatt in 1934, who could induce experimental hypertension in dogs by clipping one or both renal arteries. These observations led to a renewed interest in a renal pressor substance. In 1939, it was shown that renin was not itself a pressor substance, but an enzyme acting upon a protein to release a peptide vasoconstrictor.[1]

The juxtaglomerular apparatus (JGA) in the renal cortex represents a major structural component of the renin-angiotensin system and is one of the most important regulatory sites of renal salt and water conservation and BP regulation. Two major regulatory functions are performed by the juxtaglomerular apparatus: The high (distal tubular) NaCl concentration-induced afferent arteriolar vasoconstriction (tubuloglomerular feedback) and the low (distal tubular) NaCl concentration-induced renin release. Macula densa cells are strategically positioned in the juxtaglomerular apparatus with their apical membrane exposed to the tubular fluid, whereas their basilar aspects are in contact with cells of the mesangium and the afferent arteriole (Fig. 29.1).

The JGA consists of three components:

1. The **juxtaglomerular cells** of the afferent arteriole which synthesize and store renin, which is secreted in response to specific stimuli (e.g. low blood flow, decreased NaCl delivery in distal tubule). The juxtaglomerular cells could be considered the "effector arm" of the renin-angiotensin-aldosterone axis.

2. The **macula densa** plaque is a unique group of 15 to 20 cells in a region of the distal convoluted tubule. These cells are charac-

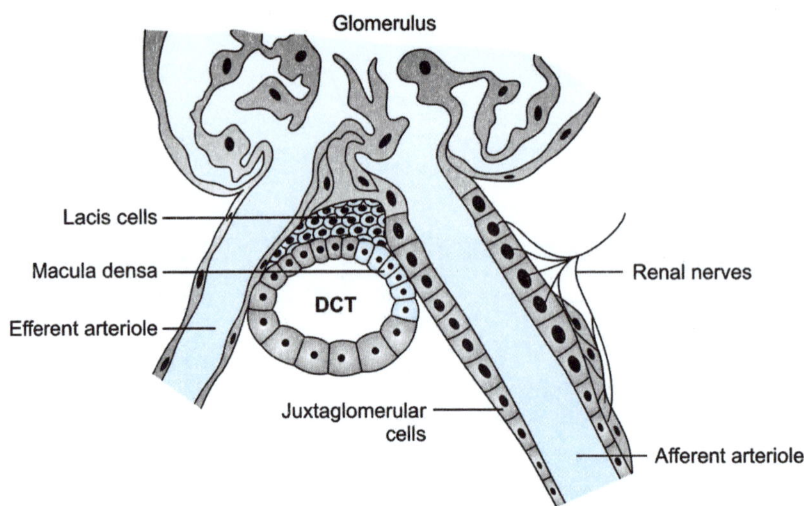

Fig. 29.1: The juxtaglomerular apparatus (DCT: Distal convoluted tubule)

terized by tubular epithelial cells which are more densely-packed than in other regions of the nephron (and thereby leading to its characteristic appearance "dense" on light microscopy). In every nephron, the distal tubule is attached to its glomerulus of origin. The macula densa cells, together with the smooth muscle cells of the afferent arterioles and the renin-secreting granular cells, form a functional unit, the juxtaglomerular apparatus. The macula densa cells play a pivotal role in sensing changes in tubular fluid composition, generating and sending signals to the other components juxtaglomerular apparatus that control renal blood flow and GFR. They are also involved in the mechanism of glomerulotubular balance.

3. The **mesangial cells**, which form connections via actin and microtubules which allow for selective vasoconstriction/vasodilation of the renal afferent and efferent arterioles.

ROLE OF JGA IN RENIN-ANGIOTENSIN-ALDOSTERONE SYSTEM

Traditionally, the renin-angiotensin-aldosterone system (RAAS) is considered as a hormonal cascade that functions in the homeostatic control of arterial blood pressure,

tissue perfusion, and extracellular volume. However, during the last decade, the view of the *renin-angiotensin-aldosterone system has been modified from a blood pressure controlling system to a ubiquitously acting system that controls a variety of biological effects in health and disease.* The present concept of the RAAS is that of a complex network that operates around the classic renin-angiotensin-aldosterone cascade. Apart from angiotensin II and aldosterone new biological effectors have been identified such as prorenin and angiotensins 1–7, which act via their specific membrane receptors.[3]

Biosynthesis of Renin

In the adult kidney, renin is synthesized by myofibroblast-like cells that are located in the media layer of renal afferent arterioles at the entrance into the glomerular capillary network. Because of their localization and cuboid-like appearance, these cells are commonly termed juxtaglomerular epithelioid cells. The cuboid form results from huge granular vesicles containing renin (Fig. 29.2). The exact nature and the origin of these cells is still unknown. Some evidence suggests that the renin-producing cells might differentiate from pericytes which are probably also precursors of preglomerular vascular cells

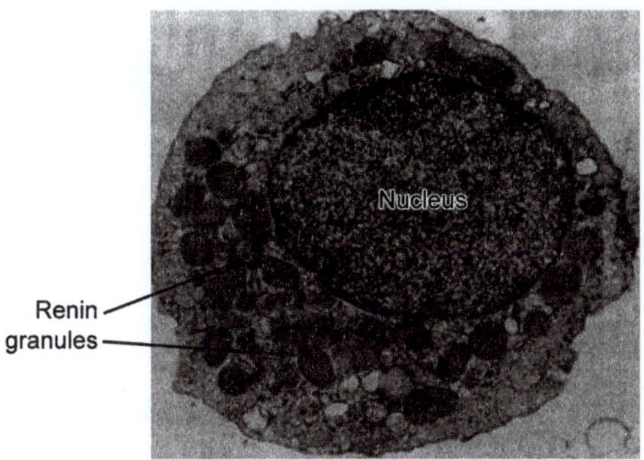

Fig. 29.2: Ultrastructure of a juxtaglomerular cell. (Modified from Kurtz A. Cellular control of renin secretion. Rev Physiol Biochem Pharmacol 1989; 113: 1–40)

and glomerular mesangial cells. In the normal adult kidney renin-producing cells appear just at the junction between these two cell types, suggesting that these cells remain in an intermediate differentiation state between vascular smooth muscle cells on the one side and mesangial cells on the other. Compatible with this idea is the capability of preglomerular vascular smooth muscle cells and extraglomerular mesangial cells to reversibly switch on and off the renin gene expression even in the adult kidney, that leads to an increase or decrease in the number of renin-producing cells[3] (*see* also below).

Biosynthesis of Renin

The renin gene transcript gives rise to pre-prorenin with a molecular mass around 48 kDa. The inactive prorenin can be follow two different pathways at the level of the Golgi apparatus (Fig. 29.3). Prorenin is either directed to (i) the constitutive secretory pathway which leads to the direct release of prorenin into circulation, or (ii) stored as renin in a prominent electron dense vesicular network.

Prorenin: Up to 80% of the produced prorenin can be directly released into circulation in men. Normally, the concentration of plasma prorenin exceeds that of plasma renin.[4] Circulating prorenin can bind to a specific membrane protein named prorenin receptor. By binding to the receptor prorenin can gain some enzymatic activity and cleave angiotensin I from angiotensinogen, a process that is normally restricted to renin. Further, in several conditions, such as pregnancy and diabetes mellitus with microvascular disease, prorenin in plasma is remarkably increased.[5] The significance of these observations remains

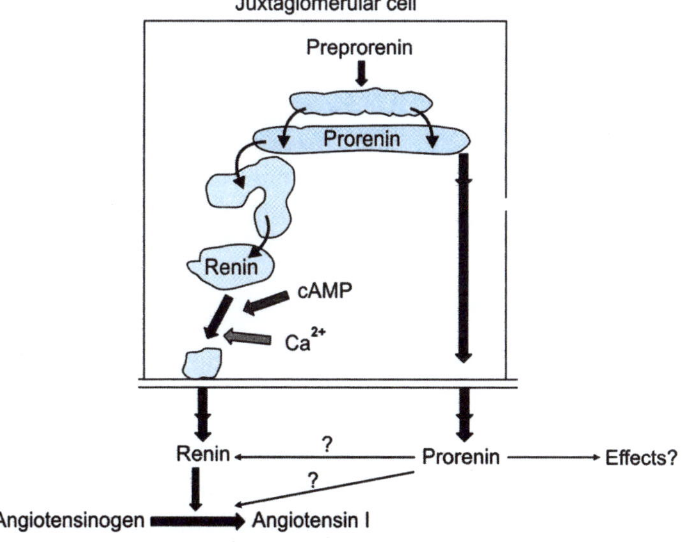

Fig. 29.3: Biosynthesis of renin (modified from Kurtz[3])

to be determined, and a little is known about factors that regulate prorenin secretion.

Renin: Only glycosylated prorenin can be directed to the vesicular network in juxtaglomerular cells. Within the vesicles, a prorenin is proteolytically cleaved to yield enzymatically active renin which has a molecular mass around 41 kDa. Renin remains stored in the vesicles awaiting controlled release. Renin is released from the vesicles by exocytosis. Increase in intracellular cAMP stimulates renin release, whereas increase in intracellular Ca^{2+} is inhibitory.[3, 6]

REGULATION OF RENIN SECRETION

Renin secretion by the juxtaglomerular renin-generating cells (JG cells) is controlled by both local as well as multiple systemic factors.[7]

Local factors
- Vascular
- Tubular
- Neural

Systemic factors
- Blood pressure
- Angiotensin II
- Salt intake
- ANP
- Aldosterone

A. Local Factors

a. Vascular Control of Renin Secretion: Baroreceptor Action

The term baroreceptors is based on the observations that intrarenal perfusion pressure and renin secretion are inversely related. Many studies have demonstrated the renin release by the JG cells is dependent on transmural pressure. However, the exact structure acting as baroreceptors have not been identified. It appears that adenosine/ATP released by the smooth muscle in response to increased intrarenal pressure is involved in the inhibition of renin release. In addition, mechanical stretching of JG cells also releases ATP (Fig. 29.4). In isolated JG

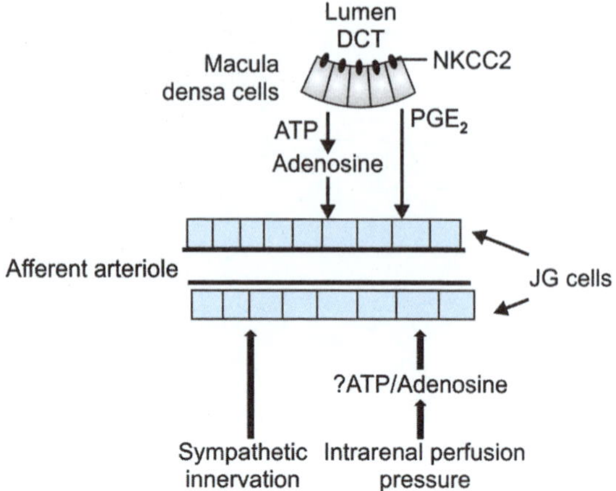

Fig. 29.4: Mechanisms of local regulation of renin secretion

cells, the increased stretching leads to proportionate decrease in both basal and post-stimulation renin release.[8]

b. Tubular Control of Renin Secretion

Renin secretion from the granular cells of the afferent arteriole is locally influenced by the tubular system of the corresponding nephron. There is a close linkage between the tubular and vascular systems within JGA. Specifically, the thick ascending limb of loop of Henle emerging from the renal medulla comes into close contact with its parent glomerulus. This anatomical link between the tubular system and the vasculature of the afferent arterioles enables macula densa (MD) to influence renin secretion. In this local control system, changes in the tubular NaCl concentration are translated into inverse changes in renin secretion. In early experiments, the most striking and direct evidence was provided by the following *in vitro* experiment: Rabbit JGAs were isolated, and ascending limbs of loop of Henle were perfused with artificial tubular fluid containing various concentrations of NaCl. Simultaneously, renin release was determined by microassay. In this experimental setup, a *decrease* in the tubular NaCl concentration resulted in a prompt *stimulation of the renin release rate"*.[9] This fundamental finding sparked interest in the signaling pathways that link the tubular NaCl concentration and renin secretion. The transport activity of the Na^+-K^+-$2Cl^-$ cotransporter-2 (NKCC2) in the apical membrane of these cells is thought to be the primary mechanism by which the MD cells detect the tubular NaCl concentration.[10] The MD cells subsequently transmit information about the tubular NaCl concentration to the granular cells of the afferent arteriole to influence renin secretion (Fig. 29.4).

Mediators of MD-dependent Control on Renin Release

Two molecules have been proposed to serve as mediators of the MD-dependent control of renin secretion: Prostanoids, and adenosine/ATP (Fig. 29.4). These molecules meet the key criteria to serve as signal mediators between MD cells and JG cells of the afferent arteriole: (1) MD cells have the capacity to generate each molecule, (2) their formation depends on the tubular salt concentration, and (3) there is solid experimental evidence to suggest that they directly influence renin secretion.[7]

a. Prostaglandins

Prostaglandin E_2 (PGE_2) and prostacyclin (PGI_2) have been shown to directly stimulate renin secretion in isolated JG cells.[11] A *decrease in salt concentration* in the tubular fluid increases local production of prostaglandins. Prostanoids like PGE_2 and PGI_2 are formed by tubular MD cells, and their generation is dependent on tubular salt concentration. PGE_2 and PGI_2 are potent stimulators of renin secretion. Their relevance as stimulatory mediators of a communication pathway between the MD cells and JG cells *in vitro* (i.e. in the absence of other regulatory mechanisms) is well established.

b. Adenosine/ATP

Adenosine meets the key requirements to serve as a mediator of the local tubular control of the renin system. First, there is solid evidence that adenosine is generated in the JGA. Second, the formation of adenosine in the JGA depends on the tubular salt concentration in the MD segment of loop of Henle. Third, adenosine exerts a direct effect on renin secretion from JG granular cells of the afferent arteriole. The function of adenosine as a paracrine factor in the JGA has been the subject of intense research over the past several years.[7, 12, 13] Adenosine receptors are expressed in the granular JG cells. A dose-dependent inhibition of renin secretion by adenosine was also demonstrated in isolated JG cells and renal cortical slices.[14]

c. Neural (Sympathetic) Control of Renin Secretion

The presence of β_1-adrenergic receptors in the JG granular cells has been established. There is a convincing evidence that neurally released norepinephrine acts directly on β_1-adrenoceptors located on the juxtaglomerular granular cells to cause renin release. The activation of β_1-receptors by catecholamines results in an augmented formation of cAMP, the central intracellular stimulus of renin secretion. The pharmacological inhibition of β_1-receptors results in reduced renin secretion.[15]

With respect to the afferent nerves from the carotid baroreceptors, carotid sinus hypotension results in a consistent and significant increase in renin secretion rate even with constant renal perfusion pressure. The results from these and other studies have clearly shown that there is an interaction among afferent neural input from arterial (carotid, aortic) and cardiopulmonary baroreceptors in the reflex neural control of renin secretion rate. Renin secretion rate increases when there is a decrease in the inhibitory input from any one of the three peripheral baroreceptor stations (carotid, aortic, or cardiopulmonary).

B. Systemic Control of Renin Synthesis and Secretion

1. Blood Pressure

The renin-angiotensin system is a key regulator of blood pressure as well as salt and water homeostasis. The inhibitors of angiotensin converting enzyme (ACE) or angiotensin II (ANG II) receptor antagonists are potent blood pressure-lowering drugs. In addition, renin inhibitors have been shown to lower blood pressure to similar levels as ANG II receptor blockers or ACE inhibitors.

Due to changes in the formation of the vasoconstrictor ANG II, alterations in renin release can also cause changes in the arterial blood pressure, such that an increased rate of renin release increases blood pressure and vice versa. As part of a negative-feedback loop, blood pressure in turn affects the synthesis and release of renin from the JG cells of the kidney: an increase in the arterial blood pressure inhibits, whereas a decrease stimulates, the synthesis and secretion of renin. This pressure-dependent mechanism, which controls renin synthesis and release, is called the "long loop negative-feedback" control on renin secretion (Fig. 29.5). Increases in the systemic blood pressure can inhibit renin release via intrarenal as well as extrarenal mechanisms. Both mechanisms have been suggested to be involved in the pressure-dependent regulation of the renin system. With regard to intrarenal mechanisms, high blood pressure (i) induces pressure-dependent natriuresis, which increases the NaCl load at the macula densa, and (ii) increases the renal perfusion pressure and therefore influences the renal baroreceptor mechanism.[7]

2. ANG II

Circulating ANG II functions in a direct negative-feedback loop ("short-loop negative-feedback control") to inhibit renin synthesis and secretion (Fig. 29.5). Numerous *in vivo* studies have shown that inhibiting the effects of ANG II with ANG II receptor antagonists or reducing the level of ANG II with ACE inhibitors causes a strong increase in renin synthesis and release. Likewise, the exogenous infusion of ANG II decreases renin synthesis and renin release, suggesting that ANG II may exert direct negative effects on this system.[16]

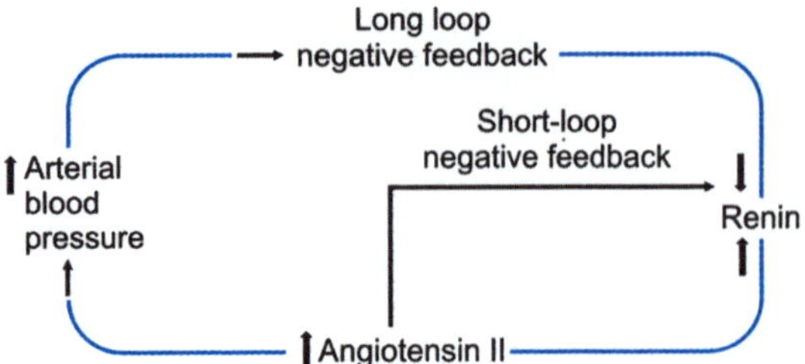

Fig. 29.5: The long loop and short-loop feedback control on renin secretion

3. Salt Intake

Oral salt intake is a well-known important regulator of RAAS activity, such that low salt intake increases and high salt intake decreases renin synthesis and secretion.[17] At present, however, it is still unclear which factor(s) or mechanism(s) mediates the observed salt-induced changes in renin secretion.

4. Atrial Natriuretic Peptide (ANP)

The cardiovascular and renal effects of ANP (hypotension, natriuresis, and diuresis) counteract the effects of ANG II. ANP can modulate the renin release through direct and indirect mechanisms. Although it is generally believed that the tubular action of ANP is predominantly exerted in the more distal segments of the nephron, ANP may also indirectly reduce the renin release via a possible effect on the proximal tubular salt reabsorption, which may lead to an increased NaCl load at the MD. In contrast, ANP can indirectly stimulate renin release by lowering the blood pressure. Indeed, somewhat conflicting data have been reported regarding the role of ANP in renin release, although most of these studies, including *in vitro* studies using isolated JG cells, suggest that ANP is an inhibitor of renin release.[7, 18]

5. Aldosterone

The stimulation of renin release increases the plasma levels of ANG II, which subsequently stimulates the secretion of aldosterone from the adrenal glands. The stimulation of angiotensin II and aldosterone formation increases the sodium reabsorption and consequently increases the body sodium content, which, in turn, inhibits the renin secretion (Fig. 29.6).

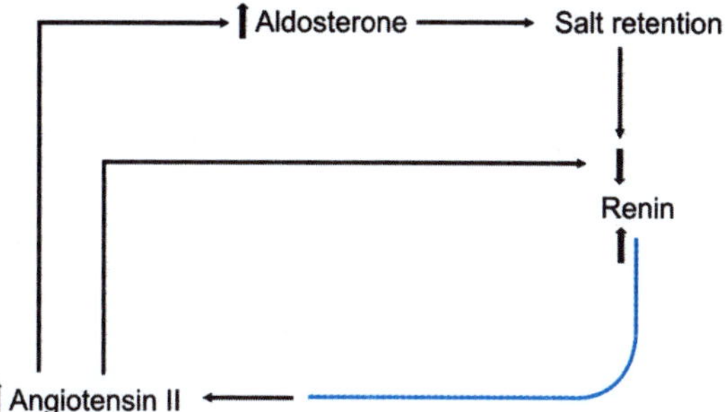

Fig. 29.6: Direct and indirect inhibition of renin secretion by angiotensin II

Physiology of Angiotensins

Angiotensinogen, a glycoprotein, is the only substrate of renin. It is synthesized in the liver and circulates in the plasma as α_2 globulin. Renin splits angiotensinogen to form a physiologically inactive decapeptide, angiotensin I. Angiotensin I is further split by an enzyme called **angiotensin converting enzyme (ACE)** into a physiologically active octapeptide, angiotensin II (Fig. 29.7). The converting enzyme is widely distributed in the endothelial cells of the vascular beds, especially in the lungs. Hence, most of the conversion of angiotensin I to angiotensin II occurs in the pulmonary circulation. Angiotensin II has half-life of 1–2 minutes. It is degraded by an enzyme angiotensinase, present in the red blood cells and many other tissues, into angiotensin III. Of many actions of angiotensin II, only aldosterone stimulating activity seems to be retained by angiotensin III. Angiotensins II and III are degraded into inactive metabolites.

Actions of Angiotensin II (Fig. 29.7)

- **Increased aldosterone secretion:** Angiotensin II acts on the zona glomerulosa cells of adrenal cortex and stimulates the secretion of aldosterone. Consequently, there is retention of NaCl (and water) in the kidneys. Renin-angiotensin II-aldosterone axis constitutes the most important mechanism for the maintenance of ECF volume.
- **Vasoconstriction:** Angiotensin II is very potent vasoconstrictor. Weight for weight it, has 4 to 8 times greater vasopressor activity than norepinephrine. However, its vasopressor activity is reduced in states of sodium depletion. In such circumstances, it is more effective as a salt retaining hormone than a vasoconstrictor hormone. Its pressor effect seems to be more important in hypotensive conditions like hemorrhage.

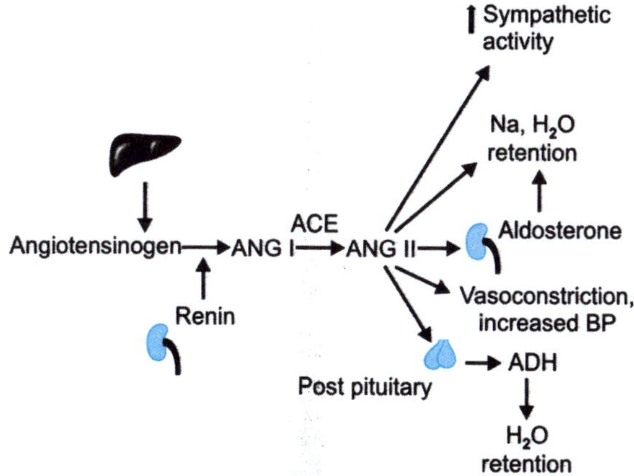

Fig. 29.7: Renin-angiotensin-aldosterone axis and its actions

- **Thirst:** Angiotensin II acts on circumventricular organs of the brain to increase thirst, thereby promoting water intake.
- **ADH secretion:** It promotes secretion of ADH by the posterior pituitary gland.
- **NaCl reabsorption:** It enhances NaCl reabsorption by the proximal convoluted tubules of the kidney.
- **Decreased GFR:** It decreases GFR by causing contraction of the mesangial cells of the renal glomeruli.
- **Norepinephrine release:** It facilitates the release of norepinephrine by a direct action on the postganglionic sympathetic nerves.

ROLE OF JUXTAGLOMERULAR APPARATUS IN AUTOREGULATION OF RBF AND GFR

One of the most striking characteristics of the renal circulation is the ability of the kidney to maintain a constant renal blood flow (RBF) and glomerular filtration rate (GFR) as renal perfusion pressure is altered. The dual regulation of both RBF and GFR is achieved by proportionate changes in the preglomerular resistance and is believed to be mediated by two mechanisms, the renal myogenic response and tubuloglomerular feedback (TGF).

a. Myogenic Response

As early as 1902, Bayliss[19] observed that the renal vasculature exhibits a profound vasoconstriction when subjected to elevated pressure (Fig. 29.8). Bayliss viewed the renal response as an example of the "myogenic" response of vascular beds. In regard to the purpose of this general response, he suggested that *"The peripheral powers of reaction possessed by the arteries is of such a nature as to provide as far as possible for the maintenance of a constant flow of blood through the tissues supplied by them, whatever may be the height of the general blood pressure ..."*.[19]

The concept that renal vascular responses to pressure might also serve to regulate function in the kidney was further advanced by the observation of Forster and Maes in 1947[20] that not only RBF but also GFR remained constant with acute elevations in

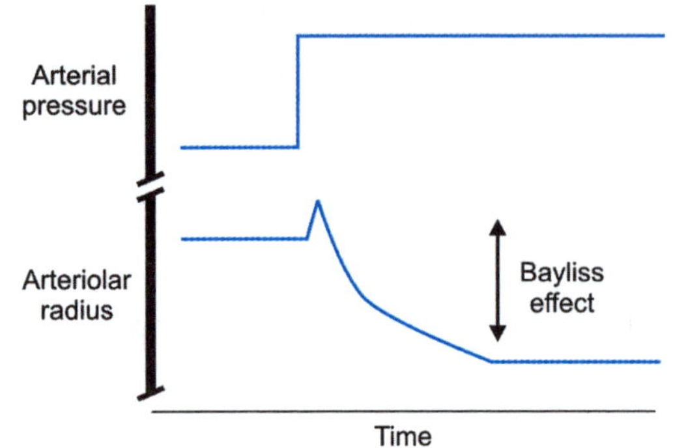

Fig. 29.8: The Bayliss response

BP. From the outset, it was recognized that the dual regulation of GFR and RBF could only be achieved if pressure-induced vasoconstriction was restricted only to preglomerular resistance vessels.

Myogenic response or Bayliss response in vascular smooth muscle cells is a response to stretch. This is especially applicable to arterioles of the body. When blood pressure is increased in the blood vessels and the blood vessels distend, they react with a constriction; this is the Bayliss effect. Stretch of the muscle membrane opens a stretch-activated ion channel. The cells then become depolarized and this results in a Ca^{2+} signal and triggers muscle contraction. It is important to understand that no action potential is necessary here; the level of entered calcium affects the level of contraction proportionally and causes tonic contraction. The contracted state of the smooth muscle depends on the grade of stretch and plays an important part in the regulation of blood flow. Myogenic response can be observed in many organs, but is particularly pronounced in the heart, brain and kidney. With the aid of the myogenic mechanism, the RBF and glomerular filtration rate remain, *within limits*, insensitive to changes in human blood pressure.[21]

b Tubuloglomerular Feedback

Autoregulation of GFR is fundamentally a phenomenon whose primary purpose is to insulate renal sodium and volume regulation from fluctuations in BP. It was first proposed by Guyton et al in 1964[22], that the autoregulation of GFR and RBF involved a unique mechanism in the kidney whereby preglomerular vasoconstriction was triggered by increased distal tubular fluid delivery. This concept is consistent with the prevailing view that, in addition to a general myogenic response, the differing physiologic and metabolic requirements of different tissues needed to be achieved by organ-specific vascular regulatory mechanisms. Since many decades, there is a consensus that both tubuloglomerular feedback and myogenic vasoconstriction are essential for normal autoregulation of renal blood flow and GFR.[23] However, their relative contributions have been a matter of debate. It is now believed that under normal conditions, both contribute equally to the maintenance of RBF and GFR.

In the kidney, the need for volume preservation requires that the capacity of the tubules to reabsorb the filtrate is not overwhelmed by excessive glomerular filtration rates. Specifically, the delivery of filtrate to the distal nephron segment, which has a limited reabsorptive capacity, needs to be precisely regulated. The unique close contact between the early distal nephron and its glomerular vascular pole of the origination glomerulus provides a potential site for such regulation.[22] By the time tubular fluid reaches the distal tubule, about 80% of the filtered water the filtered Na^+ has been already been reabsorbed. The macula densa cells play a pivotal role in sensing changes in tubular fluid composition, generating and sending signals to the juxtaglomerular apparatus that control renal blood flow and GFR. Functionally, the macula densa serves as a sensor of luminal NaCl concentration. Through the Na-K chloride cotransporter 2 (NKCC2) in the apical membrane, the macula densa cells initiate a sequence of events that regulates afferent arteriolar tone and thereby filtration rate.[23] As pointed out earlier, the NaCl signal from MD cells also controls the rate of renin secretion from juxtaglomerular granular cells. An increase in luminal NaCl concentration results in afferent arteriolar vasoconstriction and a reduction in GFR, whereas decreased salt concentration causes vasodilatation and an increase in GFR. This homeostatic feedback loop minimizes variations of salt concentration in the fluid entering the capacity-limited distal portions of the nephron. It thereby prevents filtration rate-dependent changes in urinary Na^+ excretion that could result in dysregulation of the extracellular fluid volume.[23]

The macula densa-tubular glomerular feedback (MD-TGF) relates the GFR to the NaCl delivery to MD cells. MD NaCl delivery, in turn, is related to the filtered load of NaCl and reabsorption in nephron segments upstream of the MD, all of which therefore can affect MD-TGF. Accordingly, MD-TGF integrates tubular and vascular function and is thus a unique renal mechanism of autoregulation.[24]

Biological Significance of Autoregulation of GFR

Ultrafiltration of plasma in the glomerulus and the subsequent absorption of water and solutes across the tubular epithelium have evolved as the fundamental processes responsible for formation of urine by the vertebrate kidney. In mammals, this filtration/absorption process is remarkable for its magnitude. The entire extracellular fluid volume is filtered through the renal glomeruli 15 times per day in humans, 50 times per day in rats, and 100 times per day in mice. Almost all the filtered salt and water is absorbed by the renal tubules, so that the volume of excreted urine is usually less than one hundredth of the filtrate volume.

It is evident that this process carries inherent vulnerability: Salt excretion and, therefore, maintenance of extracellular fluid volume, circulatory volume, and organ perfusion become critically dependent upon a perfect match between the rate of ultrafiltration and tubular absorption.[25] Catastrophic salt losses and salt gains by the kidneys are prevented by a set of mechanistic interactions between filtration and absorption. Changes in filtration lead to changes in absorption, and changes in tubular transport cause changes in filtration. These relationships are described as glomerular-tubular balance and tubuloglomerular feedback (TGF), respectively. Salt control at the macula densa is the ingredient which permits the kidney to enjoy the benefits of a very high filtration rate without an undue risk of inappropriate salt loss or salt retention.[25, 26] The tubuloglomerular feedback mechanism has already been discussed in detail. The phenomenon of glomerulotubular balance is briefly discussed below.

Glomerulotubular Balance

Glomerulotubular balance refers to the phenomenon whereby a constant fraction of the filtered load of the nephron is reabsorbed across a range of GFR. In other words, if the GFR spontaneously increases, the rate of water and solute reabsorption in the proximal tubule proportionally increases, thus maintaining the same fraction the filtered load being reabsorbed. The proximal tubule appears to be the principal segment in which glomerulotubular balance operates and approximately 80% of the filtered load of solutes and water is reabsorbed in the proximal tubule regardless of the value of the GFR. *The mechanistic basis of glomerulotubular balance is poorly understood. It* is likely an intrinsic property of the nephron itself. Glomerulotubular balance is a critical mechanism which protects distal segments of the nephron from being overloaded in contexts of short-term increases in GFR. Distal segments of the nephron have a very limited capacity to increase tubular reabsorption of water and solutes; consequently, a large increase in distal flow rates would result in a catastrophic loss of fluid in the urine.[27]

References

1. Montani JP, Van Vliet BN. General Physiology and Pathophysiology of the Renin–Angiotensin. https://www.springer.com/cda/content/document/cda.../9783540406402-c1.pdf?
2. https://www.renalfellow.org/2008/10/25/review-juxtaglomerular-apparatus/
3. Kurtz A.Control of Renin Synthesis and Secretion. American Jour Hypertens. 2012; 25: 839–47.
4. Toffelmire EB, Slater K, Corvol P, et al. Response of plasma prorenin and active renin to chronic and acute alterations of renin secretion in normal humans. J Clin Invest. 1989; 83: 679–87.
5. Danserb AH, Deinum J. Renin, prorenin and the putative (pro) renin receptor. Hypertension. 2005; 46: 1069–76.
6. Pratt RE, Carleton JE, Richie JP, et al. Human renin biosynthesis and secretion in normal and ischemic kidneys. Proc Natl Acad Sci USA. 1987; 84: 7837– 40.
7. Castrop H, Hocherl K, Kurtz A, et al. Physiology of kidney renin. Physiol Rev. 2010; 90: 607–73.
8. Carey RM, Grawth HE, Pentz ES, et al. Biomechanical coupling of renin-releasing cells. J Clin Invest. 1997; 100: 1566–74.
9. Skott O, Briggs JP. Direct demonstration of macula densa-mediated renin secretion. Science. 1987; 237: 1618–20.
10. Lapointe JY, Bell PD, Cardinal J. Direct evidence of apical Na^+-K^+-Cl^- cotransport in macula densa cells. Am J Physiol Renal Fluid Electrolyte Physiol. 1990; 258: F1466–69.
11. Jensen BL, Schmid C, Kurtz A. Prostaglandins stimulate renin secretion and renin mRNA in mouse juxtaglomerular apparatus. Am J Physiol Renal Fluid Electrolyte Physiol. 1996; 271: F659–69.
12. Bell PD, Lapointe JY, Sabirov R, et al. Macula densa cell signaling involves ATP release through a maxi-anion channel. Proc Natl Acad Sci, USA. 2003; 100: 4322–27.
13. Komlosi P, Peti-Peterdi J, Fuson AL, et al. Macula densa basolateral ATP release is regulated by luminal (NaCl) and dietary salt intake. Am J Physiol Renal Physiol. 2004; 286: F1054–58.
14. Churchill PC, Churchill MC. A1 and A2 adenosine receptor activation inhibits and stimulates renin secretion of rat renal cortical slices. J Pharmacol Exp Ther. 1985; 232: 589–94.
15. Keeton TK, Campbell WB. The pharmacologic alteration of renin release. Pharmacol Rev. 1980; 32: 81–227.
16. Schunkert H, Ingelfinger JR, Jacob H, et al. Reciprocal feedback regulation of kidney angiotensinogen and renin mRNA expressions by angiotensin II. Am J Physiol Endocrinol Metab. 1992; 263: E863–86.
17. Wagner C, Kurtz A. Regulation of renal renin release. Curr Opin Nephrol Hypertens. 1998; 7: 437–41.
18. Harris PJ, Thomas D, Morgan TO. Atrial natriuretic peptide inhibits angiotensin-stimulated proximal tubular sodium and water reabsorption. Nature. 1987; 326: 697–98.

19. Bayliss WM. On the local reaction of the arterial wall to changes in intraluminal pressure. J Physiol. 1902; 28: 220–31.
20. Forster RP, Maes JP. Effect of experimental neurogenic hypertension on renal blood flow and glomerular filtration rates in intact denervated kidneys of unanesthetized rabbits with adrenal glands demedullated. Am J Physiol. 1947; 150: 534–40.
21. Aukland K. Myogenic mechanisms in the kidney. Journal of Hypertension Supplement. 1989; 7: S71–76.
22. Guyton AC, Langston JB, Navar G. Theory for renal autoregulation by feedback at the juxtaglomerular apparatus. Circ Res. 964; 15(Suppl): 187–97.
23. Schnermann J, Briggs JP. The macula densa is worth its salt. J Clin Invest. 1999; 104: 1007–09.
24. Aukland K, Oien AH. Renal autoregulation: models combining tubuloglomerular feedback and myogenic response. Am J Physiol. 1987; 252: F768–83.
25. Carlstrom M, Wilcox M, Arendshorst WJ. Renal autoregulation in health and disease. Physiol Rev. 2015; 95: 405–51.
26. Loutzenhiser R, Griffin K, Williamson G, et al. Renal autoregulation: New perspectives regarding the protective and regulatory roles of the underlying mechanisms. Am J Physiol Regul Integr Comp Physiol. 2006; 290: R1153–67.
27. http://www.pathwaymedicine.org/glomerulotubular-balance

Chapter 30

Natriuretic Peptides

INTRODUCTION

Krish[1], in 1956, demonstrated on electron microscopy, certain characteristics in mammalian atrial cardiomyocytes that were previously associated with polypeptide hormone-producing cells. These features included abundant rough endoplasmic reticulum, highly developed Golgi complex and numerous dense granules (Fig. 30.1). This observation was the first milestone in ANP research. However, since the function of these intracellular organelles was not evident, these findings remained ignored for over two-decades.

In 1979, deBold[2] found that the number of atrial specific granules is decreased by water deprivation and increased by salt loading, suggesting that atrial granules contain a biologically active substance involved in volume regulation. Further, in 1981, deBold et al[3], demonstrated that extracts of rat atrial tissue induced a prompt, short-lived but profound natriuresis and diuresis in the rats. By 1983, a new peptide with 28-amino acid residues was isolated by the same researcher from rat atrial tissue[4], and by Matsuo and Kangawa from human atrial tissues.[5] The peptide, designated as atrial natriuretic peptide (ANP), exhibited diuretic, natriuretic (Fig. 30.2) and vasodilating activities. Following the discovery of ANP, Matsuo and Kangawa also isolated brain natriuretic peptide, or B-type natriuretic peptide (BNP) from the porcine brain in 1988[6], and C-type natriuretic peptide (CNP) from the porcine brain in 1990.[7] Both ANP and BNP are synthesized in the heart, mainly in the atria. The expression of ANP and BNP genes is increased in cardiac hypertrophy in response to atrial or ventricular wall stress, as well as in heart failure. In cardiac ventricles, BNP is not stored in granules but is constitutively released by myocytes suggesting that it has an

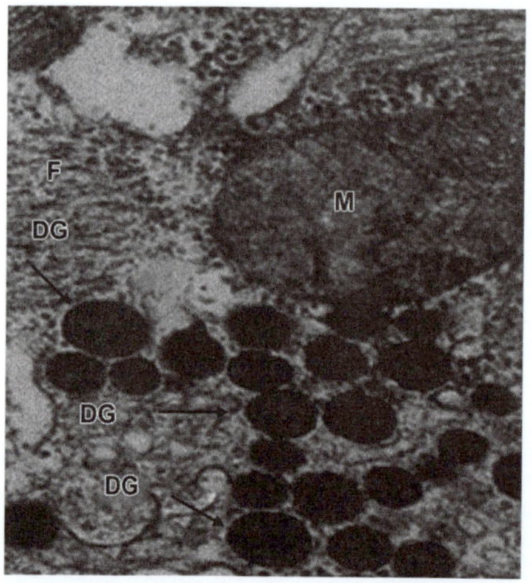

Fig. 30.1: Electron dense granules in atrial myocyte (DG). F: Myofibrils; M. Mitochondrion. (Modified from Ndungu JM, et al, Parasitol Res. 1992; 78:553–56.)

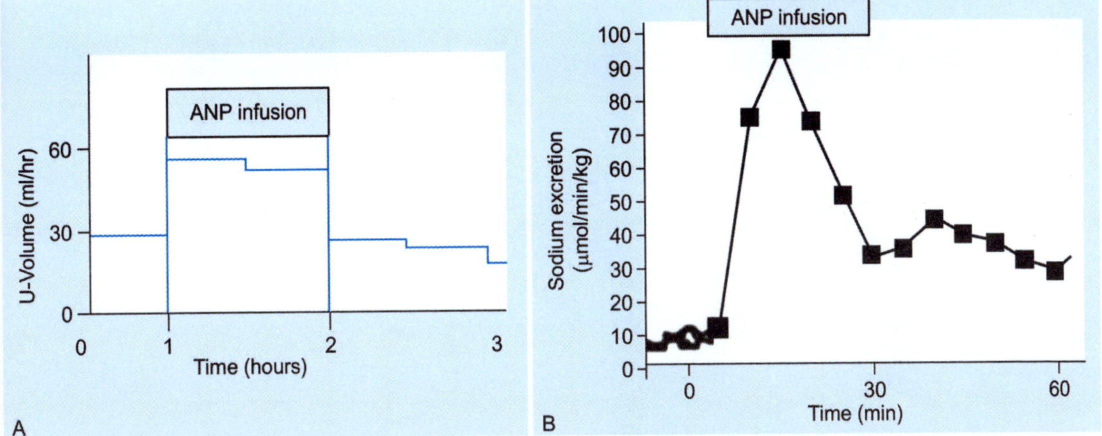

Fig. 30.2: Effect of ANP infusion on urinary water in humans (A) and sodium excretion in the rat (B)

important paracrine effect in the regulation of ventricular mass. Plasma levels of BNP are nowadays used as an important biochemical marker to detect ventricular hypertrophy and heart failure in humans.

C-type natriuretic peptide (CNP) is widely distributed in the mammalian organism, being present mostly in brain, bone, and in vascular endothelial cells. CNP is not secreted to blood but seems to have an important local paracrine actions in the bone vasculature, and central nervous system.

The greater part of ANP and BNP gene expression in mammals is found in the atria of the heart. Although it is often mentioned that BNP is a ventricular hormone while ANP is viewed as an atrial hormone, this concept needs to be reconsidered. The biochemical evidence also supports an overwhelming atrial source of ANP and BNP. The ANP concentration in normal subjects in atrium and ventricle is 9600 and 37 pmol/g respectively, while the BNP concentration of normal subjects in atrium and ventricle is 250 and 18 pmol/g respectively. Even taking into account the differences in mass between the atrial and ventricular chambers, it is not possible to conclude that BNP is a ventricular hormone.[8]

CHEMISTRY OF NATRIURETIC PEPTIDES

ANP is a 28-amino acid peptide with a ringed structure formed by intramolecular disulfide linkages. BNP is a 32-amino acid peptide, structurally similar to ANP with a 17 common amino acid ring structure. This ring structure seems to be essential for their biological activity. While ANP is stored as a 126-amino acid prohormone (proANP), BNP is stored as a mature hormone. Both ANP and BNP are released from the heart in response to blood pressure and blood volume overload. In addition, ANP and BNP secretion from the ventricular myocytes increases in association with a number of ventricular dysfunctions, and plasma levels of these peptides is increased in proportion to the severity of pathology. This observation led to the use of plasma ANP and BNP levels as diagnostic tests for heart failure. CNP consists of 22-amino acids. The ring structure of CNP is highly homologous with ANP and BNP but it uniquely lacks the carboxy-terminal extension[9, 10] (Fig. 30.3).

MECHANISM OF RELEASE OF ANP

ANP and BNP are continuously secreted from the atria under basal conditions. Mechanical stretch of the cardiac atria seems to be the most important stimulus for the release of cardiac natriuretic peptides. Initially, it was observed that inflation of a balloon in the left atrium of dogs led to marked increase in urine flow.[11] Later, it was demonstrated that atrial stretch by volume expansion caused release of ANP

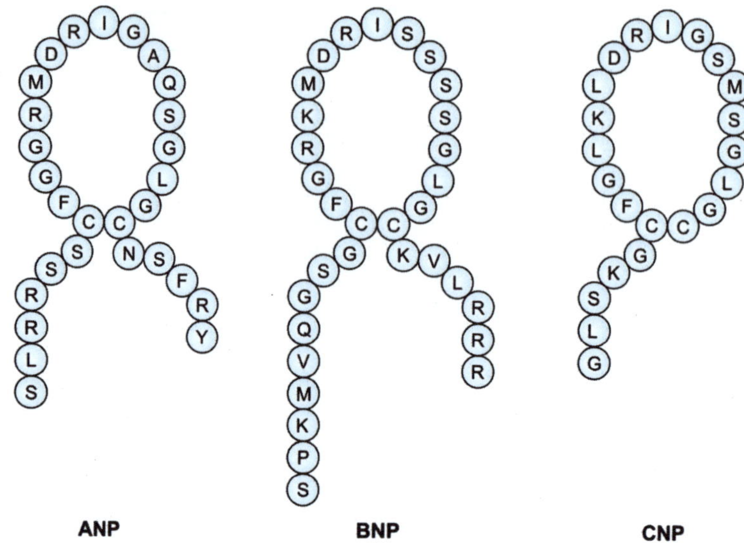

ANP **BNP** **CNP**

Fig. 30.3: Chemical structure of various natriuretic peptides

in isolated rat heart-lung preparations.[12] Subsequently, many reports confirmed the release of ANP by acute or chronic volume loading in rat as well as central hypervolemia induced by head-down tilt or head-out water immersion in humans.[13] Both, atria and ventricles increase synthesis and secretion of ANF and BNP during chronic hemodynamic overload as occurs in chronic arterial hypertension or chronic congestive heart failure. Main source of cardiac natriuretic peptides however, remains the cardiac atria, despite a significant increase in their content in the ventricles.[8]

ACTIONS OF NATRIURETIC PEPTIDES

A. Renal Actions: Natriuresis and Diuresis

Atrial natriuretic peptide infusion results in marked natriuresis and diuresis, especially when administered in supraphysiologic doses (Fig. 30.2). When administered in physiologic concentrations ANP produces less marked responses[14, 15], probably because of the concurrent action of counterregulatory neurohumoral factors. Therefore, it is logical to ask whether ANP is involved in physiological regulation of ECF volume. In this context, an experiment by Mizelle et al[16] in conscious dogs with indwelling catheters in the two kidneys is illuminating. ANP was infused in one kidney in physiologic concentration, and the vehicle was infused in the other kidney. Urine was collected from the two kidneys separately by split bladder technique. There was clear increase in salt and water excretion from the ANP-infused kidney. Since the counterregulatory factors were acting on both the kidneys, the natriuretic effect of ANP in physiologic concentrations was beyond doubt. This conclusion is also supported by the observation that administration of ANP degradation blocker results in modest increase in salt and water excretion.

Mechanism(s) of ANP-induced natriuresis and diuresis (Fig. 30.4)

A number of mechanisms seem to be involved in ANP-induced natriuresis and diuresis:

- Increased GFR
- Proximal tubular Na^+ reabsorption suppression
- Collecting duct Na^+ reabsorption suppression
- ADH-induced H_2O reabsorption suppression
- Increased blood flow through vasa recta

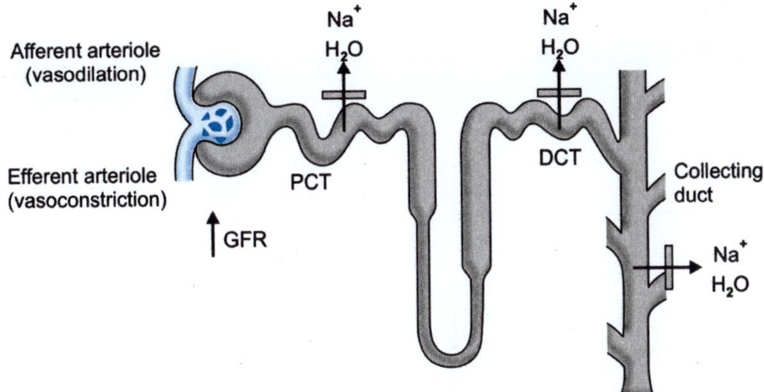

Fig.30.4: Renal actions of ANP

- Decreased renin secretion
- Decreased aldosterone secretion

1. Increased GFR

i. **Increased GFR through increased RBF:** ANP increases GFR, possibly through a number of mechanisms. Initially some reports indicated increased renal blood flow rate as a result of ANP infusion. However, after many contradictory reports, it has now been accepted that ANP does not increase renal blood flow.

ii. **Increased filtration fraction:** ANP dilates preglomerular (afferent) arterioles as well as constricts postglomerular (efferent) arterioles of the kidney, resulting in increased hydrostatic pressure in the glomerular capillaries (Pc).[17] An additional factor is increased ultrafiltration coefficient (Kf) mainly as a result of relaxation of mesangial cells.[18] As a result of increased filtration fraction filtered Na$^+$ load is increased.

2. Decreased Proximal Tubular Na$^+$ Reabsorption

ANP decreases proximal tubular Na$^+$ reabsorption. This inference has been drawn from the reports that ANP produced natriuresis and diuresis in many experiments in which GFR had not increased. This was particularly true with low dose infusion of ANP.[19]

3. Direct Decreased Salt and Water Reabsorption in Collecting Ducts

Many reports have suggested an addition mechanism contributing to natriuresis and diuresis: Direct decrease in salt and H$_2$O reabsorption in the collecting ducts. Nonoguchi and Knepper[20] observed dose related inhibition of vasopressin action at the level of inner medullary collecting ducts in the rat (Fig. 30.5). Stanton[21] reported a direct inhibition of reabsorption of sodium by ANP in the inner medullary collecting ducts.

4. Decreased Renin Secretion

Infusion of ANP markedly lowers rate of renin secretion (Fig. 30.6). As regards the

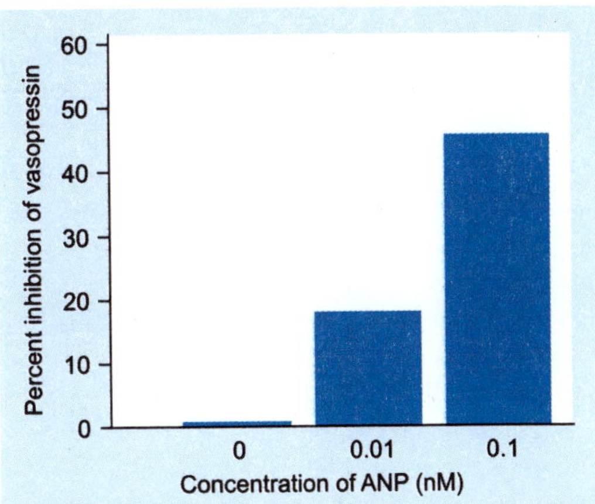

Fig. 30.5: Dose-related inhibition of vasopressin by ANP, after Nonoguchi and Knepper[20]

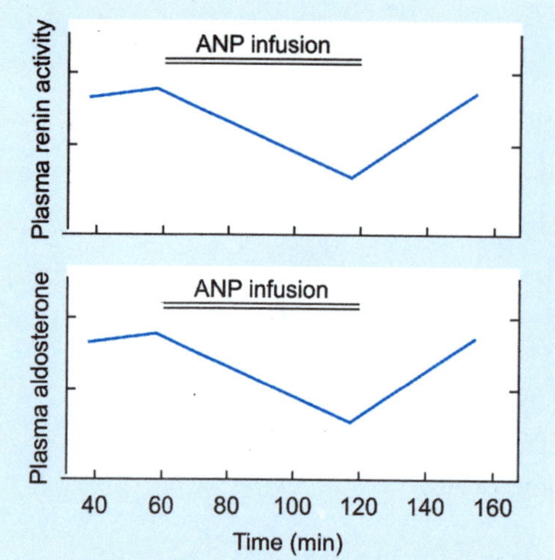

Fig. 30.6: Plasma renin and aldosterone levels in response to ANP infusion

mechanism of inhibition, Opgenorth et al[22] have reported that the effects of ANP on renin secretion occurs only in kidney undergoing filtration, but has no effect in non-filtering kidney. This observation suggests that the effect of ANP on renin secretion is mediated through macula densa when distal tubular Na^+ concentration is increased by actions 1 and 2 discussed above. In addition, ANP may also inhibit renin secretion by inhibition of sympathetic discharge.

5. Inhibition of Aldosterone Secretion

ANP inhibits aldosterone secretion (Fig. 30.6), not only through inhibition of renin-angiotensin-aldosterone axis but also direct inhibition of aldosterone-secreting zona glomerulosa cells.[23]

6. Inhibition of ADH Secretion

Besides inhibition of vasopressin action on collecting ducts discussed above, ANP has been reported to directly inhibit ADH secretion from the posterior pituitary gland.[23]

7. Increased Blood Flow Through Vasa Recta

ANP increases blood flow through the vasa recta[24], which washes the solutes (sodium chloride, and urea) out of the medullary interstitium. The consequent lower osmolality of the medullary interstitium leads to decreased power of urinary concentration and hence diuresis.

B. Cardiovascular Actions of Natriuretic Peptides (Fig. 30.7)

The effects of ANP on cardiocirculatory hemodynamics are complex and depend on the baseline status of the cardiovascular system. In normotensive experimental animals and humans, ANP decreases blood

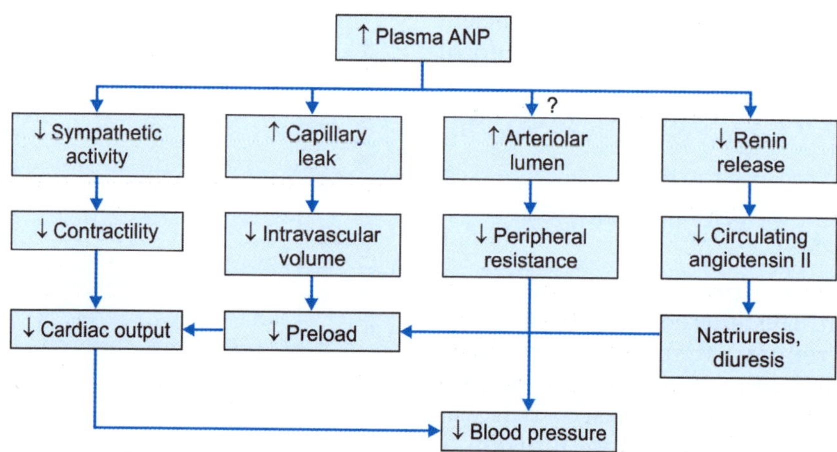

Fig. 30.7: Cardiovascular effects of ANP

pressure slightly but consistently. In several models of hypertension in experimental animals, ANP decreases blood pressure markedly.[25] ANP lowers blood pressure by several mechanisms.

1. Shift of Fluid from Intravascular to Extravascular Compartment

A reduction in blood pressure and increased hematocrit on infusion of ANP in bilaterally nephrectomised rats has been reported. The plasma volume decreased by an average of 14%. Simultaneously 7% increase in hematocrit, suggesting that increased capillary permeability leads to efflux of fluid from the circulatory system.[4, 26] The reports of increased rate of clearance of radioiodinated albumin from the circulatory system of mice infused with ANP[27], lend further support to the view that ANP causes shift of fluid from the intravascular to extravascular compartment by increasing capillary permeability. A decrease in plasma volume contributes to a reduction in both cardiac output and blood pressure.

2. Inhibition of Sympathetic Activity

ANP decreases sympathetic outflows by modulating ganglionic neurotransmission.[28] The decrease in sympathetic activities contributes to a decrease in systemic vasoconstriction and thus a decrease in systemic blood pressure. Administration of ANP in intact animals markedly reduces mean arterial pressure without augmenting heart rate or renal sympathetic activity. In contrast, when similar reduction in blood pressure was induced by nitroprusside, marked stimulation of heart rate and renal sympathetic activity occurred. Thus, unlike, nitroprusside, ANP causes hypotension and suppresses the stimulatory effect of baroreceptor activity on sympathetic nervous system.[29]

3. Functional Antagonism of the Renin-angiotensin-aldosterone System

As discussed earlier, ANP inhibits renin secretion by the kidney and aldosterone synthesis by the adrenal gland, resulting in a decrease in plasma levels of these antinatriuretic hormones. ANP-mediated decrease in renin secretion is due to the increase in sodium load to the macula densa brought about by an increase in GFR and possibly a decrease in proximal tubule reabsorption of sodium. ANP inhibits aldosterone synthesis by a direct action on adrenal zona glomerulosa cells, and indirectly by decreasing plasma renin. ANP also antagonizes all of the known effects of angiotensin II, including its peripheral vasoconstriction, growth promoting activity in vascular smooth muscle cells, stimulation of proximal renal tubular sodium reabsorption, and central dipsogenic effect. The inhibition of the renin-angiotensin-aldosterone system by administered ANP in humans is striking. At infusion rates that barely increased plasma levels of the hormone and only slightly increased fluid and electrolyte excretion by the kidney, ANP markedly decreased plasma aldosterone levels and plasma renin in humans.[25]

4. Arteriolar Vasodilation

ANP increases NO synthesis capability and NO production. The NO pathway could be an intercellular messenger in the ANP-mediated endothelium-dependent vasorelaxation mechanism.[30] According to another point of view, natriuretic peptides are not vasorelaxant *per se* but are functional antagonists of vasoconstriction.[25]

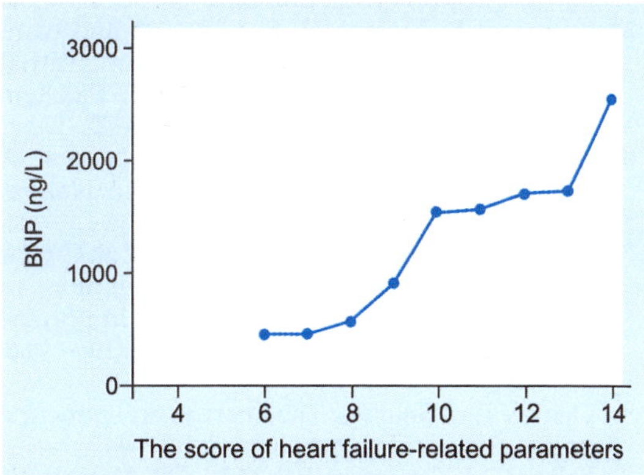

Fig. 30.8: BNP levels in relation to the severity of congestive heart failure

Diagnostic Value of Plasma BNP Estimation

Plasma BNP level is markedly elevated in patients with acute and chronic congestive heart failure. Determining plasma BNP concentration is an effective diagnostic tool in patients presenting with acute dyspnea. Studies have demonstrated that patients diagnosed with dyspnea secondary to heart failure have significantly elevated plasma BNP concentration compared to those with noncardiogenic dyspnea (due to respiratory disease). BNP analysis has also been shown to have a superior diagnostic accuracy in heart failure compared to other parameters like high venous pressure recruitment, increase of LV dimension and interstitial edema detected using chest X-ray. Hence BNP is not only a good marker for the diagnosis of heart failure but is also an excellent indicator of its severity. In patients who present with acute dyspnea, BNP testing provides useful information to physicians in clinical practice[31] (Fig. 30.8).

References

1. Kisch B. Electron microscopy of the atrium of the heart. Exp Med Surg. 1956; 14: 99–112.
2. deBold AJ. Heart atria granularity effects of changes in water-electrolyte balance.Proc Soc Exp Biol Med. 1979; 161: 508–11.
3. deBold AJ, Borenstein HB, Veress AT, et al. A rapid and potent natriuretic response to intravenous injection of atrial myocyte extracts in rat. Life Sci. 1981; 28: 89–94.
4. deBold AJ, Flynn TG. Cardionatrin I—a novel heart peptide with potent diuretic and natriuretic properties. Life Sci.1983; 33: 297–302.
5. Kangawa K, Matsuo H. Purification and complete amino acid sequence of alpha-human atrial natriuretic polypeptide (alpha-hANP). Biochem Biophys Res Commun. 1984; 118: 131–39.
6. Sudoh T, Kangawa K, N. Minamino N, et al. A new natriuretic peptide in porcine brain. Nature. 1988; 332: 78–81.
7. Sudoh T, Minamino N, Kangawa K, et al. C-type natriuretic peptide (CNP): a new member of natriuretic peptide family identified in porcine brain. Biochem Biophys Res Commun. 1990; 168: 863–70.
8. Ogawa T, deBold AJ. The heart as an endocrine organ. Endocr Connect. 2014; 3: R31–44.
9. Felker GM, Petwerson JW, Mark DB. Natriuretic peptides in the diagnosis and management of heart faiure. CMAJ. 2006; 175: 611–17.
10. Komatsu Y, Nakao K, Suga S, et al. C-type natriuretic peptide (CNP) in rats and humans. Endocrinology. 1991; 129: 1104–06.
11. Goetz KL, Wang BC, Greer PG, et al. Atrial stretch receptors and renal function. Am J Physiol. 1970; 219: 1417–23.
12. Dietz JR. Release of natriuretic factor from rat heart-lung preparation by atrial distension. Am J Physiol. 1984; 247: R1093–96.
13. Epstein MR, Loutzenhiser R, Friedland E, et al. Relationship of increased atrial natriuretic factor and renal sodium handling during immersion-induced hypervolemia in normal humans. J Clin Invest. 1987; 79: 738–45.
14. Anderson JV, Donckier J, Payne NM, et al. Atrial naturetic peptide: evidence of action as natural hormone at physiologic plasma concentrations. Clin Sci (London). 1987; 72: 305–12.
15. Burnett JC, Grangless JP, Opegenorth J. Effects of synthetic atrial natriuretic peptide on renal function and renin release. Am J Physiol (Renal Physiology). 1984; 16: F863–66.
16. Mizelle HL, Hall SE, Hilderbrant DA, et al. Renal hemodynamics and electrolyte excretion during chronic infusion of atrial natriuretic peptide. FASEB J. 1989; 3: A697.
17. Dunn BR, Ichikawa I, Pfeffer JM, et al. Renal and systemic hemodynamic effects of synthetic atrial natriuretic peptide in anesthetized rat. Circ Res. 1986; 59: 237–46.
18. Fiied TA, McCoy RN, Osgood RW, et al. Effects of atriopeptin II on determinants of glomerular filtration rate *in vitro* perfused dog glomerulus. Am J Physiol (Renal Physiol). 1986; 19: F1119–22.
19. Murray RD, Itoh S, Inagami T, et al. Effects of synthetic natriuretic factor in the isolated perfused rat kidney, Am J Physiol (Renal Physiol). 1985; F603–09.
20. Nonoguchi HJ, Knepper MA. Atrial natriuretic factor inhibits vasopressin-stimulated osmotic water permeability in rat inner medullary collecting duct. J Clin Invest. 1988; 82: 1383–90.
21. Stanton BA. Molecular mechanisms of ANP inhibition of renal sodium transport. Can J Physiol Pharmacol. 1991; 69: 1546–52.
22. Opgenorth TA, Burnett JC, Granger JP, et al. Effects of atrial natriuretic peptide on renin secretion in non-filtering kidney. Am J Physiol (Renal Physiol). 1986; 19: F798–801.
23. Gerstberger R, Schütz H, Luther-Dyroff D, et al. Inhibition of vasopressin and aldosterone release by atrial natriuretic peptide in conscious rabbits. Exp Physiol. 1992; 77: 587–600.
24. Kiberd BA, Larson TS, Robertson CR, et al. Effect of atrial natriuretic peptide on vasa recta blood flow in the rat. Am J Physiol. 1987; 252: F1112–17.

25. Maack T. The broad homeostatic role of natriuretic peptides. *Arq Bras Endocrinol Metab* [online]. 2006; 50: 198–207.
26. Almeida FA, Suzuki M, Maack T. Atrial natriuretic factor increases hematocrit and decreases plasma volume in nephrectomized rats. Life Sci. 1986; 39: 1193–99.
27. Chen W, Gassner B, Börner S, et al. Atrial natriuretic peptide enhances microvascular albumin permeability by the caveolae-mediated transcellular pathway. Cardiovasc Res 2012; 93: 141–51.
28. Floras JS. Inhibitory effect of atrial natriuretic factor on sympathetic ganglionic neuro-transmission in humans. Am J Physiol Regul Integr Comp Physiol. 1995; 269: R406–12.
29. Brenner BM, Ballermann BJ, Gunning ME, et al. Diverse biological actions of natriuretic peptide. Physiolo Rev. 1990; 70: 665–99.
30. Costa MD, Bosc LV, Majowicz MP, et al. Atrial natriuretic peptide modifies arterial blood pressure through nitric oxide pathway in rats. Hypertension. 2000; 35: 1119–23.
31. Chopra S, Cherian D, Verghese PP, et al. Physiology and clinical significance of natriuretic hormones. Indian J Endocrinol Metab. 2013; 17: 83–90.

Index

AAS 147
Actin 93
Adenosine 29, 40
Adiponectin 136
Agouti-related peptide 139
Alpha cells 183
Anemia of chronic diseases 9
Angiotensin II
   actions 247
Angiotensins 247
Anorexigenic neurons 134
ANP 246, 252
ANP release mechanism 253
Aphagia 132
Apoferritin 6
Appetite control 72
   brainstem control 140
   central control 137
   peripheral signals
   reward system 141
Appetite regulation 132
Arachnoid CSF blood barrier 168
ARAS 146
ARC 134, 135, 138
Arcuate nucleus 134, 135
Ascending
   activating system 147
   reticular activating system 146
Asplenia 21
Astrocyte
   glutamate-glutamine cycle 105
   neuron lactate shuttle 106
   tripartite synapse 106
Astrocytes 104
   fibrous 104
   foot plates 163
   glycogen store 108
   ion transport 105
   morphology 104
   oxidative stress 108
   protoplasmic 104
Atrial natriuretic peptide 246, 252
Auerbach's plexus 57
Autoregulation of GFR 248
   biological significance 249
   myogenic mechanism 248
Autoregulation of RBF 248

Barrett's esophagus 75
Basal ganglia 112
   afferents 114
   cognitive function 120
   efferents 115
   executive function 120
   functions 117
   intrinsic connections 115
   limbic circuit 120
   motor circuit 118
   oculomotor circuit 118
   prefrontal circuits 120
BAT 128
   catecholamines 130
   effect cold exposure 129
   location adults 128
   sex hormones 130
BBB 162, 163
Beta cells 183
Blood–brain barrier 162
   amino acid transport 165
   anatomy 163
   clinical implication 169
   enzymatic 165
   functions 169
   glucose transport 164
   ion transport 165
   macromolecules transport 165
   physiology 164
   water transport 164
Blood–CSF barrier 166
   ion transport 167
Blood–testis barrier 208
BNP 252
Bone 202
   endocrine function 202
   glucose metabolism 203
   insulin 203
     metabolism 202
   structure 202
Brain natriuretic peptide 252
Brown adipose tissue 128

Caldesmon 99
Calponin 100
CART 134, 135, 140
Caudate nucleus 113

CCK releasing peptide 69
CCK 69, 86, 137
   actions 70
   trophic effect 70
CCK-RP 69
Central command 36, 44, 52
Cerebral capillaries 163
Cholecystokinin 69
Choroid
   epithelium 166
   plexus 166
Circadian oscillator
   central 153
   peripheral 153
Circadian
   photoreception 154
   rhythms 152
CNP 252
   cascade model 16
   cell-based theory 16
   factors 15
   tests 18
   waterfall model 16
Coagulation 15
Cocain- and amphetamine-regulated
      transcript neurons 134
Cold-sensitive neurons 127
Cords of Billroth 22
Core body temperature 158
   biological markers 154
   blood pressure 159
Core-shell temperature 122
   clinical implication 123
Coronary
   autoregulation 28
   circulation
      autonomic control 31
      parasympathetic control 32
      sympathetic control 31
Cortex
   basal ganglia and
   cerebral cortex loop 115
      direct pathway 116
      indirect pathway 116
Cortisol 159
   human performance 160

D cells 68
Delta cells 183
    role 191
DHEA 228
Diaphragmatic sphincter 78
    contraction 79
    relaxation 80
Diuresis 254
Divalent metal transporter-1 6
DMH 139
Dopamine 234
    withdrawal 237
Dorsal vagal nucleus 78
Dual center hypothesis 132

ECL cells 66, 68
EEC 64, 73
End-feet 104
ENS 56
    defense responses 63
    fluid exchange 62
    motor input 60
    sensory output 61
ENS–CNS interaction 60
Enteric
    ganglia 57
    interneurons 58
    motor neurons 58
    secretomotor 59
    nervous system 56
    neuron plexuses 56
    neurons 58
        chemical nature 58
        classification 58
Enterochromaffin cells 64, 73
    closed type 65
    open type
Enterochromaffin-like cells 68
Entero-enteric reflexes 63
EPO 3
    chronic diseases 3
    history 3
    kidney disease 3
    regulation 4
EPOR 2
Ergogenic phase 159
Erythroid regulator 8
Erythropoiesis 1
    angiotensin II 4
    stages 1
Erythropoietin 1
Executive functions 172
    development 175
Exercise hyperemia 38
    mechanical factors 38
    metabolic factors 39
    sympathetic innervation 39
Exercise hyperpnea 51
    blood pressure regulation 44
    cardiac output 36
    central chemoreceptors 52
    central command 52

hypothalamic locomotor center 53
    mechanisms, redundancy 54
    motor cortex 53
    peripheral chemoreceptors 51
    peripheral neural reflex 54
Exercise mechanoreflex 38
Exercise pressor reflex 37, 45
Exercise 35, 44, 51
    baroreflex resetting 46
    cardiopulmonary baroreflex 46
    cardiovascular adjustments 35
    metaboreflex 37
    splanchnic blood flow 42
    sympatholysis 39
Exocrine pancreas 82
    bicarbonate secretion 83
    enzyme secretion 83
    innervation 82
    structure 82
Exocrine pancreatic secretion 83
    neural control 86
    phases 84
Extravascular compression 26
Extrinsic pathway 16

F-actin 95
Feeding center 132
Ferritin 6
Ferroportin 7
Ferroxidase 7
Fetal
    cortisol 228
    CRH 227
    prostaglandins 229
Food intake control 72

G-actin 95
Gastric endocrines 67
Gastrin release 68
Gastrin releasing peptide 67
Gastrins 67
Gastrin
    mucosal growth 69
Gastroesophageal reflux disease 75
Gastrointestinal
    endocrines 64, 67
    mucosal cells, turnover 65
G cells 67
GERD 75
Ghrelin 72, 137
GHRH 177
GIP 70
Gliotransmitters 106
Globus pallidus 113
Glomerulotubular balance 250
GLP-1 71, 88, 137, 140
Glucagon 184, 187
Glucagon-like peptide-1 71
Glucagon secretion 187
    amino acids 190
    autonomic regulation 190
    glucose 188

insulin 189
    mechanism 188
    regulation 188
Glucocorticoids 178
Glucose-dependent insulinotropic
        peptide 70
Glucostatic hypothesis 133
GnRH neurons
    negative feedback 222
    positive feedback 222
Growth hormone 176, 181
    actions 179
    age 181
    ghrelin 177
    glucose metabolism 179
    inhibitors 177
    lipid metabolism 180
    metabolic effects 179
    protein metabolism 180
    pulsatile secretion 176
    secretion regulation 177
    sex steroids 177
    stimulators 177
Growth hormone release 172
Growth promotion 179

H$_2$O$_2$ 30
HCP-1 6
Hedonic factors 141
Heme carrier protein-1 6
Heme oxygenase-1 7
Hemochromatosis 9
Hemoglobin, high altitude 3
Hemosiderin 7
Hepcidin 4, 7
Hephaestin 6
HIFs 4
Histamine
    gastric 68
5-HT 73
HTLR 53
5-hydroxytryptamine 73
Hyperphagia 132
Hyperprolactinemia, infertility 238
Hypoxia-inducible factors 4
Hypoxia 29

I cells 69
ICC 61
    slow wave activity 61
IGF-1 actions 178, 179
Ileal-brake 71
Insulin 136, 184
Insulin secretion
    autonomic innervation 186
    GIP 186
    GLP-1 186
    glucose 185
    incretins 186
    mechanism 185

phases 187
proteins 186
regulating factors 185, 187
Insulin-like growth factor-1 178
Interneurons
  ascending 58
  descending 58
Interstitial cells of Cajal 61
Intestinal endocrines 69
Intrinsic pathway 16
Intrinsic primary afferent neuron 58
Intrinsically photosensitive retinal
      ganglion cells 154
IPAN 58
ipRGCs 154
Iron stores 7
Iron transport
  enterocyte 6
  plasma 6
Iron
  absorption 5
  export 7
Islets of Langerhans 183
  A cells 183
  B cells 183
  D cells 183
  F cells
  PP cells 183

JGA 241
JGA, RAAS 242
Juxtaglomerular
  apparatus 241
  cells 241

K cells 70
Kisspeptin
  adults
    feedback control 221
    GnRH 221
    neurons 218
    puberty regulation 219
    terminology 218
Kisspeptins 217
  role in adults 221
KNDy
  hypothesis 222
  neurons 222

Lactate 40
Lactation 235
  hormones required 235
Lactotrophs 233
Latch effect 101
Lateral
  hypothalamic area 135
  hypothalamus 132
  parabrachial nucleus 124
Law of the intestines 56
L cells 71
Lentiform nucleus 113
Leptin 135

LES 75
  relaxation 79
LHA/PFA 139
Light-dark cycles 156
Lipostatic hypothesis 134
Local nervous system 56
Lower esophageal sphincter 75
  electrophysiology 77
  neural control 77
  relaxation 79
  tone 76

Macula densa 241
Malabsorption
  physiological 90
Maternal CRH 227
  positive feed forward 229
M cells 73
Medial hypothalamus 132
Megakaryocytes 13
Meissner's plexus 57
Melanopsin 154
Melatonin 155, 156
  antioxidant 167
  immune system 157
Mesangial cells 242
Metabolic theory 29
Microglia 109
MLCK 94, 98
Motilin 73
Mucosal
  block hypothesis 6
  chemosensors 58
  mechanoreceptors 58
Myenteric plexus 57
Myogenic theory 28
Myometrial activation 227
Myosin light chain kinases 94
  essential 94
  regulatory 94
Myosin 94
  heavy chains 94
  hinge points 94
  papain site 94
  trypsin site 94

Natriuresis 254
Natriuretic peptides 252
  actions 254
  cardiovascular actions 256
  chemistry 253
  diagnostic value 258
  filtration fraction 255
  GFR 255
Neuroglia 103
Neuropeptide Y 139
Neurotensin 73
Nigrostriatal projection 117
Nitric oxide 41
NREM sleep 144
  neurobiology 148
  stages 144

NREM-REM cycles 144
Nutrition and puberty 220

Ob/Ob mice 220
Oligodendrocytes 109
Orexigenic neurons 134
Orexin 148
Osteocalcin 203
  carboxylated 203
  glucose homeostasis 205
  testicular function 205
  testosterone 206
  undercarboxylated 204
Oxyntomodulin 137
Oxytocin, endometrial 231

PALS 21
Pancreatic enzyme ratio 89
Pancreatic exocrine response to diet
      89
  gastric acid 88
  protein diet 89
Pancreatic exocrine secretion 84
  feedback control 90
  termination 90
Pancreatic polypeptide 88, 136
Paneth cells 66
Parturition 225
  CRH 227
  estrogens 229
  oxytocin 231
  progesterone 229
  prostaglandins 231
  timing 226
Peptide YY 72, 87, 137
Peristaltic reflex circuitry 59
Peritubular interstitial fibroblasts 4
Phasic blood flow 26
Pineal gland 155
Pinealocytes 155
Platelet activation 14
Platelets 11
  granules 13
  ultrastructure 12
Platelet plug 15
POA 125
Polycythemia vera 3
Polycythemia
  secondary 3
POMC 134, 135, 140
Post-exercise hypotension 46
  baroreflex resetting 48
  environmental factors 49
  histamine receptors 49
  magnitude 47
  mechanisms 47
  vasodilatory mechanisms 48
Postpartum hemorrhage 231
Prefrontal cortex 172
  executive functions 171
Prefrontal lobotomy 171
Preoptic area 125

Primary hemostasis 11
Progesterone withdrawal
    functional 231
Prolactin 233
    CNS action 238
    kisspeptin 239
    metabolic functions 237
    orgasm 238
    role in males 239
Prolactin secretion 233
    during lactation 236
    estrogens 235
    hypothalamic control 233
    lactation 235
    pregnancy 234
Proopiomelanocortin 134
Proplatelet 13
Prorenin 243
Prostaglandins 41
Puberty 219
    obesity 220
Putamen 113
PVN 138
PYY 72

Red pulp 22
REM sleep 145, 160
    effector neurons 150
    neurobiology 150
Renin granules 242
Renin secretion
    adenosine
    aldosterone 246
    ANG II 246
    ANP 246
    baroreceptors 244
    blood pressure 245
    macula densa 244
    prostaglandins 245
    regulation 244
    salt intake 246
    sympathetic control 245
Renin
    biosynthesis 242
Resistin 136
Retinohypothalamic tract 154

Saccadic eye movement 119
Satiety center 132
S cells 70
Schwann cells 110
Secondary hemostasis 15
Secretin 70, 86
Seminiferous tubules 208
    adluminal compartment 210
    basal compartment 210
Sertoli cells 208
    development 209
    proliferation 209

Sertoli cell barrier 211
    anatomical 212
    germ cell transport 241
    physiologic 212
Sham feeding 84
Skeletal muscle
    relaxation 97
    ultrastructure 92
Sleep
    flip-flop switch 149
    physiological changes 146
    promoting area 148
Sleep-wakeful cycles 143
Sliding filament mechanism 96
Smooth muscle 97
    actin:myosin ratio 97
    contraction 92, 98, 100
    excitation contraction coupling 97
    G-protein 98
    IP3 98
    power generation 101
    relaxation 101
    ultrastructure 97
Somatopause 181
Somatostatin 71, 88
    gastric 67
Spermatogenesis
    stages 210
Spinal LPB–POA pathway 125
Spinothalamocortical pathway 124
Spleen 20
    closed circulation 23
    culling function 23
    follicles 22
    functions 23
    iron metabolism 24
    marginal zone 22
    open circulation 23
    reservoir function 23
    sinuses 22
    structure 21
Splenectomy 20
Stretch responsive neurons 58
Striatum 113
Subendocardial blood flow 27
Subepicardial blood flow 27
Submucosal plexus 57
Substantia nigra 113
Subthalamic nucleus 113
Suckling 236
Suprachiasmatic nucleus 153

TA cells 66
Temperature regulation 122
    afferent pathways 124
    anterior hypothalamus 123
    autonomic responses 123
    control 126
    effector systems 127

posterior hypothalamus 123
    set point 126
    somatic responses 123
TRP channels 124
Temperature transduction 124
Testis
    immunological barrier 213
TFPI 18
Thermometer 122
Thick filaments 95
Thin filament activation 96
Thin filaments 95
Thrombocytes 11
Thrombopoietin 14
TIDA neurons 234
Tissue factor pathway inhibitor 18
TLESR 80
Transferrin 6
Transient LES relaxation 80
Trophotropic phase 159
Tropomyosin 95
Troponin (TnI, TnC, TnT) 95
Tubuloglomerular feedback 248

UCP-1 128
Uncoupling protein-1 128
Uterine quiescence 226

Vagovagal reflex 60
Vasodilatation
    conducted 41
Vasodilator metabolites 39
    criteria 40
Ventral striatum 113
Ventromedial hypothalamus 135
Vitamin D hormone 193
    actions
        bone 197
        intestine 196
        receptor 196
    cutaneous production 193
    hepatic metabolism 194
    immune function 198
    insulin secretion 198
    kidney 197
    parathyroid function 199
    psoriasis
    renal metabolism 195
    reproductive function 199
VLPO 148
VMH 135, 139
von Willebrand factor 14

Wakefulness 146
Warm-sensitive neurons 127
WAT 128
Weight control 132
White adipose tissue 128
White pulp 21